T0184055

Lecture Notes in Artificial Intelligence   11160

Subseries of Lecture Notes in Computer Science

LNAI Series Editors

Randy Goebel
University of Alberta, Edmonton, Canada
Yuzuru Tanaka
Hokkaido University, Sapporo, Japan
Wolfgang Wahlster
DFKI and Saarland University, Saarbrücken, Germany

LNAI Founding Series Editor

Joerg Siekmann
DFKI and Saarland University, Saarbrücken, Germany

More information about this series at http://www.springer.com/series/1244

Francisco Herrera · Sergio Damas
Rosana Montes · Sergio Alonso
Óscar Cordón · Antonio González
Alicia Troncoso (Eds.)

Advances in Artificial Intelligence

18th Conference of the Spanish Association
for Artificial Intelligence, CAEPIA 2018
Granada, Spain, October 23–26, 2018
Proceedings

 Springer

Editors
Francisco Herrera ⓘ
Andalusian Research Institute on Data
  Science and Computational Intelligence
  (DaSCI)
University of Granada
Granada, Spain

Sergio Damas ⓘ
Andalusian Research Institute on Data
  Science and Computational Intelligence
  (DaSCI)
University of Granada
Granada, Spain

Rosana Montes ⓘ
Andalusian Research Institute on Data
  Science and Computational Intelligence
  (DaSCI)
University of Granada
Granada, Spain

Sergio Alonso ⓘ
Andalusian Research Institute on Data
  Science and Computational Intelligence
  (DaSCI)
University of Granada
Granada, Spain

Óscar Cordón ⓘ
Andalusian Research Institute on Data
  Science and Computational Intelligence
  (DaSCI)
University of Granada
Granada, Spain

Antonio González ⓘ
Department of Computer Science
  and Artificial Intelligence
University of Granada
Granada, Spain

Alicia Troncoso ⓘ
School of Engineering
Pablo de Olavide University
Seville, Spain

ISSN 0302-9743   ISSN 1611-3349 (electronic)
Lecture Notes in Artificial Intelligence
ISBN 978-3-030-00373-9   ISBN 978-3-030-00374-6 (eBook)
https://doi.org/10.1007/978-3-030-00374-6

Library of Congress Control Number: 2018939448

LNCS Sublibrary: SL7 – Artificial Intelligence

This Springer imprint is published by the registered company Springer Nature Switzerland AG
The registered company address is: Gewerbestrasse 11, 6330 Cham, Switzerland

Preface

This volume contains the selected papers presented at CAEPIA 2018, XVIII Conference of the Spanish Association for Artificial Intelligence, during October 23–26, 2018, held in Granada, Spain. The CAEPIA series of conferences is a biennial event that started in 1985. Previous editions took place in Alicante, Málaga, Murcia, Gijón, San Sebastián, Santiago de Compostela, Sevilla, La Laguna, Madrid, Albacete, and Salamanca.

CAEPIA is a forum open to researchers from all over the world to present and discuss their latest scientific and technological advances in Artificial Intelligence (AI). Authors were kindly requested to submit unpublished original papers describing relevant achievements on AI topics. Either formal, methodological, technical, or applied research were welcome.

Several federated congresses and workshops related to relevant AI tracks took place within CAEPIA: XIX Spanish Congress on Fuzzy Logic and Technologies (ESTYLF); XIII Spanish Congress on Metaheuristics, Evolutionary and Bioinspired Algorithms (MAEB); IX Symposium of Theory and Applications of Data Mining (TAMIDA); V Congress of the Spanish Society for Video Game Science (CoSECiVi); IV Symposium on Information Fusion and Ensemble Learning (FINO), II Workshop in Big Data and Scalable Data Analytics (BigDADE); I Workshop in Deep Learning (DEEPL); I Workshop on Applications of Artificial Intelligence for Industry 4.0 (IndustrIA 4.0); I Workshop on Data Science in Social Media (CiDReS); and I Workshop of Spanish AI Research Groups in Biomedicine (IABiomed).

Within CAEPIA 2018, the Doctoral Consortium (DC) was also organized. It is a forum for PhD students to interact with other researchers by discussing their PhD advances and plans. With the aim of highlighting the practical importance of AI, the 3rd Competition on Mobile Apps with AI Techniques was held at CAEPIA as well as the AI Dissemination Video Competition. CAEPIA 2018 aimed to maintain the high quality standards of previous editions.

Apart from the presentation of technical full papers and the DC, the scientific program of CAEPIA 2018 included an app contest and a track on outstanding recent papers (Key Works: KW) already published in renowned journals or forums.

CAEPIA aims to be recognized as a flagship conference in AI. This implies to achieve high-quality standards in the review process. In particular, the total number of submissions to CAEPIA 2018 was 240 (neither DC nor KW submissions were included in those 240 contributions since their review process was different). Only 36 outstanding manuscripts were selected for this volume after a thorough review process that involved at least 3 reviews per submission. This involved a lot of tough work by the CAEPIA 2018 Program Committee (PC) that was really appreciated. The reviewers judged the overall quality of the submitted manuscripts, together with the quality of the methodology employed, the soundness of the conclusions, the significance of the topic, the clarity and the organization, among other evaluation fields. The reviewers stated their confidence in the subject area in addition to detailed written comments. On the basis of the reviews, the PC Chairs proposed the final decisions that were finally made by both the CAEPIA 2018 general chairman and the president of the Organizing Committee.

Likewise, CAEPIA 2018 invited five internationally renowned researchers for a plenary talk. Pedro Larrañaga (Technical University of Madrid, Spain) presented "Bayesian Networks in Action." Sergio Guadarrama (Google Research, USA) introduced "AI in Google: latest advances and tendencies." Carlos A. Coello (CINVESTAD-IPN, Mexico) focused on "Trends in Evolutionary Multiobjective Optimization Research." Joao Gama (University of Porto, Portugal) discussed on "Real-Time Data Mining." Finally, Humberto Bustince (Public University of Navarra, Spain) gave the talk "From Integrals to Pre-aggregations: Applications in Classification, the Computational Brain and Image Processing."

AEPIA and the organization of CAEPIA 2018 recognized the best PhD work submitted to the DC with a prize, as well as the best student and conference paper presented at CAEPIA 2018. Furthermore, CAEPIA 2018 also aimed to promote the presence of women in AI research. As in previous editions, the Frances Allen award recognized the two best AI PhD thesis defended by a woman during the last two years.

The editors of this volume would like to thank many people who contributed to the success of CAEPIA 2018: authors, members of Scientific and Program Committees, invited speakers, etc. We would like to especially thank the tireless work of the Organizing Committee, our local sponsors (Andalusian Research Institute on Data Science and Computational Intelligence – DaSCI and the University of Granada), the Springer team, and AEPIA for their support.

Last but not least, on behalf of the CAEPIA 2018 participants, Francisco Herrera (CAEPIA 2018 General Chair) and Sergio Damas (CAEPIA 2018 President of the Organizing Committee), would like to especially thank Prof. Enric Trillas (CAEPIA 2018 Honorary Chair) for his pioneering work in AI, his numerous initiatives to promote AI in Spain, and his friendship.

July 2018

Francisco Herrera
Sergio Damas
Rosana Montes
Sergio Alonso
Óscar Cordón
Antonio González
Alicia Troncoso

Organization

Honorary Chair

Enric Trillas   University of Oviedo, Spain

General Chair

Francisco Herrera   University of Granada, Spain

Chair of Organizing Committee

Sergio Damas   University of Granada, Spain

Chair of the Awards Committee

Enrique Alba   University of Málaga, Spain

Chair of the Tutorials and Workshops

Luís Martínez   University of Jaén, Spain

Chairs of the Federated Conferences and Workshops

ESTYLF 2018

Antonio González   University of Granada, Spain
Enrique Herrera-Viedma   University of Granada, Spain

MAEB 2018

Oscar Cordón   University of Granada, Spain
Rafael Martí   University of Valencia, Spain

TAMIDA 2018

José Riquelme   University of Sevilla, Spain
Alicia Troncoso   University Pablo de Olavide, Spain
Salvador García   University of Granada, Spain

CoSECiVi 2018

Pedro Antonio González   Complutense University of Madrid, Spain
David Camacho   Autonomous University of Madrid, Spain

FINO 2018

Emilio Corchado   University of Salamanca, Spain
Mikel Galar   Public University of Navarra, Spain
Bruno Baruque   Burgos University, Spain
Alberto Fernández   University of Granada, Spain

BigDADE 2018

Amparo Alonso   University of Coruña, Spain
Maria José del Jesus   University of Jaén, Spain
Francisco Herrera   University of Granada, Spain

DEEPL 2018

Siham Tabik   University of Granada, Spain
Antonio Bahamonde   University of Oviedo, Spain
Juan Manuel Górriz   University of Granada, Spain

IndustrIA 4.0

Javier Del Ser   TECNALIA and University of the Basque Country,
                           Spain
Jose Antonio Lozano   University of the Basque Country, Spain

CiDReS 2018

David Camacho   Autonomous University of Madrid, Spain
Victoria Luzón   University of Granada, Spain

IABiomed 2018

Jose M. Juarez   University of Murcia, Spain
Mar Marcos   Universitat Jaume I, Spain
David Riaño   Universitat Rovira i Virgili, Spain

Doctoral Consortium

José Riquelme   University of Seville, Spain
Humberto Bustince   Public University of Navarra, Spain

Apps Competition

Alberto Bugarín   University of Santiago de Compostela, Spain
Jose Antonio Gámez   University de Castilla–La Mancha, Spain

Organizing Committee

Secretary

Pedro Villar   University of Granada, Spain

Publishing and Program

Sergio Alonso   University of Granada, Spain
Jorge Casillas   University of Granada, Spain
Julián Luengo   University of Granada, Spain

Publicity

M. Dolores Pérez   University of Jaén, Spain
Pedro González   University of Jaén, Spain

Logotype

Manuel Parra   University of Granada, Spain

Web

Rosana Montes   University of Granada, Spain

Technical Secretaries

ESTYLF
Raul Pérez   University of Granada, Spain
Javier Cabrerizo   University of Granada, Spain

MAEB
Daniel Molina   University of Granada, Spain
Maria Isabel García   University of Granada, Spain

CoSECiVi
Marco Antonio Gómez   Complutense University of Madrid
Antonio Mora   University of Granada, Spain

CiDRES
Eugenio Martínez   University of Granada, Spain

Area Chairs

Uncertainty in A.I.

Pedro Larrañaga Múgica   Technical University of Madrid, Spain

Natural Language Processing

Patricio Martínez Barco   Alicante University, Spain

Representation of Knowledge, Reasoning and Logic

Pedro Meseguer González   IIIA-CSIC, Spain

Restrictions, Searching, and Planning

Eva Onaindia De La   Polytechnic University of Valencia, Spain
  Rivaherrera

Multi-agent Systems

Juan Pavón Mestras   Complutense University of Madrid, Spain

Intelligent Web and Information Retrieval

Juan Manuel Fernández   University of Granada, Spain
  Luna

Computer Vision and Robotics

Petia Radeva   University of Barcelona, Spain

Open Data and Ontologies

Asunción Gómez Pérez   Technical University of Madrid, Spain

Ambient Intelligence and Smart Environments

Jesús García Herrero   Carlos III University of Madrid, Spain

Creativity and Artificial Intelligence

Miguel Molina Solana   Imperial College London, UK

Foundations, Models, and Applications of A.I.

Serafín Moral Callejón   University of Granada, Spain

Program Committee

Jesús S. Aguilar-Ruiz   University Pablo de Olavide, Spain
Enrique Alba   University of Málaga, Spain
Cristina Alcalde   University of the Basque Country, Spain
Rafael Alcalá   University of Granada, Spain
Jesús Alcalá-Fdez   University of Granada, Spain
Amparo Alonso   University of A Coruña, Spain
José María Alonso   University of Santiago de Compostela, Spain
Sergio Alonso   University of Granada, Spain
Amparo Alonso-Betanzos   University of A Coruña, Spain
Ada Álvarez   Universidad Autónoma de Nuevo León, Mexico
Martín Álvarez   W3C, Spain

Jorge Casillas	University of Granada, Spain
José Luis Casteleiro-Roca	University of A Coruña, Spain
Robert Castelo	Universitat Pompeu Fabra, Spain
Pedro A. Castillo	University of Granada, Spain
Juan Luis Castro	University of Granada, Spain
José Jesús Castro-Sánchez	University of Castilla–La Mancha, Spain
Manuel Chica	European Centre for Soft Computing, Spain
Francisco Chicano	University of Málaga, Spain
Miguel Chover	Universitat Jaume I, Spain
Francisco Chávez	University of Extremadura, Spain
Manuel Jesus Cobo Martin	University of Cádiz, Spain
Carlos A. Coello	CINVESTAV-IPN, Spain
José Manuel Colmenar	King Juan Carlos University, Spain
Ángel Corberán	University of Valencia, Spain
Emilio Corchado	University of Salamanca, Spain
Juan Corchado	University of Salamanca, Spain
Oscar Corcho	Universidad Politécnica de Madrid, Spain
Rafael Corchuelo	University of Seville, Spain
Oscar Cordón	University of Granada, Spain
Ulises Cortés	Universitat Politècnica de Catalunya, Spain
Carlos Cotta	University of Málaga, Spain
Inés Couso	University of Oviedo, Spain
Susana Cubillo	Polytechnic University of Madrid, Spain
Leticia Curiel	University of Burgos, Spain
Sergio Damas	University of Granada, Spain
Rocío de Andrés	University of Salamanca, Spain
Luis M. de Campos	University of Granada, Spain
Andre de Carvalho	University of São Paulo, Brazil
María Jesús De La Fuente	University of Valladolid, Spain
Luis De La Ossa	University of Castilla–La Mancha, Spain
Juan José Del Coz	University of Oviedo, Spain
Maria Jose Del Jesus	Universidad de Jaén, Spain
Javier Del Ser	University of the Basque Country, Spain
Miguel Delgado	University of Granada, Spain
Bernabé Dorronsoro	University of Cádiz, Spain
Jose Dorronsoro	Autonomous University of Madrid, Spain
Abraham Duarte	King Juan Carlos University, Spain
Richard Duro	University of A Coruña, Spain
Susana Díaz	University of Oviedo, Spain
Francisco Javier Díez	National Distance Education University, Spain
José Egea	Polytechnic University of Cartagena, Spain
Jorge Elorza	University of Navarra, Spain
Juan Manuel Escaño	Universidad Loyola Andalucía, Spain
Francisco Escolano	University of Alicante, Spain
Luis Espinosa-Anke	Universitat Pompeu Fabra, Spain
Francesc Esteva	Technical University of Catalonia, Spain

Javier Faulín	Public University of Navarra, Spain
Francisco Fernández De Vega	University of Extremadura, Spain
Alberto Fernández	University of Granada, Spain
Antonio J. Fernández	University of Málaga, Spain
Francisco Fernández	University of Extremadura, Spain
Francisco Javier Fernández	Public University of Navarra, Spain
Juan Carlos Fernández	University of Córdoba, Spain
Antonio J. Fernández Leiva	University of Málaga, Spain
Mariano Fernández López	San Pablo CEU University, Spain
Jesualdo Tomás Fernández-Breis	University of Murcia, Spain
Antonio Fernández-Caballero	University of Castilla–La Mancha, Spain
Juan M. Fernández-Luna	University of Granada, Spain
Juan Fernández-Olivares	University of Granada, Spain
Francesc J. Ferri	University of Valencia, Spain
Aníbal R. Figueiras	Carlos III University of Madrid, Spain
Aníbal R. Figueiras-Vidal	Carlos III University of Madrid, Spain
Ramón Fuentes-González	Public University of Navarra, Spain
Maribel G. Arenas	University of Granada, Spain
Alfredo G. Hernández-Díaz	Pablo de Olavide University, Spain
Mikel Galar	University of Navarra, Spain
Mikel Galar	Public University of Navarra, Spain
José Manuel Galán	University of Burgos, Spain
Jose Gámez	University of Castilla–La Mancha, Spain
Jesus García	Carlos III University of Madrid, Spain
Nicolás García	University of Córdoba, Spain
Carlos García	University of Córdoba, Spain
Salvador García	University of Granada, Spain
Pablo García Bringas	University of Deusto, Spain
Pablo García Sánchez	University of Granada, Spain
Raúl García-Castro	Universidad Politécnica de Madrid, Spain
José Luis García-Lapresta	University of Valladolid, Spain
César García-Osorio	University of Burgos, Spain
Nicolás García-Pedrajas	University of Córdoba, Spain
Rafael M. Gasca	University of Seville, Spain
Pablo Gervás	Complutense University of Madrid, Spain
María Ángeles Gil	University of Oviedo, Spain
Ana Belén Gil González	University of Salamanca, Spain
Lluis Godo	IIIA-CSIC, Spain
Koldo Gojenola	University of the Basque Country, Spain
Juan Gómez Romero	University of Granada, Spain
Juan A. Gómez-Pulido	University of Extremadura, Spain
Antonio González	University of Granada, Spain
Pedro González	University of Jaén, Spain

Contents

Data Mining

Applications

Artificial Intelligence

Neighbor Selection for Cold Users
in Collaborative Filtering
with Positive-Only Feedback

Alejandro Bellogín[1]([⊠]), Ignacio Fernández-Tobías[2], Iván Cantador[1],
and Paolo Tomeo[3]

[1] Universidad Autónoma de Madrid, 28049 Madrid, Spain
{alejandro.bellogin,ivan.cantador}@uam.es
[2] NTENT, 08018 Barcelona, Spain
ifernandez@ntent.com
[3] Politecnico di Bari, 70125 Bari, Italy
paolo.tomeo@poliba.it

Abstract. Recommender systems heavily rely on the availability of historical user preference data, struggling to provide relevant suggestions for new users. The cold start user scenario is thus recognized as one of the most challenging problems in the recommender systems research area. Previous work has focused on exploiting additional information about users and items –e.g., user personality and item metadata– to mitigate the lack of user feedback. However, it is still unclear how to approach the worst scenario where no side information is available to a recommender system. Addressing this problem, in this paper we focus on new users of memory-based collaborative filtering methods with positive-only feedback, and conduct a comprehensive study of a number of neighbor selection strategies. Specifically, we present empirical results on several datasets analyzing the effects of choosing adequately the user similarity, the set of candidate neighbors, and the size of the user neighborhoods. In particular, we show that even few but reliable neighbors lead to better recommendations than large neighborhoods where cold start users belong to.

Keywords: Recommender systems · Collaborative filtering
Cold start · Neighbor selection

1 Introduction

Recommender systems are designed to predict the most potentially relevant unknown items for a particular user considering her historical information, such as her interaction with the system in the form of clicks, likes, ratings and so on. This task is highly difficult with new users, since for them the available information is not enough to properly understand their preferences. In fact, this problem, known as the *user cold start*, has been recognized as one of the most

© Springer Nature Switzerland AG 2018
F. Herrera et al. (Eds.): CAEPIA 2018, LNAI 11160, pp. 3–12, 2018.
https://doi.org/10.1007/978-3-030-00374-6_1

challenging problems in the recommender systems research area, and there is not a unique solution that can be generically applied [5,9].

Previous work has focused on acquiring and exploiting additional information about users and items to mitigate the lack of user feedback, such as side information [1,6], item metadata [11], cross-domain data [4], elicited user preferences [7], and user personality [5]. In this work, we face the worst scenario where no side information is available to the recommender system, which has to rely only on historical information about the users.

Within the context of cold start, in [9] Kluver and Konstan presented an analysis of the behavior of different recommendation algorithms, including a user-based nearest-neighbor (kNN) method with a neighborhood size of 30. Their empirical comparison showed that kNN produces poor quality recommendations for new users, resulting the worst method in that case. Therefore, the authors proposed to better investigate this issue as future work.

In this paper, we address the problem of low recommendation quality of user-based kNN for cold users when positive-only feedback (e.g., likes, purchases, views) is available. Hence, a novelty of our study arises in the combination of such issues: cold start and positive-only feedback. Several authors have already studied nearest-neighbor algorithms [2,8], but none of them have dealt explicitly with cold start situations. Moreover, in general, they have used data consisted of numeric ratings. In many social media, however, this type of user feedback is not common, and is replaced by unary or binary ratings, such as the *likes* of Facebook, Twitter and Instagram, and the *thumbs up/down* of YouTube, or by implicit user feedback, such as product views and purchases in Amazon, and music play counts in Spotify and Last.fm. Motivated by this situation, we aim to provide some guidelines about how user-based kNN algorithms should be exploited in cold start scenarios with positive-only feedback.

More specifically, on datasets from Facebook and Last.fm, we present a comprehensive study showing that neighborhoods of cold start users may be composed of other cold start users who negatively impact on the recommendation quality. Considering this, we investigate a number of strategies to select candidate neighbors, based on how the performance changes when considering all the users, only cold start users, and only warm (i.e., not cold start) users. For each strategy, we also evaluate several user similarity metrics and neighborhood sizes.

Our empirical results indicate that a compromise should be met between large neighborhoods of cold users (to promote diversity) and small neighborhoods of warm users (to achieve high accuracy). Moreover, the results show that some user similarity metrics (like Jaccard coefficient and Cosine) are more sensitive to these experimental configurations than others, namely the Overlap and Log-Likelihood metrics. We provide an analysis on the rationale behind these effects, and propose solutions to deal with them.

2   Case Study

The main intuition behind this work is that there are cases where the number of neighbors chosen as input parameter for a kNN approach should not be the only

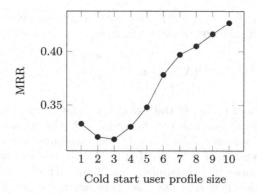

Fig. 1. User-based kNN with Jaccard and $k = 100$ (Facebook movie dataset).

variable to consider in order to produce positive variations on item relevance prediction results. We argue that a proper selection of "good quality" neighbors may also positively affect the performance of a recommender system, especially when dealing with cold users having very few (even one) ratings in their profiles.

Let us consider the case represented in Fig. 1, which shows the performance results (in terms of Mean Reciprocal Rank, MRR, that measures the inverse of the rank position of the first relevant item recommended) of a user-based kNN algorithm using the Jaccard coefficient as similarity metric, and 100 users as neighbors, on a dataset from Facebook with positive-only feedback. The performance of the algorithm decreases when we move from cold start user profiles containing 1 item to those with 2 and 3 items, whereas it increases with larger profiles. We observed this behavior in other datasets, such as those included in our experiments; the performance improves when one (music domain) or two (movie domain) additional interactions are introduced into the users' profiles after a performance drop that occurs when one interaction exists in the profiles.

We hypothesize that this effect on recommendation performance comes from a deficiency in how the used similarity metric selects the neighbors in cold start cases. For presentation purposes, we show the rationale behind our intuition by means of an analytic discussion about the Jaccard coefficient. The observations also hold for other similarity metrics used in kNN approaches.

Let u and v be two user profiles represented as vectors of positive-only feedback components in the item space, and let $|u|$ be the number of elements of u. The Jaccard coefficient (similarity) between u and v is defined as follows:

$$J(u,v) = \frac{|u \cap v|}{|u \cup v|} = \frac{|u \cap v|}{|u| + |v| - |u \cap v|} \tag{1}$$

As typical similarity metrics in the field, the Jaccard coefficient depends on the overlap $|u \cap v|$ between the two users involved.

It can be shown that for a target user, the larger a neighbor's profile, the more likely the probability of overlap X between the two users will be positive.

This comes from the fact that the number of items common to the users can be modeled as a hypergeometric distribution as:

$$P(X = k) = \frac{\binom{K}{k}\binom{N-K}{n-k}}{\binom{N}{n}} \tag{2}$$

being N the number of items, K the size of the target (cold start) user's profile, $n \leq N$ the size of the neighbor's profile, and $k \leq K$ the size of a target overlap.

For instance, if we take $k = K = 1$, then the probability that a random neighbor has an overlap of k is $\frac{n}{N}$, that is, the larger the neighbor's profile, the higher her chances of overlapping one item with the cold start user. In general, this applies when K grows, as there are more possibilities of having some (non-empty) overlap ($1 \leq k \leq K$), and those probabilities also depend on the neighbor's profile size.

One can argue that selecting larger neighbors is, in general, a good practice, since it provides more confidence on the resulting similarity, and thus allows for better recommendations; in fact, ad-hoc factors have been introduced into rating prediction to take this confidence into account when computing user similarity [8]. However, the Jaccard coefficient combines the overlap with the union of the users' profiles, hence providing a trade-off between the confidence on the similarity and the ratio of the overlap that belongs to each user; in other words, it does not promote larger neighbors anymore, but something in between, in order not to bias recommendations coming from heavy users. This can be shown by the following derivation:

$$J(u,v) = \frac{|u \cap v|}{|u| + |v| - |u \cap v|} = \frac{1}{\frac{|u|}{|u \cap v|} + \frac{|v|}{|u \cap v|} - 1} \tag{3}$$

In the context of cold start recommendation (where $|u \cap v| \leq |u|$ is very small), Eq. 3 lets us conclude that those users v with large training profile sizes, and higher overlap probabilities (Eq. 2), would generate very low values of the Jaccard similarity, whereas smaller users having some overlap with u would get higher similarity values. Actually, the optimal neighbor according to this metric is the one where $|u \cap v| = |u \cup v|$, since in that case $J(u,v) = 1$ (directly from Eq. 1). However, these neighbors do not contribute with new items for the target user in recommendation time, and thus are useless in practice.

Because of this, and in order to assess our hypothesis that the observed drop in recommendation performance comes from the way the neighbors are selected by the similarity metric, in the next section, we present three strategies specifically tailored to the cold start scenario previously introduced. Furthermore, in the following sections, we analyze the impact of these strategies using different similarity metrics with small and large neighborhoods.

Unless stated otherwise, every experiment presented in this paper follows the methodology introduced in [9] to evaluate cold start scenarios, whose main characteristic is that every user being tested has the same number of items in training; in our experiments, from 1 to 10 items.

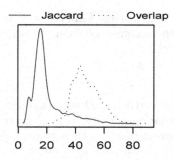

Fig. 2. Distributions of the average neighbor size using Jaccard coefficient and the overlap size (Facebook movie dataset). The former mostly selects neighbors with a very small profile, evidenced by the peak around 20 likes, whereas the latter shows a distribution with a moderate peak around 40 likes.

3   Strategies for Neighbors Selection

In the previous section, we hypothesized that a user-based kNN recommender using the Jaccard coefficient as similarity metric does not perform optimally due to the way it selects the neighbors. We noted that, although larger users have a higher probability of being selected because of their overlap, the Jaccard coefficient shows an inverse dependency on the neighbors' profile size. To further validate this hypothesis, in Fig. 2 we show the distribution of the average neighbor size obtained when using the Jaccard coefficient as similarity metric, and compare it against the one obtained when pure overlap is used.

We observe that the average neighbor size when overlap is used as a similarity metric is much larger than when Jaccard coefficient is used. This is related to the previous observation that optimal neighbors for Jaccard are those small users who will not contribute with new items for test, and hence, no improvements can be expected from using such a similarity metric.

To mitigate this effect, we propose three strategies that impose constraints on the type of users that will be used in the neighborhoods:

All. All users are equally considered to build the neighborhoods; this strategy is equivalent to the standard nearest-neighbor algorithm.

Cold. Only cold start users are used to build the neighborhoods, that is, users with a profile larger than a threshold are not considered as potential neighbors.

Warm. Cold start users are deleted from the neighborhoods, and only users with enough items (more than a given threshold) in their profiles are considered as neighbors.

The use of these neighbor selection strategies will allow us to analyze the effect that the type of neighbor (in terms of profile size) has in the final performance of the collaborative filtering method. Besides, by applying these strategies with different similarity metrics, we study whether some similarities are more sensitive to these effects than others. Finally, we study the impact that the neighborhood size has on the recommendation performance for each strategy.

4   Experiments

In this section we detail the settings and results of the experiments conducted to evaluate the recommendation quality in cold start situations of the user-based kNN method on three datasets with positive-only feedback, namely two Facebook datasets used in [5] with page *likes* on the movie and music domains, and the HetRec 2011 dataset[1] with Last.fm music listening records.

4.1   Experimental Settings

The conducted evaluation is based on a modified user-based 5-fold cross-validation methodology proposed in [9] for the cold start user scenario. First, we selected the users with at least 16 likes, shuffled and split into five (roughly) equally sized subsets. In each cross-validation stage, we kept all the likes from four of the groups in the training set, whereas the likes from the users in the fifth group were randomly split into three subsets to properly select the best configuration of the algorithms: training set (10 likes), validation set (5 likes), and testing (the remaining likes, at least 1). In order to simulate different user profile sizes from 1 to 10 likes, we repeated the training and the evaluation processes ten times, starting with the first like in the training set, and incrementally increasing it one by one feedback. This evaluation setting allowed us to evaluate each profile size with the same test set, avoiding potential biases in the evaluation, since some accuracy metrics have been proven to be sensitive to the test set size [9].

After preprocessing, the Facebook music dataset contained $49,369$ users, $5,748$ music bands and artists, and $2,084,462$ likes; the Facebook movie dataset contained $26,943$ users, $3,901$ movies, and $876,501$ likes. The Last.fm dataset contained $1,892$ users, $17,632$ artists, and $92,834$ relations between users and listened band/artist. Note that, although the Last.fm dataset includes listening counts, these are ignored and the information from the three datasets is considered as *unary feedback*, that is, only the signal that a user interacted with an item is used by the recommendation algorithms.

4.2   Results

In order to assess the ability of the strategies proposed in Sect. 3 to select good neighbors in collaborative filtering, and how this aspect affects the recommendation performance, we evaluated a user-based kNN method with a fixed value for the neighborhood size ($k = 100$), and varying its similarity metric and neighbor selection strategy.

---

[1] https://grouplens.org/datasets/hetrec-2011.

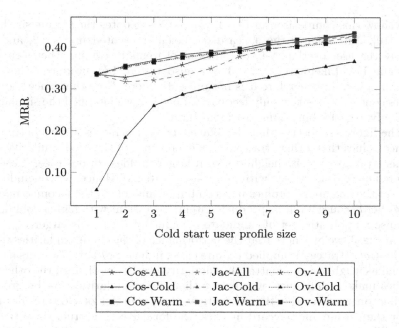

Fig. 3. MRR results in the Facebook movie dataset when $k = 100$.

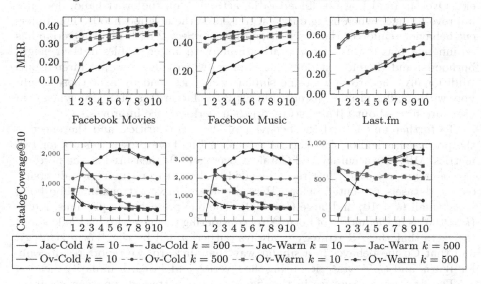

Fig. 4. Accuracy and diversity results at different neighborhood sizes (k). Cold start user profile size is represented in the X axis.

We show this comparison in Fig. 3, where **Cos** denotes the Cosine similarity metric [10], **Jac** denotes the Jaccard coefficient presented in Sect. 2, and **Ov** is simply the user overlap, i.e., the amount of items in common between both users. The Log Likelihood ratio [3] was also tested, but produced the same results than Ov, and therefore it is not presented here. Note that since the used datasets contained positive-only feedback, if needed, we modified the similarities accordingly, e.g., by binarizing the input data.

In the figure we observe that the *Warm* strategy achieves always better performance values than other strategies, independently of the used similarity metric. This strategy exploits neighbors with larger profiles (in our case, those not considered as cold-users, i.e., with more than 16 items). Hence, we conclude that the size of the neighbors' profiles plays an important role in the recommendation stage, either because it allows producing item suggestions with higher confidence, or because its potential set of candidate recommended items is larger.

From Fig. 3, we also note that the performance of the three similarities tested when the *Cold* strategy is applied is quite close, indicating that cold users are not beneficial as neighbors, no matter the similarity used. Nonetheless, the other two strategies make a difference; we observe that these similarities can be grouped depending on whether the *Warm* strategy obtains better or close results than *All*. For the Cosine and Jaccard metrics, the former case is true, since there is a clear difference between their performance, being larger for the *Warm* strategy. Overlap (and Log Likelihood ratio, actually), on the other hand, does not improve so much when using only warm users in the neighborhoods. This different behavior reinforces the observation raised in Sect. 2 with Fig. 1: Overlap has an inherent bias towards incorporating users with larger profiles in their neighborhoods, and therefore it does not improve when cold users are filtered out. Additionally, Jaccard and Cosine similarity metrics tend to favor those neighbors with smaller profiles (as noted in Eq. 3 for Jaccard), and, hence, when cold users are filtered out (*Warm* strategy) their performance improves.

To further understand the behavior of these similarities, and the effect of the proposed neighbor selection strategies, in Fig. 4 we present results for two metrics, MRR for accuracy and CatalogCoverage@10 –measured as the percentage of items a method has recommended at least to one user– for diversity, on the datasets presented before. Here, we analyze two extreme neighborhood sizes: small ($k = 10$) and large ($k = 500$). The one presented in previous figures ($k = 100$) is in the middle of both. Also, considering the observations made about the facts that the *All* strategy is always outperformed by the *Warm* strategy, and that Cosine behaves in a similar way than Jaccard, we report results only for *Warm* and *Cold* strategies, using the Overlap and Jaccard similarity metrics.

The first issue to notice in this figure is that a tradeoff between accuracy and diversity is very clear in these experiments: worst recommenders in terms of MRR achieve the highest values of CatalogCoverage@10. In terms of neighborhood sizes, the tradeoff is also present, although less obvious in principle: small neighborhoods of large neighbors (*Warm* with $k = 10$) always outperform large neighborhoods of small neighbors (*Cold* with $k = 500$) in terms of accuracy.

However, depending on the dataset and the size of the cold user, the same effect occurs in terms of diversity.

For the Facebook movies and music datasets, large neighborhoods of small neighbors are more diverse than small neighborhoods of large neighbors when the target user is very cold (less than 3 observations in her profile). For the Last.fm dataset, in contrast, the situation is the other way around: recommendations are more diverse when more interactions are available in the cold start user profile. A possible explanation for this result could be attributed to the different nature of the datasets: whereas Facebook movies and music likes are explicit one-class user preferences (someone likes a page –and this is the only allowed interaction), Last.fm listening counts represent an implicit user feedback (the user's frequency shows the interest towards that item, but it is relative to the user's whole listening history). To some extent, this different nature may also affect how a cold user is defined in each system: someone with 10 page likes in Facebook is probably better represented than someone with 10 listening records in Last.fm. In any case, it is interesting to observe that the same gap in performance between warm and cold neighbors is found in both systems.

Finally, we also note that, independently of the neighbor selection strategy, the user similarity metric, and the size of the cold start user profile, the larger the neighborhood is, the better the accuracy the recommendation algorithm achieves. Hence, if computational resources (memory and processing time to generate the recommendations) are not a problem, the simplest solution to solve the accuracy problem in cold start scenarios would be to increase the size of the computed neighborhoods.

5   Conclusions and Future Work

In this paper, we have investigated the impact of some parameters in user-based nearest-neighbor algorithms when applied to cold start users with positive-only feedback. We have compared three strategies to build user neighborhoods – considering only cold users, excluding cold users from the neighborhood, and without any constraint–, and have observed that neighborhoods based on cold users are usually worse for accuracy, but may allow for more diversified recommendations, depending on the domain. Furthermore, our experiments showed that small neighborhoods of large neighbors outperform large neighborhoods of small neighbors, which might be used as a guideline for deploying kNN algorithms in cold start scenarios with positive-only feedback.

As future work, we plan to explore the behavior of the proposed neighbor selection strategies in a general setting where the ratio of cold users is unknown. Besides, according to our analysis on how the Jaccard coefficient finds optimal neighbors, we aim to study other formulations to improve its performance by, for example, filtering those neighbors with similarities equal or very close to 1.0, since they cannot contribute with new items for the target user.

Acknowledgments. This research was supported by the Spanish Ministry of Economy, Industry and Competitiveness (TIN2016-80630-P).

References

1. Barjasteh, I., Forsati, R., Masrour, F., Esfahanian, A.H., Radha, H.: Cold-start item and user recommendation with decoupled completion and transduction. In: Proceedings of the 9th ACM Conference on Recommender Systems, RecSys 2015, pp. 91–98. ACM, New York (2015). https://doi.org/10.1145/2792838.2800196
2. Bellogín, A., Castells, P., Cantador, I.: Neighbor selection and weighting in user-based collaborative filtering: a performance prediction approach. ACM Trans. Web **8**(2), 12:1–12:30 (2014). https://doi.org/10.1145/2579993
3. Dunning, T.: Accurate methods for the statistics of surprise and coincidence. Comput. Linguist. **19**(1), 61–74 (1993)
4. Enrich, M., Braunhofer, M., Ricci, F.: Cold-start management with cross-domain collaborative filtering and tags. In: Huemer, C., Lops, P. (eds.) EC-Web 2013. LNBIP, vol. 152, pp. 101–112. Springer, Heidelberg (2013). https://doi.org/10.1007/978-3-642-39878-0_10
5. Fernández-Tobías, I., Braunhofer, M., Elahi, M., Ricci, F., Cantador, I.: Alleviating the new user problem in collaborative filtering by exploiting personality information. User Model. User-Adap. Inter. **26**(2–3), 221–255 (2016). https://doi.org/10.1007/s11257-016-9172-z
6. Gantner, Z., Drumond, L., Freudenthaler, C., Rendle, S., Schmidt-Thieme, L.: Learning attribute-to-feature mappings for cold-start recommendations. In: Proceedings of the 2010 IEEE International Conference on Data Mining, ICDM 2010, pp. 176–185. IEEE Computer Society, Washington, DC (2010). https://doi.org/10.1109/ICDM.2010.129
7. Graus, M.P., Willemsen, M.C.: Improving the user experience during cold start through choice-based preference elicitation. In: Proceedings of the 9th ACM Conference on Recommender Systems, RecSys 2015, pp. 273–276. ACM, New York (2015). https://doi.org/10.1145/2792838.2799681
8. Herlocker, J., Konstan, J.A., Riedl, J.: An empirical analysis of design choices in neighborhood-based collaborative filtering algorithms. Inf. Retrieval **5**(4), 287–310 (2002). https://doi.org/10.1023/A:1020443909834
9. Kluver, D., Konstan, J.A.: Evaluating recommender behavior for new users. In: Proceedings of the 8th ACM Conference on Recommender Systems, RecSys 2014, pp. 121–128. ACM, New York (2014). https://doi.org/10.1145/2645710.2645742
10. Ning, X., Desrosiers, C., Karypis, G.: A comprehensive survey of neighborhood-based recommendation methods. In: Ricci, F., Rokach, L., Shapira, B. (eds.) Recommender Systems Handbook, pp. 37–76. Springer, Boston, MA (2015). https://doi.org/10.1007/978-1-4899-7637-6_2
11. Tomeo, P., Fernández-Tobías, I., Cantador, I., Noia, T.D.: Addressing the cold start with positive-only feedback through semantic-based recommendations. Int. J. Uncertain. Fuzziness Knowl.-Based Syst. **25**(Suppl.-2), 57–78 (2017). https://doi.org/10.1142/S0218488517400116

Crowd Learning with Candidate Labeling:
An EM-Based Solution

Iker Beñaran-Muñoz[1]([✉]), Jerónimo Hernández-González[2], and Aritz Pérez[1]

[1] Basque Center for Applied Mathematics, Al. Mazarredo 14, Bilbao, Spain
{ibenaran,aperez}@bcamath.org
[2] University of the Basque Country UPV/EHU,
P. Manuel de Lardizabal 1, Donostia, Spain
jeronimo.hernandez@ehu.eus

Abstract. Crowdsourcing is widely used nowadays in machine learning for data labeling. Although in the traditional case annotators are asked to provide a single label for each instance, novel approaches allow annotators, in case of doubt, to choose a subset of labels as a way to extract more information from them. In both the traditional and these novel approaches, the reliability of the labelers can be modeled based on the collections of labels that they provide. In this paper, we propose an Expectation-Maximization-based method for crowdsourced data with candidate sets. Iteratively the likelihood of the parameters that model the reliability of the labelers is maximized, while the ground truth is estimated. The experimental results suggest that the proposed method performs better than the baseline aggregation schemes in terms of estimated accuracy.

Keywords: Supervised classification · Crowdsourced labels
Weak supervision · Candidate labeling
Expectation-maximization based method

1 Introduction

Nowadays, due to the ever-increasing use of the Internet, there is a huge amount of data available. Among the machine learning community, the use of crowdsourcing has become popular as a means of gathering labels at a relatively low cost. In the crowdsourcing context, **crowd labeling** is the process of getting noisy labels for the instances in the training set from a set of various non-expert **annotators (or labelers)** A. In this sense, an annotator $a \in A$ can be seen as a classifier which provides labels with a certain amount of noise. In the traditional crowdsourcing scenario, referred to as **full labeling** throughout this paper, every annotator is asked to select a single label for each instance.

Crowd learning consists of learning a classifier from a dataset with crowdsourced labels. A straightforward approach would separate this learning task into two stages: (i) label aggregation (to determine the ground truth label of each

© Springer Nature Switzerland AG 2018
F. Herrera et al. (Eds.): CAEPIA 2018, LNAI 11160, pp. 13–23, 2018.
https://doi.org/10.1007/978-3-030-00374-6_2

instance of the training set) and (ii) learning (to learn a model using the aggregated labels and standard supervised classification techniques). In this paper, we assume this approach for crowd learning and focus on its first stage.

Probably the most popular aggregation technique is majority voting (MV), which labels every instance with the label that most annotators have selected. In weighted voting [11], the vote of each annotator is weighted according to their reliability. Many aggregation methods that also model the reliability of annotators were derived from the expectation-maximization (EM) strategy [6], which was first applied to crowdsourcing by Dawid and Skene [5], and has been widely used since then [4,12,14,17,18,20]. An extensive review of different label aggregation and crowd learning techniques can be found in [19].

In crowd learning, annotators are usually assumed to be non-experts. In this scenario, it may seem reasonable to consider a more relaxed request than to force them to provide a single label. Some approaches are more flexible with labelers by allowing annotators to (i) express how sure they are about the provided labels [9,15], or (ii) state that they do not know the answer [7,16,21]. Recently, **candidate labeling** [2], inspired by weak supervision [10], allows annotators to select a subset of labels, called the **candidate set**, instead of just one. It has been shown that this type of labeling can extract more information from labelers than full labeling, especially with few annotators or difficult instances. It could also lead to faster and/or less costly labeling [1].

In social sciences, a similar problem has been extensively studied under the name of **approval voting** (AV) [3,8,13]. Without ground truth, the objective is to identify popular (approved) options. When a single option needs to be selected, aggregation is usually carried out as follows (using machine learning terminology): Given an instance x, the label included in most candidate sets is chosen. This approach is used as a baseline for comparison in this paper.

In this paper, an EM-based technique is proposed to aggregate the candidate labels of crowdsourcedly annotated examples. Firstly, the candidate labeling framework is set. In Sect. 3, annotator modeling is explained, and maximum likelihood estimates for the parameters are provided. Next, the proposal is presented and its performance is discussed in Sect. 5 based on experiments with synthesized data. Finally, conclusions are drawn and future work is discussed.

2 Candidate Labeling

This work deals with multi-class classification problems, where each instance of the training set belongs to one of $r > 2$ possible classes. Two types of random variables are considered: (i) the features X, that take values in the space Ω_X, and (ii) the class variable C, which takes r distinct values in the space $\Omega_C = \{1, \ldots, r\}$. We assume that the multidimensional random variable (X, C) is distributed according to an (unknown) probability distribution $p(X, C)$. In this work it is assumed that there is a single **ground truth** label for each unlabeled instance x, denoted by c_x. A crowd learning problem with candidate labeling [2] is also considered. In this scenario, for each instance x in the training set, each annotator $a \in A$ provides a candidate set, denoted by $L_x^a \subseteq \Omega_C$.

We focus on estimating the ground truth for an unlabeled dataset $\mathcal{D}$ given the candidate sets provided by the labelers. The goal is to maximize the **estimated accuracy** (or, for the sake of brevity, accuracy) function:

$$acc(\phi) = \frac{1}{|\mathcal{D}|} \sum_{x \in \mathcal{D}} \mathbb{1}(c_x = \phi(x)) \tag{1}$$

where ϕ is a classifier, i.e., a function that maps Ω_X into Ω_C, and $\mathbb{1}(b)$ is a function which returns 1 if the condition b is *true* and 0 otherwise.

For estimating the ground truth using data with candidate sets, the candidate voting (CV) function [2] is defined, given an instance x and the set of candidate sets $\mathcal{L}_x = \{L_x^a\}_{a \in A}$ gathered for it, as follows,

$$\omega(\mathcal{L}_x) = \underset{c}{\operatorname{argmax}}\, w_x(c), \tag{2}$$

where

$$w_x(c) = \frac{1}{|A|} \sum_{a \in A} \frac{\mathbb{1}(c \in L_x^a)}{|L_x^a|} \tag{3}$$

is the **candidate voting estimate**. CV can be understood as a weighted voting function where the weights depend on the sizes of the provided candidate sets. It works as a generalization of the MV strategy from the full labeling to the candidate labeling context. It can be easily observed that CV behaves as MV when annotations are obtained by means of full labeling.

As the trustworthiness of the labelers is not homogeneous, having information about their reliability can be of great advantage to aggregate the labels that they provide. In the next section, a model of the annotators based on their reliability is proposed and the maximum likelihood estimates of its parameters are obtained.

3   Modeling Annotators and Maximum Likelihood Estimate

In order to aggregate the candidate sets gathered through candidate labeling, the contribution of each labeler can be weighted according to their reliability. In this section, a model for the behavior of annotators is described, with parameters that control their reliability and the way the candidate sets are generated. Then, the maximum likelihood estimates are inferred for those parameters. Finally, a procedure for estimating the labels is presented.

In the presented framework, the candidate set L_x^a is assumed to be generated by asking annotator a one question of the kind "Do you consider that the given instance x might belong to class c?" for each $c \in \Omega_C$. Let α_c^a denote the probability that annotator a includes label c in the candidate set for instances which really belong to class c. Let us also define β_c^a as the probability that annotator a includes any label $c' \neq c$ ($c' \in \Omega_C$) in the candidate set when annotating instances which really belong to class c. Note that we assume that, given an instance of a certain class, the rest of class labels have the same probability of being mistakenly selected. The parameters α_c^a and β_c^a provide us insights into the behavior of annotator a when labeling instances that really belong to class c.

Assuming the process of generation of the candidate sets described above, the likelihood of the parameters given a candidate set L_x^a is:

$$
\begin{aligned}
Pr(L_x^a|\boldsymbol{\alpha}, \boldsymbol{\beta}, c_x) = (\alpha_{c_x}^a)^{\mathbb{1}(c_x \in L_x^a)} \cdot (1 - \alpha_{c_x}^a)^{1-\mathbb{1}(c_x \in L_x^a)} \\
\cdot (\beta_{c_x}^a)^{|L_x^a|-\mathbb{1}(c_x \in L_x^a)} \cdot (1 - \beta_{c_x}^a)^{r-|L_x^a|+\mathbb{1}(c_x \in L_x^a)},
\end{aligned}
\tag{4}
$$

where the set of probabilities for annotators of selecting the (unknown) correct label c_x is $\boldsymbol{\alpha} = \{\alpha_c^a\}_{a \in A, c \in \Omega_C}$ and the set of probabilities for annotators of selecting each incorrect label is $\boldsymbol{\beta} = \{\beta_c^a\}_{a \in A, c \in \Omega_C}$.

Assuming that annotators provide the candidate sets independently and that all instances are i.i.d. according to $p(X, C)$, the likelihood given a dataset $\mathcal{D}$ where each instance is annotated with a set of candidate sets is:

$$
Pr(\{\mathcal{L}_x\}_{x \in \mathcal{D}}|\boldsymbol{\alpha}, \boldsymbol{\beta}) = \prod_{x \in \mathcal{D}} \prod_{a \in A} Pr(L_x^a|\boldsymbol{\alpha}, \boldsymbol{\beta}, c_x)
\tag{5}
$$

From this expression, the maximum likelihood estimates of both alpha and the beta parameters are:

$$
\hat{\alpha}_c^a = \frac{\sum_{x \in \mathcal{D}} \mathbb{1}(c_x = c)\mathbb{1}(c \in L_x^a)}{\sum_{x \in \mathcal{D}} \mathbb{1}(c_x = c)}
\tag{6}
$$

$$
\hat{\beta}_c^a = \frac{\sum_{x \in \mathcal{D}} \mathbb{1}(c_x = c) \cdot (|L_x^a| - \mathbb{1}(c \in L_x^a))}{r \cdot \sum_{x \in \mathcal{D}} \mathbb{1}(c_x = c)}
\tag{7}
$$

The estimate $\hat{\alpha}_c^a$, given by maximum likelihood, is the number of instances of class c for which annotator a included class label c in the candidate set over the total number of instances of class c. On the other hand, the estimate $\hat{\beta}_c^a$, given by maximum likelihood, is the number of mistaken class labels that annotator a included in the candidate sets of all the instances of class c over the whole set of possible class labels for the total number of instances of class c.

The estimates in Eqs. (6) and (7) can be computed when the true class labels are known for all instances. Conversely, if the true labels are not known, they can be estimated by means of the $\boldsymbol{\alpha}$ and $\boldsymbol{\beta}$ parameters. Using Bayes' Theorem, it follows that:

$$
Pr(c|\mathcal{L}_x, \boldsymbol{\alpha}, \boldsymbol{\beta}) \propto Pr(c) \cdot Pr(\mathcal{L}_x|\boldsymbol{\alpha}, \boldsymbol{\beta}, c)
\tag{8}
$$

Using Eq. (4) for the case that $c_x = c$ and estimating the marginal probability as $Pr(c) = \frac{\sum_{x \in \mathcal{D}} \mathbb{1}(c_x = c)}{|\mathcal{D}|}$, Eq. (8) can be rewritten as:

$$
\begin{aligned}
Pr(c|\mathcal{L}_x, \boldsymbol{\alpha}, \boldsymbol{\beta}) \propto \frac{\sum_{x \in \mathcal{D}} \mathbb{1}(c_x = c)}{|\mathcal{D}|} \cdot \prod_{a \in A} \left((\alpha_c^a)^{\mathbb{1}(c \in L_x^a)} \cdot (1 - \alpha_c^a)^{1-\mathbb{1}(c \in L_x^a)} \right. \\
\left. \cdot (\beta_c^a)^{|L|-\mathbb{1}(c \in L_x^a)} \cdot (1 - \beta_c^a)^{r-|L|+\mathbb{1}(c \in L_x^a)} \right)
\end{aligned}
\tag{9}
$$

In this way, the probability that a given instance x belongs to each possible class label can be computed by means of the parameters $\boldsymbol{\alpha}$ and $\boldsymbol{\beta}$, and the candidate sets. This probability distribution could be considered as an estimate for the ground truth. In practice, neither the true labels nor the values of the parameters $\boldsymbol{\alpha}$ and $\boldsymbol{\beta}$ are known. A method based on the EM strategy [6] that estimates all of them jointly is proposed.

4   EM-Based Method for Candidate Labeling Aggregation

The EM strategy attempts to gather maximum likelihood estimates when there is missing data. Two steps are iterated: (i) Expectation (E-step), where the expected values of the missing data are computed using the current parameter estimates and (ii) Maximization (M-step), where the parameters are updated with the new maximum likelihood estimates given the current expected data. This method is guaranteed to converge to a local maximum.

In the crowdsourcing context, the true class labels of the training instances are the missing data. Our method is based on the Dawid-Skene approach [5], which is implemented as follows: First, an initial estimate of the ground truth labels is obtained. After that, the method consists of two steps: (i) **M-step**: The parameters that model the reliability of the annotators are updated with estimates that maximize or, at least, improve the likelihood achieved in the previous E-step; and (ii) **E-step**: Given an estimate of the parameters, the expected values of the ground truth labels are obtained for every instance, given the expected labels. The M and E steps are carried out iteratively until convergence.

Our proposal is an adaptation of this strategy to the candidate labeling scenario. Firstly, let us define $q(c|x)$ as the estimate of the probability $Pr(c|\mathcal{L}_x, \boldsymbol{\alpha}, \boldsymbol{\beta})$ described in Eq. (8), that is, the probability that x belongs to class c. In Eqs. (4), (6) and (7), the $q(c|x)$ estimates can substitute the expression $\mathbb{1}(c_x = c)$, switching from two discrete values (0 or 1) to any possible value in the continuous interval $[0, 1]$. Note that the true label c_x is unknown and this modification allows this approach to work with the probabilistic estimates of the ground truth. Our method works in the following way:

After a first step where the estimates $q(c|x)$ are initialized for all $x \in \mathcal{D}$ and $c \in \Omega_C$, the M and E steps of the proposed method are as follows:

- **M-step.** For every $a \in A$ and $c \in \Omega_C$, the estimates $\hat{\alpha}_c^a$ and $\hat{\beta}_c^a$ are computed given q by means of Eqs. (6) and (7), using the estimates $q(c|x)$ instead of $\mathbb{1}(c_x = c)$.
- **E-step.** For every $x \in D$ and $c \in \Omega_C$, Eq. (9) is used to compute the probability distributions $q(c|x)$ given $\hat{\alpha}_c^a$ and $\hat{\beta}_c^a$. As in the E-step, the terms $\mathbb{1}(c_x = c)$ are substituted by the previous estimates $q(c|x)$.

In the next section, the performance of the previously described method is tested using artificial data.

5   Experiments

In this section, the performance of the presented method is evaluated in different scenarios. In order to have insights into its performance: (i) the accuracy is computed for different scenarios, varying the numbers of annotators, classes, and instances, and the values of the α and β parameters, (ii) the method is compared with candidate voting [2], approval voting [3] and the **privileged aggregation** (where all α and β parameters are known), and (iii) the evolution is observed through each iteration of the method.

5.1   Experimental Setting

To the best of our knowledge, there is not any publicly available dataset for crowd learning with candidate labeling. Thus, artificial data has been used as a means of obtaining experimental results. Simulated data is also useful to control the settings and explore different scenarios.

In order to generate different situations, the following experimental parameters are set to different values: number of instances (n), number of annotators (m), number of classes (r), minimum and maximum values of the α parameters ($\underline{\alpha}$ and $\overline{\alpha}$) and minimum and maximum β parameters ($\underline{\beta}$ and $\overline{\beta}$). The parameters $\underline{\alpha}$ and $\overline{\beta}$ have both been fixed to 0.5, so that there always can be annotators of minimum expertise and adversarial annotators are not generated.

The method itself has two additional parameters:

- The convergence threshold δ. If $\frac{|\overline{\alpha}_{(it)}-\overline{\alpha}_{(it-1)}|}{\overline{\alpha}_{(it-1)}} < \delta$ or $\frac{|\overline{\beta}_{(it)}-\overline{\beta}_{(it-1)}|}{\overline{\beta}_{(it-1)}} < \delta$, where $\overline{\alpha}_{(it)}$ ($\overline{\beta}_{(it)}$) is the mean value of $\alpha_{(it)}$ ($\beta_{(it)}$) at iteration it, it is considered that the EM has converged. It has been set to $\delta = 0.05$.
- The smoothing parameter γ. There are two factors that lead to undesirable results, such as the likelihood equal to 0: (i) There is a large number of parameters to be estimated ($2 \cdot m \cdot r$) and there is not always sufficient information, and (ii) sometimes, the parameter estimates can get close to 0 or to 1, leading to error. An additive smoothing is used for the $\hat{\alpha}_c^a$ estimates:

$$\hat{\alpha}_c^a = \frac{\gamma + \sum_{x \in \mathcal{D}} \mathbb{1}(c_x = c)\mathbb{1}(c \in L_x^a)}{2 \cdot \gamma + \sum_{x \in \mathcal{D}} \mathbb{1}(c_x = c)} \tag{10}$$

In this way, all possible values are reached at least once, that is, there is at least one instance of class c such that $c \in L_x^a$ and another instance of class c such that $c \in L_x^a$. In these experiments, Eq. (10) is used instead of Eq. (6) with $\gamma = 1$.

Datasets are simulated as follows: The ground truth class labels are distributed uniformly among all instances, that is, there are $\frac{n}{r}$ instances belonging to each class. Next, the α and β parameters are generated. In order to have annotators with different types of knowledge, a maximum ($\overline{\alpha}$) value of α and a minimum value for β ($\underline{\beta}$) are set. All the parameters are sampled uniformly from the intervals $[0.5, \overline{\alpha}]$ and $[\underline{\beta}, 0.5]$. By means of the α and β parameters, candidate sets are generated following the interpretation explained at the beginning of Sect. 3. That is, given an instance that belongs to class c, annotator a includes class c in the candidate set with probability α_c^a and each of the classes $c' \neq c$ with probability β_c^a.

Once the candidate sets are generated, 4 different schemes are used to aggregate them: (i) our EM-based method, (ii) CV (Eq. (2)), (iii) AV and (iv) privileged aggregation (PA). The PA is obtained by computing the estimate from Eq. (9), using the original parameters and the ground truth class labels.

As mentioned above, EM is ensured to converge to a local maximum, so various initializations should be carried out to achieve desirable results. In order to obtain different initializations from the same candidate sets, we initialize the estimates $q(c|x)$ for each instance x in the following way: First, the candidate voting estimates $w_x(c)$ (Eq. (3)) are computed for all $c \in \Omega_C$, using an additive smoothing of $\frac{1}{r}$ for each one. The $q(\cdot|x)$ are normalized so that $0 \leq q(c|x) \leq 1$ and $\sum_{c \in \Omega_C} q(c|x) = 1$. Next, to initialize $q(\cdot|x)$, a Dirichlet distribution with hyperparameters $r \cdot w_x(c_1), \ldots, r \cdot w_x(c_r)$ is sampled: $q(\cdot|x) \sim Dir(r \cdot w_x(c_1), \ldots, r \cdot w_x(c_r))$.

30 initializations are carried out and the values of the final $q(c|x)$ estimates that maximize the likelihood are used to infer the labels: each instance x takes the class label c that maximizes $q(c|x)$. The process is repeated 30 times and the expected accuracy is approximated by computing the mean of the obtained accuracy estimates.

5.2   Experimental Results

Experiments with artificial data have been performed, varying a number of parameters to compare our method and previous approaches in different scenarios.

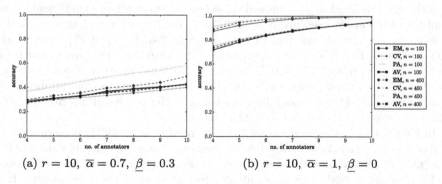

(a) $r = 10$, $\overline{\alpha} = 0.7$, $\underline{\beta} = 0.3$ (b) $r = 10$, $\overline{\alpha} = 1$, $\underline{\beta} = 0$

Fig. 1. Graphical description of the accuracy obtained by annotations simulated with different numbers of annotators.

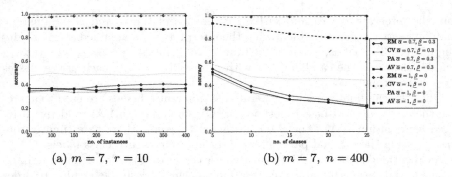

Fig. 2. Graphical description of the accuracy obtained by annotations simulated with different numbers of instances and classes.

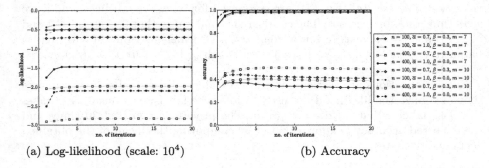

Fig. 3. Graphical description of the log-likelihood and the accuracy obtained throughout different iterations, with $r = 10$.

Except for the graphics where their evolution is examined, standard values have been chosen for the parameters. The number of annotators varies from 4 to 10, although it is fixed to its standard value ($m = 7$) in different experiments. The numbers of instances used are $n = \{100, 400\}$. In the case $n = 100$, $r = \{5, 10\}$ class labels are considered, and in the case $n = 400$, $r = \{10, 20\}$ class labels are considered. Regarding the expertise of annotators, two scenarios have been studied: (i) $\underline{\beta} = 0.3$ and $\overline{\alpha} = 0.7$, where the average expertise is low, and (ii) $\underline{\beta} = 0$ and $\overline{\alpha} = 1$, where the expertise of the annotators ranges from minimum to maximum values. Due to space limitations, only the results of a representative subset of experiments are shown in this paper.

In Figs. 1 and 2, the accuracy of the presented method (EM) is compared to that of the CV, the AV and the PA, in scenarios where the number of annotators (m, Fig. 1), the number of instances (n, Fig. 2a) and the number of classes (r, Fig. 2b) are varied. The experimental results suggest that, in general, EM outperforms CV and AV in terms of the accuracy (Eq. (1)). The accuracies are similar only in the case where the average expertise is low and the number of classes is high with respect to the number of instances (see Fig. 2 with $\underline{\beta} = 0.3$

and $\overline{\alpha} = 0.7$). Moreover, in the case that $\beta = 0$ and $\overline{\alpha} = 1$ (Fig. 1b), the proposed method reaches the accuracy of the PA. In other words, in the presence of annotators that are experts in a subset of classes, our EM-based strategy can reach the highest possible accuracy. Note as well that the accuracy of the EM approach decreases at a smoother pace than that of CV or AV as the number of annotators is reduced.

As can be seen in Fig. 2a, the number of instances (n) does not seem to affect the differences between the accuracies of the different methods, when it ranges between 100 and 400 (Fig. 2a). On the other hand, the number of classes (r) has a negative effect on the accuracy of all the methods (Fig. 2b). The only exception is that when the expertise of the annotators ranges from minimum to maximum values (Fig. 2b, $\overline{\alpha} = 1$, $\beta = 0$), our EM approach outperforms the baselines.

The evolution of the log-likelihood and the accuracy in each iteration of the EM can be seen in Fig. 3. In Fig. 3b, the accuracy in iteration number 0 is the one reached using the initial q estimates. As could be expected, generally, the log-likelihood increases monotonically and remains stable after some point (Fig. 3a). The accuracy increases in the first iterations as well, and then remains stable in most cases (Fig. 3b), but decreases in one case ($n = 100$, $\overline{\alpha} = 0.7$, $\beta = 0.3$). This decline may be due to overfitting, since scarce data (each annotator labels 100 instances) is used to estimate many parameters (20 per annotator).

To sum up, according to the experiments, EM seems to outperform CV and AV in most scenarios, especially when the expertise of the annotators is varied. In favorable settings, EM can reach a high accuracy - as if the real α and β parameters were known (PA).

6   Conclusions and Future Work

In this work, a crowd learning problem is approached with candidate labeling. A model for the reliability of the annotators is proposed. An EM-based method is presented, as an extension to the traditional methods for aggregating crowdsourced labels into the candidate labeling scenario. Experimental results obtained with artificial data suggest that the presented method has an enhanced performance, in terms of estimated accuracy, compared with the baseline methods. Particularly, it stands out when few annotators are available and when they show different levels of expertise.

For future work, more realistic data could be used, not following the assumption that, given an instance of a certain class, the rest of labels can be selected with the same probability. Also, a real-world dataset with candidate sets could be gathered in order to test the presented method, as well as other aggregation and learning schemes. The presented method could also be refined by reducing the number of parameters or, similar to [14], learning a classification model from data as the crowd-modeling parameters are estimated.

Acknowledgments. IBM and AP are both supported by the Spanish Ministry MINECO through BCAM Severo Ochoa excellence accreditation SEV-2013-0323 and the project TIN2017-82626-R funded by (AEI/FEDER, UE). IBM is also supported by the grant BES-2016-078095. AP is also supported by the Basque Government through the BERC 2014-2017 and the ELKARTEK programs, and by the MINECO through BCAM Severo Ochoa excellence accreditation SVP-2014-068574. JHG is supported by the Basque Government (IT609-13, Elkartek BID3A) and the MINECO (TIN2016-78365-R).

References

1. Banerjee, S.O.A., Gurari, D.: Let's agree to disagree: a meta-analysis of disagreement among crowdworkers during visual question answering. In: GroupSight Workshop at AAAI HCOMP, Quebec City, Canada (2017)
2. Beñaran-Muñoz, I., Hernández-González, J., Pérez, A.: Weak Labeling for Crowd Learning. arXiv e-prints (2018)
3. Brams, S.J., Fishburn, P.C.: Approval voting. Am. Polit. Sci. Rev. **72**(3), 831–847 (1978)
4. Côme, E., Oukhellou, L., Denoeux, T., Aknin, P.: Learning from partially supervised data using mixture models and belief functions. Pattern Recognit. **42**(3), 334–348 (2009)
5. Dawid, A.P., Skene, A.M.: Maximum likelihood estimation of observer error-rates using the EM algorithm. J. Roy. Stat. Soc. Ser. C **28**(1), 20–28 (1979)
6. Dempster, A.P., Laird, N.M., Rubin, D.B.: Maximum likelihood from incomplete data via the EM algorithm. J. Roy. Stat. Soc. Ser. B **39**(1), 1–38 (1977)
7. Ding, Y.X., Zhou, Z.H.: Crowdsourcing with unsure option. Mach. Learn. **107**(4), 749–766 (2018)
8. Falmagne, J.C., Regenwetter, M.: A random utility model for approval voting. J. Math. Psychol. **40**(2), 152–159 (1996)
9. Grady, C., Lease, M.: Crowdsourcing document relevance assessment with mechanical turk. In: NAACL HLT 2010 Workshop, pp. 172–179 (2010)
10. Hernández-González, J., Inza, I., Lozano, J.A.: Weak supervision and other non-standard classification problems: a taxonomy. Pattern Rec. Lett. **69**, 49–55 (2016)
11. Karger, D.R., Oh, S., Shah, D.: Iterative learning for reliable crowdsourcing systems. In: NIPS, pp. 1953–1961 (2011)
12. López-Cruz, P.L., Bielza, C., Larrañaga, P.: Learning conditional linear gaussian classifiers with probabilistic class labels. In: Bielza, C., et al. (eds.) CAEPIA 2013. LNCS (LNAI), vol. 8109, pp. 139–148. Springer, Heidelberg (2013). https://doi.org/10.1007/978-3-642-40643-0_15
13. Procaccia, A.D., Shah, N.: Is approval voting optimal given approval votes? In: NIPS, pp. 1801–1809 (2015)
14. Raykar, V.C., et al.: Learning from crowds. J. Mach. Learn. Res. **11**, 1297–1322 (2010)
15. Smyth, P., Fayyad, U.M., Burl, M.C., Perona, P., Baldi, P.: Inferring ground truth from subjective labelling of venus images. In: Proceedings of NIPS 7, pp. 1085–1092 (1994)
16. Venanzi, M., Guiver, J., Kohli, P., Jennings, N.R.: Time-sensitive bayesian information aggregation for crowdsourcing systems. J. Artif. Intell. Res. **56**, 517–545 (2016)

17. Welinder, P., Branson, S., Belongie, S., Perona, P.: The multidimensional wisdom of crowds. In: Proceedings of NIPS 23, pp. 2424–2432 (2010)
18. Whitehill, J., Ruvolo, P., Wu, T., Bergsma, J., Movellan, J.R.: Whose vote should count more: optimal integration of labels from labelers of unknown expertise. In: Proceedings of NIPS 22, pp. 2035–2043 (2009)
19. Zhang, J., Sheng, V.S., Wu, J., Wu, X.: Multi-class ground truth inference in crowdsourcing with clustering. IEEE Trans. Knowl. Data Eng. **28**(4), 1080–1085 (2016)
20. Zhang, Y., Chen, X., Zhou, D., Jordan, M.I.: Spectral methods meet EM: a provably optimal algorithm for crowdsourcing. In: Advances in Neural Information Processing Systems, pp. 1260–1268 (2014)
21. Zhong, J., Tang, K., Zhou, Z.H.: Active learning from crowds with unsure option. In: Proceedings of 24th IJCAI, pp. 1061–1068 (2015)

Comparing Deep Recurrent Networks Based on the MAE Random Sampling, a First Approach

Andrés Camero[✉], Jamal Toutouh, and Enrique Alba

Departamento de Lenguajes y Ciencias de la Computación,
Universidad de Málaga, Málaga, Spain
andrescamero@uma.es, {jamal,eat}@lcc.uma.es

Abstract. Recurrent neural networks have demonstrated to be good at tackling prediction problems, however due to their high sensitivity to hyper-parameter configuration, finding an appropriate network is a tough task. Automatic hyper-parameter optimization methods have emerged to find the most suitable configuration to a given problem, but these methods are not generally adopted because of their high computational cost. Therefore, in this study we extend the MAE random sampling, a low-cost method to compare single-hidden layer architectures, to multiple-hidden-layer ones. We validate empirically our proposal and show that it is possible to predict and compare the expected performance of an hyper-parameter configuration in a low-cost way.

Keywords: Deep learning · Recurrent neural network
MAE random sampling

1 Introduction

In recent years, Machine Learning (ML) approaches have gained significant interest as a way of building powerful applications by learning directly from examples, data, and experience. Increasing data availability and computer processing power have allowed ML systems to be efficiently trained on a large pool of examples. Deep learning (DL) is a specific branch of ML that focuses on learning features from data through multiple layers of abstraction by applying deep architectures, i.e., Deep Neural Networks (DNNs) [15]. DL has improved dramatically state-of-the-art in many pattern recognition and prediction applications [17,18].

Deep feedforward networks have incorporated feedback connections between layers and neurons in order to capture long-term dependency in the input. These special case of DNNs are known as Recurrent Neural Networks (RNNs). Thus, RNNs have successfully been applied to address problems that involve sequential modeling and prediction, such as natural language, image, and speech recognition and modeling [15]. However, RNNs present a limitation on their learning process due to two main issues: the *vanishing* and the *exploding* gradient [3,20].

© Springer Nature Switzerland AG 2018
F. Herrera et al. (Eds.): CAEPIA 2018, LNAI 11160, pp. 24–33, 2018.
https://doi.org/10.1007/978-3-030-00374-6_3

A promising alternative to mitigate the problems related to the learning process in DNNs is to select/optimize the hyper-parameters of a network. By selecting an appropriate configuration of the parameters of the DNN (e.g. the activation functions, the number of hidden layers, the kernel size of a layer, etc.), the network is adapted to the problem and by this mean the performance is improved [4,6,12]. DNN hyper-parameter optimization methods can be grouped into two main groups: the manual exploration-based approaches, usually lead by expert knowledge, and the automatic search-based methods (e.g., grid, evolutionary or random search) [19].

The hyper-parameter optimization of DNN implies dealing with a high-dimensional search space. However, most methods (manual and automatic) are based on *trial-and-error*, i.e., each hyper-parameter configuration is trained and tested to evaluate its numerical accuracy. Thus, the high-dimensional search space and the high cost of the evaluation limit the results of this methodology.

Some authors have explored different approaches to speed up the evaluation of DNN architectures in order to improve the efficiency of the automatic hyper-parameter optimization algorithms [8,9].

A promising early approach to evaluate one-hidden-layer stacked RNN architectures is the *MAE (mean absolute error) random sampling* [8]. The main idea behind this method, inspired by the linear time-invariant theory (LTI), is to infer the numerical accuracy of a given network without actually training it. Given an input, they generate sets of random weights and analyze the output in terms of the MAE. Then, they estimate the probability of finding a set of weights whose MAE is below a predefined threshold.

In this study, we propose to extend the *MAE random sampling* to multiple-hidden-layer networks and to study the suitability of this method to deal with deeper RNNs. Particularly, we implemented our proposal and empirically tested it using stacked RNNs with up to three hidden layers. The results show that as the RNN gets deeper (more complex), the proposal is capable of given even better results. The reminder of this paper is organized as follows: the next section outlines the related work. Section 3 introduces the multiple-hidden-layer extension of the MAE random sampling. Section 4 presents the results, and finally, Sect. 5 discusses the conclusions drawn from this study and presents the future work.

2 Related Work

An RNN incorporates recurrent (or feedback) edges that may form cycles and self connections. This approach introduces the notion of time to the model. Thus, at a time t, a node connected to a recurrent edge receives input from the current data point x^t and also from the hidden node h^{t-1} (the previous state of the network). The output y^t at each time t is computed according to the hidden node values at time t (h^t). Input at time $t-1$ (x^{t-1}) can determine the output at time t (y^t) and later by way of recurrent connections [16].

Most of DL approaches to train a network are based on gradient-based optimization procedures, e.g., using a local numerical optimization such as stochastic gradient descent or second order methods. However, these methods are not suitable for RNNs. This is mainly because they keep a vector of activations, which makes RNNs extremely deep and aggravates the exploding and the vanishing gradient problems [3,14,20].

More recently, Long Short-Term Memory (LSTM) have emerged as a specific type of RNN architectures, which contain special units called memory blocks in the recurrent hidden layer [11]. LSTM mitigates gradient problems, and therefore, they are easier to train than standard RNNs. Not just the network architecture affects the learning process. The weight initialization procedure determines the learning rate, the convergence rate, and the probability of classification error. Ramos et al. [21] analyzed different weight initialization strategies and provided a quantitative measure for each one of them.

A promising research line in DL proposes to define a specific hyper-parameter configuration for a neural network to improve its numerical accuracy, instead of using a generalized one [4,6,12]. The idea is to select the most suitable number of layers, number of hidden unit per layer, activation function, kernel size of a layer, etc. for a given dataset.

When dealing with hyper-parameter configuration, human experts are able to discard hyper-parameterizations without requiring their evaluation by using their expertise. However, intelligent automatic hyper-parameter configuration procedures searches more efficiently through the high-dimensional hyper-parameter space. Even though the intelligent methods are more competitive than the humans, they are not generally adopted because they require high computational resources. This is mainly because they require fitting a model and evaluating its performance on validation data (i.e., they are data driven), which can be an expensive process [2,4,22].

Therefore, few methods have been proposed to address this issue by speeding up the evaluation of the proposed hyper-parameterization. Domhan et al. [9] analyzed an approach that detects and finishes the neural networks evaluations that under-perform a previously computed one. This solution was able to reduce the hyper-parameterization search time up to 50%. More recently, Camero et al. [8] presented the MAE random sampling, a novel low-cost method to compare one-hidden layer RNN architectures without training them. MAE random sampling evaluates RNN architecture by generating a set of random weights and evaluating their performance.

In line with the latter approach, we propose to extend MAE random sampling to evaluate RNNs with multiple-hidden-layers. Therefore, it will be suitable to evaluate deeper RNNs.

3   Proposal

We start our discussion with an inspiring fact: changing the weights of a neural network affects its output [10]. In spite of the simplicity (and even triviality) of this fact, it might hide some important clues to characterize the behavior of a net. Camero et al. [8] introduced a novel approach to compare RNN architectures based on this fact: the *MAE random sampling*. They showed its usefulness for comparing the expected performance (in terms of the error) of RNNs with a single-hidden-layer.

In this study we propose to extend the MAE random sampling [8] to multiple-hidden-layers, aiming to validate its usefulness for comparing deeper RNNs. Given an input time series, the idea is to take an arbitrary number of samples of the output of a specific RNN architecture, whose weights are normally initialized independently every time a sample is taken. Then, we propose to fit a truncated normal distribution to the MAE values sampled and estimate the probability p_t of finding a set of weights whose error is below an arbitrary *threshold*. Finally, we propose to use p_t as a predictor of the performance (in terms of the error) of the analyzed architecture.

Figure 1 depicts the MAE random sampling originally introduced by Camero et al. [8] extended to multiple-hidden-layers RNNs. The distribution of the sampled errors is used to estimate the probability of finding a *good solution*.

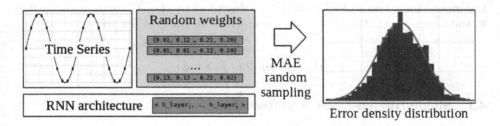

Fig. 1. MAE error sampling based on a random weight initialization

Algorithm 1 presents the adaptation of the MAE random sampling [8] to multiple hidden layers. Given an architecture ($ARCH$), encoded as a vector whose terms represents the number of LSTM cells in the correspondent hidden layer, a number of time steps or look back (LB), and a user defined time series (data), the algorithm takes $MAX\_SAMPLES$ samples of the MAE by initializing the weights with a normal distribution. After the sampling is done, a truncated normal distribution is fitted to the MAE values sampled, and finally p_t is estimated for the inputed $THRESHOLD$.

Algorithm 1. MAE random sampling of a given architecture.

1: *data* ← LoadData()
2: *rnn* ← InitializeRNN(*ARCH, LB*)
3: *mae* ← ∅
4: **while** sample ≤ *MAX_SAMPLES* **do**
5:     *weights* ← GenerateNormalWeights(0,1)
6:     UpdateWeights(*rnn, weights*)
7:     *mae*[sample] ← MAE(*rnn, data*)
8:     sample++
9: **end while**
10: *mean, sd* ← FitTruncatedNormal(*mae*)
11: p_t ← PTruncatedNormal(*mean, sd, THRESHOLD*)

---

4   Results

We implemented our proposal[1] in Python 3, using **dlopt** [7], **keras** (version 2.1) and **tensorflow** (version 1.3) [1]. Then, we tested our proposal using a standard problem: the sine wave. We selected this problem because of two main reasons: Camero et al. [8] also studied it, thus we have a baseline to compare to, and any periodic waveform can be approximated by adding sine waves [5].

Particularly, a sine wave can be noted as a function of time (t), where A is the peak amplitude, f is the frequency, and ϕ is the phase (Eq. 1). We defined the input of our test to be the sine wave described by $A = 1$, $f = 1$, and $\phi = 0$, in the range $t \in [0, 100]$ seconds (s), sampled at 10 samples per second.

$$y(t) = A \cdot sin(2\pi \cdot f \cdot t + \phi) \tag{1}$$

4.1   Single-Hidden-Layer Architectures

To begin with our experimentation we studied the MAE random sampling performance prediction in the set of RNNs with 1 to 100 LSTM cells in the hidden layer and with a look back ranging from 1 to 30. For each architecture we took 100 samples ($MAX\_SAMPLES$) and estimated $p_{0.01}$ (i.e., $THRESHOLD$=0.01). Figure 2 shows the relation between the number of hidden cells and $p_{0.01}$, each color represent a different look back. The probability rapidly increases from 1 to 25 cells, from that point $p_{0.01}$ tends to *converge*.

We selected 100 architectures (i.e. the number of LSTM cells and the look back) and trained them using Adam optimizer [13]. Then, we analyzed the relation between the estimated probability $p_{0.01}$ and the observed MAE. Table 1 presents the correlation between the MAE random sampling results (Mean, Sd, and log $p_{0.01}$) and the observed MAE after training the RNNs for a predefined number of epochs. The table also presents the mean (Mean MAE) and standard deviation (Sd MAE) values of the observed MAE.

---

[1] Code available at https://github.com/acamero/dlopt.

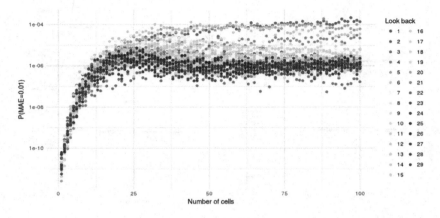

Fig. 2. Probability of finding a *good* set of weights for single-hidden-layer RNNs

Table 1. Correlation between the MAE observed after training the model and the MAE random sampling results (single hidden layer)

Epochs	Correlation			Results	
	Mean	Sd	log $p_{0.01}$	Mean MAE	Sd MAE
1	−0.447	−0.317	−0.211	0.618	0.017
10	−0.726	−0.431	−0.321	0.517	0.111
100	−0.790	−0.641	−0.650	0.133	0.146
1000	−0.668	−0.458	−0.515	0.036	0.079

There is a moderate to strong negative correlation between $p_{0.01}$ and the MAE observed, thus the results suggest that the estimated probability is useful for comparing RNN architectures. In a rough sense, given two RNNs we might select the one that has a higher probability. Note that we are not predicting the performance, instead we are predicting how *likely* would be to find a set of weights that have a *good* error performance.

4.2   Two-Hidden-Layer Architectures

The previous results suggest that the proposal is useful to compare single-hidden-layer architectures (as it is also stated in [8]), but what about multiple hidden layers? To begin with the validation in a deeper context, we studied the problem using the same sine wave and the search space defined by the stacked RNNs with 7 to 31 LSTM cells per layer, with up to two-hidden-layer, and a look back in {1, 10, 20, 30}. The selection of the number of cells and the look back was made upon the results observed in the single-hidden-layer study, the greatest variation of p_t occurs in the referred region (see Fig. 2). Note that there are 625 possible architectures, without taking into account the look back.

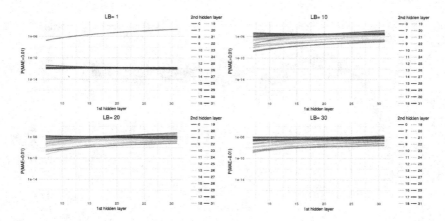

Fig. 3. Probability of finding a *good* set of weights for two-hidden-layer RNNs

We repeated the MAE random sampling for each architecture, i.e. we took 100 samples, and estimated $p_{0.01}$. Figure 3 presents the estimated $p_{0.01}$ smoothed using a logarithmic model (to ease the visualization of trends). The x-axis shows the number of cells in the first hidden layer, the y-axis represents the estimated probability, the different colors represent the number of cells in the second hidden layer, and LB stands for the look back. A value equal to 0 in the second hidden layer implies that the RNN is a single-hidden-layer one.

We uniformly selected 100 architectures and trained them using Adam optimizer [13]. Table 2 presents the correlation between the MAE random sampling results (Mean, Sd, $\log p_{0.01}$) and the MAE observed after training the model. The table also presents the mean (Mean MAE) and standard deviation (Sd MAE) values of the observed MAE.

Table 2. Correlation between the MAE observed after training the model and the MAE random sampling results (two-hidden-layer)

Epochs	Correlation			Results	
	Mean	Sd	$\log p_{0.01}$	Mean MAE	Sd MAE
1	−0.086	−0.135	−0.171	0.617	0.009
10	−0.450	−0.632	−0.635	0.591	0.018
100	−0.709	−0.827	−0.905	0.147	0.145
1000	−0.695	−0.843	−0.922	0.106	0.147

Again, the results show a string negative correlation between the estimated probability and the MAE observer after training. In this case, two things gain our attention: first, the single-hidden-layer architectures outperforms the two-hidden-layer ones on average in terms of the observed MAE, and second the correlation is higher for two-hidden-layer architectures (refer to Tables 1 and 2). Intuitively we suspect that both insights are related and both can be explained by the fact that two-hidden-layer architectures are more complex than single-layer ones. Therefore, the optimizing algorithm has to struggle harder to find a *good* set of weights as the number of hidden layers increases.

4.3 Three-Hidden-Layer Architectures

We added a third hidden layer to the search space (with 7 to 31 LSTM cells on it). We performed the MAE random sampling for all the architectures (15625 architectures, without considering the look back variations) and estimated $p_{0.01}$.

Afterwards, we uniformly selected 100 three-hidden-layer architectures and trained them using Adam optimizer [13]. Table 3 presents the correlation between the MAE random sampling results (Mean, Sd, log $p_{0.01}$) and the MAE observed after training the model. The table also presents the mean (Mean MAE) and standard deviation (Sd MAE) values of the observed MAE.

Table 3. Correlation between the MAE observed after training the model and the MAE random sampling results (three-hidden-layer)

Epochs	Correlation			Results	
	Mean	Sd	log $p_{0.01}$	Mean MAE	Sd MAE
1	−0.334	−0.447	−0.475	0.616	0.003
10	−0.546	−0.724	−0.745	0.605	0.010
100	−0.720	−0.869	−0.906	0.180	0.159
1000	−0.130	−0.873	−0.911	0.143	0.164

The results follow the same trends highlighted in the two-hidden-layer study. The correlation between the observed MAE and $p_{0.01}$ suggests that for this specific scenario the probability acts as a proxy of the expected performance. Moreover, the results indicate that as the problem gets more complex, the probability is even more correlated to the actual performance.

4.4 Memory and Time Comparison

Finally, after showing the usefulness of the MAE random sampling to compare stacked RNN architectures in terms of the expected performance, we studied the time and memory needed to perform the referred sampling.

We analyzed the execution logs and extracted the memory and time consumed by each process. It is important to notice that every process was executed using similar hardware and software configurations. Table 4 summarizes the time and memory usage. Despite the simplicity of the input, there is notable difference between the time needed to train an RNN and to perform a MAE random sampling. On the other hand, there is only a small difference in the memory required. Due to the *low-cost* of sampling an RNN and the usefulness of the approach for comparing architectures, we believe that using the MAE random sampling is worthwhile.

Table 4. Time and memory usage comparison

	Mean time [s]	Sd time	Mean mem [MB]	Sd mem
Training 1000 epochs	996	0.006	127	6.338
MAE random sampling	6	0.001	150	98.264

5   Conclusions and Future Work

In this study we extend the MAE random sampling technique [8] to multiple-hidden-layer architectures. Given an RNN architecture and an input time series, we generate a set of normally distributed weights and compute the MAE (the *samples*). Then, we fit a truncated normal distribution to the MAE samples and estimate the probability of finding a set of weights whose MAE is below an arbitrary threshold.

We test our proposal on stacked RNN architectures with up to three-hidden-layers, using a sine wave. The results show that there is a strong negative correlation between the estimated probability and the MAE measured after training the model using Adam optimizer. Moreover, as we add hidden-layers to the RNN, the correlation between the probability and the MAE measured increases. We think that this might be explained in part by the increasing complexity of the training process, however further analysis is required to explain this observation.

Overall, the results suggest that the MAE random sampling is a *"low-cost, training-free*, rule of thumb" method to compare deep RNN architectures.

As future work we propose to validate the results of this study using other time series and to explore a broader hyper-parameter search space.

Acknowledgements. This research was partially funded by Ministerio de Economía, Industria y Competitividad, Gobierno de España, and European Regional Development Fund grant numbers TIN2016-81766-REDT (http://cirti.es) and TIN2017-88213-R (http://6city.lcc.uma.es). Universidad de Málaga, Andalucía TECH.

References

1. Abadi, M., et al.: Tensorflow: a system for large-scale machine learning. In: OSDI, vol. 16, pp. 265–283 (2016)
2. Albelwi, S., Mahmood, A.: A framework for designing the architectures of deep convolutional neural networks. Entropy **19**(6), 242 (2017)
3. Bengio, Y., Simard, P., Frasconi, P.: Learning long-term dependencies with gradient descent is difficult. IEEE Trans. Neural Netw. **5**(2), 157–166 (1994)

4. Bergstra, J.S., Bardenet, R., Bengio, Y., Kégl, B.: Algorithms for hyper-parameter optimization. In: Shawe-Taylor, J., Zemel, R.S., Bartlett, P.L., Pereira, F., Weinberger, K.Q. (eds.) Advances in Neural Information Processing Systems 24, pp. 2546–2554. Curran Associates, Inc. (2011)
5. Bracewell, R.N., Bracewell, R.N.: The Fourier Transform and Its Applications, vol. 31999. McGraw-Hill, New York (1986)
6. Camero, A., Toutouh, J., Stolfi, D.H., Alba, E.: Evolutionary deep learning for car park occupancy prediction in smart cities. In: Kotsireas, I., Pardalos, P. (eds.) Learning and Intelligent OptimizatioN (LION) 12, pp. 1–15. Springer, Heidelberg (2018)
7. Camero, A., Toutouh, J., Alba, E.: DLOPT: deep learning optimization library. arXiv preprint arXiv:1807.03523, July 2018
8. Camero, A., Toutouh, J., Alba, E.: Low-cost recurrent neural network expected performance evaluation. arXiv preprint arXiv:1805.07159, May 2018
9. Domhan, T., Springenberg, J.T., Hutter, F.: Speeding up automatic hyperparameter optimization of deep neural networks by extrapolation of learning curves. In: Proceedings of the 24th International Conference on Artificial Intelligence, IJCAI 2015, pp. 3460–3468. AAAI Press (2015)
10. Haykin, S.: Neural Networks and Learning Machines, vol. 3. Pearson, Upper Saddle River (2009)
11. Hochreiter, S., Schmidhuber, J.: Long short-term memory. Neural Comput. 9(8), 1735–1780 (1997)
12. Jozefowicz, R., Zaremba, W., Sutskever, I.: An empirical exploration of recurrent network architectures. In: Proceedings of the 32nd International Conference on International Conference on Machine Learning, ICML 2015, vol. 37, pp. 2342–2350. JMLR.org (2015)
13. Kingma, D.P., Ba, J.: Adam: a method for stochastic optimization. arXiv preprint arXiv:1412.6980 (2014)
14. Kolen, J.F., Kremer, S.C.: Gradient Flow in Recurrent Nets: The Difficulty of Learning LongTerm Dependencies, pp. 464–479. Wiley-IEEE Press (2001)
15. LeCun, Y., Bengio, Y., Hinton, G.: Deep learning. Nature 521(7553), 436 (2015)
16. Lipton, Z.C., Berkowitz, J., Elkan, C.: A critical review of recurrent neural networks for sequence learning. arXiv preprint arXiv:1506.00019 (2015)
17. Litjens, G., et al.: A survey on deep learning in medical image analysis. Med. Image Anal. 42, 60–88 (2017)
18. Min, S., Lee, B., Yoon, S.: Deep learning in bioinformatics. Brief. Bioinf. 18(5), 851–869 (2017)
19. Ojha, V.K., Abraham, A., Snášel, V.: Metaheuristic design of feedforward neural networks: a review of two decades of research. Eng. Appl. Artif. Intell. 60, 97–116 (2017)
20. Pascanu, R., Mikolov, T., Bengio, Y.: On the difficulty of training recurrent neural networks. In: Proceedings of the 30th International Conference on International Conference on Machine Learning, ICML 2013, vol. 28, pp. III-1310–III-1318. JMLR.org (2013)
21. Ramos, E.Z., Nakakuni, M., Yfantis, E.: Quantitative measures to evaluate neural network weight initialization strategies. In: 2017 IEEE 7th Annual Computing and Communication Workshop and Conference (CCWC), pp. 1–7 (2017)
22. Smithson, S.C., Yang, G., Gross, W.J., Meyer, B.H.: Neural networks designing neural networks: multi-objective hyper-parameter optimization. In: 2016 IEEE/ACM International Conference on Computer-Aided Design (ICCAD), pp. 1–8. IEEE (2016)

PMSC-UGR: A Test Collection for Expert Recommendation Based on PubMed and Scopus

César Albusac, Luis M. de Campos, Juan M. Fernández-Luna[✉], and Juan F. Huete

Departamento de Ciencias de la Computación e Inteligencia Artificial, ETSI Informática y de Telecomunicación, CITIC-UGR, Universidad de Granada, 18071 Granada, Spain
calbusac@ugr.es, {lci,jmfluna,jhg}@decsai.ugr.es

Abstract. A new test document collection, PMSC-UGR, is presented in this paper. It has been built using a large subset of MEDLINE/PubMed scientific articles, which have been subjected to a disambiguation process to identify unequivocally who are their authors (using ORCID). The collection has also been completed by adding citations to these articles available through Scopus/Elsevier's API. Although this test collection can be used for different purposes, we focus here on its use for expert recommendation and document filtering, reporting some preliminary experiments and their results.

Keywords: Test collection · Authors disambiguation
Expert finding · Document filtering · MEDLINE/PubMed · Scopus

1 Introduction

In most areas of science it is necessary to test the possible merits and advantages of new methods and techniques against the state-of-the-art, as well as to compare competing new methods among them. To do that it is necessary to have benchmark data available where we can compare the different alternatives in a controlled environment. This is also true in the field of recommender systems and within expert finding [2] (also known as expert search or expert recommendation), where the goal is to recommend persons (experts) given a topic of interest for a given user. A different, but formally very related problem, is that of document filtering [3,9], where the goal is to decide which users should receive (because they are interested in) a new incoming document from a document stream. In both cases we have a set of individuals (experts or users) which are characterized by some kind of profile[1], and a "query" (the topic of interest or the document to be filtered). The system must be able to decide which individuals

---

[1] Profile extracted, for example, from documents authored by this individual.

© Springer Nature Switzerland AG 2018
F. Herrera et al. (Eds.): CAEPIA 2018, LNAI 11160, pp. 34–43, 2018.
https://doi.org/10.1007/978-3-030-00374-6_4

are more related to this query, according to their profiles. In such a context, benchmark data consists in a test collection composed of documents which can be associated to their authors. In this paper we describe a new test collection, primarily based on MEDLINE/**P**ubMed scientific documents but also using the **SC**opus tool from Elsevier, PMSC-UGR, which can be used for expert finding and document filtering but also for other tasks.

In developing the collection, we have invested considerable effort in the disambiguation of the authors names. This is important for expert finding because different experts may have exactly the same name (or the same surname and the same initials), or the name of a given expert can appear in different versions. The only way to avoid ambiguities is to use a unique identifier, like ORCID. The problem is that in the PubMed articles the authors are not always identified by their ORCID. We have used Elsevier's API to complete the collection by adding validated authors (i.e. unambiguously identified by their ORCID codes) to those articles that do not have them but it is checked that they should be.

Moreover, we have enlarged the collection by adding available citations (also through the Elsevier's API) to the PubMed articles. In this way we can use these citations to create graphs of either articles or authors. These networks can be used, for example, to enlarge/improve the profile of an author au by taking into account information about either other authors or articles that cite the articles authored by au.

Our collection contains neither explicit queries nor explicit relevance judgements, which are necessary to evaluate any proposed model. We propose to use a machine learning based methodology for the evaluation of the models, splitting the dataset into training and test set. Then, given an article in the test set, our objective is to find –from the information in the training data– those researchers that might be interested in this article. As query, we propose the use of information from the article, such as the title, the abstract or the keywords. With respect to the relevance judgements, we consider that the experts who are relevant for a query are own authors of the article and/or the authors citing this article. Note that by means of this approach not everyone who is interested in the paper is in the judgements list, but everyone on it has expressed his/her interest. So, we can evaluate a model over a large number of pairs (automatic query – implicit relevance judgements).

The remainder of this paper is structured as follows: Sect. 2 outlines related work. In Sect. 3 we give details about how the PMSC-UGR test collection was built. Section 4 describes how the collection can be used in the evaluation of expert search and document filtering methods, together with some preliminary results. Section 5 outlines other possible uses of the test collection. Finally, Sect. 6 contains the concluding remarks.

2   Related Work

The are several collections suitable for expert finding. For example, some of them are LExR [10], ArnetMiner [12], CERC [1] and W3C [7].

W3C collection includes W3C working groups members as candidate experts (a total of 1,092 experts), 331,037 documents (mainly emails) and 50 queries (the topics were the own working groups, and working group membership was considered as the ground truth). The CERC collection includes 3,500 Commonwealth Scientific and Industrial Research Organisation (CSIRO) employees as candidate experts, contains 370,715 documents and 50 topics (created by SCIRO Science Communicators, which include a short list of key contacts used as ground truth). ArnetMiner is a test collection focused in an academic setting, the candidate experts are computer science researchers. The publication data come from online databases including DBLP, ACM Digital library, Citeseer, and others. ArnetMiner is a large collection, it includes 1,048,504 researcher profiles and 3,258,504 publications, and also contains citations (although the reported number of queries used for expert finding is only 13). Also from an academic setting, the most recent test collection is LExR, where candidate experts are researchers working in Brazil. Initially it contains 206,697 researchers and 11,942,014 references to publications (although only 483,222 of these references include the abstract). The number of available queries for expert finding is 235.

As we shall see in the next section, PMSC-UGR is much larger than W3C and CERC (in both number of documents and experts). Also, the number of queries we can use in our collection (as many as documents in the test set) is much larger than those in all the other collections. Therefore, the results obtained using PMSC-UGR can generate more reliable conclusions, from a statistical point of view. Finally, other important points in favor of PMSC-UGR are that it includes citations (only ArnetMiner also contains citations) and the strict disambiguation process carried out, which avoids incorrect attribution of papers to authors.

3   Building the PMSC-UGR Test Collection

This section describes in some detail the steps followed to build our test collection, starting from the MEDLINE/PubMed collection, which contains articles for biomedical literature from MEDLINE. This collection is property of the US National Library of Medicine (NLM).

3.1   MEDLINE/PubMed Collection

The initial step was to download the complete colleccion of articles[2] from PubMed[3]. The download was carried out on June 8, 2017. In this collection, there are 892 files in XML format and inside each one, there are around 30,000 articles. In this way, the initial collection contained 26,759,991 articles, being possible to find a lot of information in each article. Some of the fields are:

---

[2] Although PubMed does not contain complete articles but references to articles, called citations, we will use the term articles to refer to these citations, and reserve the name citations to refer to other articles that cite in their bibliographic references a given article.

[3] ftp://ftp.ncbi.nlm.nih.gov/pubmed/baseline/.

- *PubMedID* is a unique identifier of the article from PubMed.
- *Journal* is the name of the journal where the article was published.
- *ArticleTitle* is the complete title of the article, in English.
- *Abstract* of the article.
- *AuthorList* contains information about the authors of the article. For each one, we can find:
  - *LastName* contains the surname or the single name used by an individual.
  - *ForeName* contains the remainder of name.
  - *Identifier* is a unique identifier associated with the name. The value in the Identifier attribute Source designates the organizational authority that established the unique identifier. For our purposes we will pay special attention to the ORCID identifier.
- *MeshHeadingList* is NLM controlled vocabulary, Medical Subject Headings (MeSH). It is used to characterize the content of the article, using descriptors from this thesaurus.
- *KeywordList* contains controlled terms in Keywords that also describe the content of the article.

Formats and meanings of all fields can be consulted at the official PubMed page[4].

3.2 Disambiguation of Author Names

In a context of expert recommendation, it is important to remove possible ambiguities between authors, in order to not to attribute articles to the wrong author or to lose articles from the right author. To do it, the only and efficient way is to use a unique digital identifier for each author. We decided to use the perhaps more widely accepted identifier for authors of scientific articles, namely the Open Researcher and Contributor ID (ORCID). Therefore, we restricted our PubMed collection to include only authors with an ORCID. There are 112,546 differents authors with ORCID. The problem is that in the PubMed collection frequently occurs that in an article the author appears with his ORCID whereas in another article (possibly older), perhaps of the same author, the ORCID does not appear. Therefore we are not completely sure that both articles belong to the same author. This fact would severely limit the completeness of our document collection, because we should discard a lot of articles where the ORCIDs of the authors do not appear, thus limiting the sources of information about the interests of these authors[5]. The goal is to be able to attribute an article appearing in PubMed to an author with an associated ORCID, although the ORCID does not appear in the original PubMed record. To deal with this problem we will try to solve some ambiguities by using another information source, namely the Scopus database through the Elsevier's APIs[6].

---

[4] https://www.nlm.nih.gov/bsd/licensee/elements_descriptions.html.
[5] In fact 26,661,157 articles in PubMed have not any ORCID.
[6] The data was downloaded from Scopus API between July 3 and September 27, 2017 via http://api.elsevier.com and http://www.scopus.com.

In order to do it, we need to associate the ORCID of an author with the digital identifier used in Scopus, namely ScopusID, and then search for the articles published by this author within Scopus, finally comparing these articles with those appearing in PubMed.

For each author with ORCID in PubMed we have carried out the following query to Scopus:

```
http://api.elsevier.com/content/search/
author?query=ORCID(AuthorORCID)&apiKey=yourApiKey
```

where *AuthorORCID* is an ORCID and *yourApiKey* is a Key that can be generated once a user is registered in the Elsevier system.

This query obtains the ScopusID for each author as well as his name (givenname, surname and initials) as the result.

After this step, 21,048 different authors with ScopusID were obtained (those authors which can be simultaneously associated with an ORCID and a ScopusID). These authors will be the ones used primarily in this collection. We also included those authors with ORCID in PubMed but not appearing in Scopus (91,498) as a secondary collection[7].

In the next step, for each ScopusID, the following query was created to get information about the articles written by the corresponding author:

```
http://api.elsevier.com/content/search/
scopus?query=AU-ID(AuthorScopusID)&field=dc:identifier,dc:title,
doi,pubmed-id&count=200&apiKey=yourApiKey
```

where *AuthorScopusID* is the ScopusID previously obtained and the field parameter specifies wich article data we want to get: *dc:identifier* is the Scopus identifier for the article (ArticleScopusID); *dc:title* is the article title; *doi* is the digital object identifier of the article; *pubmed-id* is de PubMed identifier (if it exists).

In this way, for each author we have obtained all their articles (within Scopus). All this information has been stored in a csv file with the ORCID, ScopusID, ArticleScopusID, Title, doi and PubMedID.

Then, to try to assign validated authors to articles in PubMed where the ORCID of the authors does not appear, the following process was carried out: an index with the search engine library Lucene[8] was created and all the PubMed articles downloaded at the beginning were indexed. The only fields in these articles that were indexed (those which are necessary to our purposes) are Article title, Article authors and PubmedID. Next, for each author au with ScopusID and ORCID, we obtain the list l_{au} of their articles retrieved from Scopus (those safely assigned to au in Scopus) which can be used to complete the information of au in PubMed. Then, for each pair $(au, l_{au}(i))$ we perform the following process: If an article has PubMedID, then it is clear that it belongs to the PubMed collection and moreover it can be safely associated with the ORCID of the author. Otherwise, we will try to set such association algorithmically. Particularly, a query by the article title was run against the PubMed index. Then, focusing

---

[7] The reason is that for these authors we cannot obtain citations to their articles, so this secondary collection is larger but contains less information.

[8] https://lucene.apache.org.

only on the top 20 results we check whether both the title and the author field in the retrieved PubMed article match[9] the pair $(au, l_{au}(i))$. If we find such a match, we can also safely associate the ORCID of the author with this article.

After this process, there are articles and validated authors who wrote them. In total, there are 762,508 validated articles (each one stored in an independent file), which form our primary test collection. In fact we have been able to enlarge the initial set of validated (article, author) pairs from 161,609 to a total of 868,498. It is worth mentioning that 642 out of 21,048 authors being considered have not validated articles[10], so finally there are 20,406 authors in our collection. The distribution of the number of articles written by each author follows a power-law distribution, where there are lots of authors with few articles and few authors with lots of articles. For example, there are 1,033 authors who have written a single article, whereas at the other extreme there is a single author who has written 1384 articles.

3.3 Adding Citations

To complete our collection with citations (not available in MEDLINE/PubMed), we have used again the Elsevier's APIs. Starting from the PubMedID of each article in our collection, we needed to obtain an identifier called *eid* by means of the following query:

`http://api.elsevier.com/content/search/`
`scopus?query=pmid(PubMedID)&apikey=YourApiKey&field=eid`

Once obtained the eid for a PubMed article, the list of citations of this article (appearing in Scopus) can be obtained using the following query:

`https://api.elsevier.com/content/search/`
`scopus?query=refeid(eid)&apikey=YourApiKey&field=pubmed-id`

which returns the list of the PubMedIDs of these articles (we only keep those articles belonging to our collection). Finally, each file containing one article in our collection is enlarged by adding information about the articles citing it (PubMedIDs) as well as the corresponding authors (ORCIDs). In this way we can easily build a network of articles (with an arc going from article a to article b if article b is cited by article a) as well as a network of authors (with an arc from author a to author b if in one of the articles written by a is cited an article written by b, this arc can be weighted by the number of such citations).

We were able to consult the citations of 749,811 articles (the remaining articles did not possess an eid), and 509,202 of them had citations in Scopus. Therefore 66.78% of the articles in our collection have citations. The average number of validated citations per article is 7. The total number of citations found is 3,593,931. The number of citations per article also follows a typical power-law

---

[9] We do not require a perfect match, allowing an edit distance of 5 for title and 3 for author.

[10] This may happen, for example, when the articles (probably only one) in PubMed of an author (having ORCID and ScopusID) do not appear in the list of papers in Scopus written by this author.

distribution, where a few articles have many citations and many articles have few citations. For example, there are 116,365 articles having only one citation, and there is only one article having 713 citations.

4 Using the Collection for Expert Search and Document Filtering

In order to simulate a document filtering scenario, we will consider that each author in the collection is a possible user, the documents to be recommended/filtered are the articles in the collection and the queries representing them may be composed of their abstracts and/or their titles. In the expert finding scenario the experts to be recommended are the authors in the collection and the titles of the articles in the collection may be considered as the simulated topics of interest (queries). Obviously, the data used for gathering information about the authors (in order to learn about their interests and perhaps build their profiles) should be different from the data used for building the queries. In other words, a partition of the collection in training and test sets must be carried out. Given the size of the collection, using for example 80% of the articles for training and 20% for testing, we have more than 150,000 cases in the test set, which is a size large enough to allow extracting statistically significant conclusions from the experiments.

As we mentioned previously, our collection has not explicit relevance judgements stating which are the relevant authors given a query. However, we can establish two levels of implicit relevance judgements: authors and citers. For example, in the context of expert finding, it is reasonable to assume that given a topic of interest represented by an article, then the own authors of this article are relevant experts for the topic. In the context of document filtering, if the document to be filtered is an article (represented by its abstract), we can assume that the authors of this article, together with the authors of other articles citing it are the users interested in this document. It is true that these are not exhaustive relevance judgements, because probably there are more authors interested in the document than just their authors and their citers. Therefore we have an scenario where the implicitly fixed relevance judgements are correct but possibly incomplete. However, there are some previous studies [5, 6, 11] establishing that this situation of incompleteness of judgements does not represent a problem to reliably compare different systems, provided that the number of queries is large (and in our case it is quite large). For example, Carterette et al. [5] says that "evaluation over more queries with fewer or noisier judgments is preferable to evaluation over fewer queries with more judgments".

4.1 Building a Recommender/Filtering System Through an Information Retrieval System

In order to build a recommender/filtering system we can use essentially two techniques [4, 8]: either information retrieval-based (IR) methods or machine learning-based methods. Focusing on IR methods, we are going to use a rather simple approach (which could serve as baseline for other more sophisticated approaches), where we do not build an explicit profile for each expert/user. Instead, we are going to use an Information Retrieval System (IRS) indexing the (training) collection of articles. Then, following a document model-based approach [2], given a query (either the abstract or the title of a test article), the IRS will return a ranked list of the articles that better match this query. Then we replace each article in the ranking by their associated authors, in order to get a ranking of authors. As this ranking may contain duplicate authors (if two or more articles of the same author appear in the original ranking), we combine all the scores associated to the same author and rerank the list of unique authors according to the combined score. In the experiments reported in this paper we have used the maximum as the combination function.

4.2 Preliminary Results

We randomly partitioned the collection of articles into 80% for training and 20% for testing. To index the abstracts, they were previously preprocessed (removing stopwords and doing stemming). The time required to index the collection using Lucene was less than five minutes. In these experiments we have used the abstracts of the test articles as the queries, and the implementation in Lucene of the classical vector space retrieval model as the IRS.

We have used trec_eval[11], the standard tool used by the TREC community for evaluating an ad hoc retrieval run, which uses the ranked list of authors obtained by the system in response to a query and the relevance judgements corresponding to this query. Concerning this, we have evaluated the system using two different sets of relevance judgements: the authors which are relevant given a query associated to a test article are (1) the own authors of this article and (2) the authors of this article plus the authors of other articles that cite it (the citers).

We will focus on three performance measures: R-Precision (Rprec), Precision at 10 (P_10) and Recall at 10 (recall_10). Table 1 shows these measures for the two sets of relevance judgements.

Table 1. Results of the experiments.

Only authors			Authors + Citers		
Rprec	P_10	recall_10	Rprec	P_10	recall_10
0.5232	0.0927	0.7399	0.4270	0.1490	0.5713

We can observe that the results when using only the own authors as relevant are better than the case where we also use the citers (except in the case of

---

[11] http://trec.nist.gov/trec_eval/.

P_10). This fact is expected, as in the second case we have a greater number of relevant authors to identify, so that the problem is more difficult in some sense. Obviously, this is not true for P_10, as having more relevant authors also means that it is easier to find more within the top 10 results. In absolute terms, we believe that the results are not bad for a baseline approach. On the average we find one relevant author within the first 10 authors (P_10); also, around 75% of the relevant authors are found within the first 10 results (recall_10); when we recover a number of results equal to the true number of relevant authors (Rprec), we find around 50% of these relevant authors.

5   Other Use Cases of the Collection

- As all the articles have an associated journal, we could use titles and abstracts of articles to build and evaluate recommender systems of scientific journals [13], to help authors to find the more appropriate journals to publish their new papers (as for example Springer Journal Suggester and Elsevier Journal Finder do).
- Using MeshTerms (associated to articles and indirectly to authors) we could evaluate expert profiling techniques. Also (hierarchical and multilabel) text classification methods could be studied.
- As all the authors' names are unambiguously associated to the corresponding ORCIDs, we could study disambiguation methods, trying to distinguish between authors with the same names taking into account the text (title and abstract) of their articles.
- We could use the graphs connecting authors (or articles) by means of citations to enhance the profiles of authors (using information about either the authors or the articles that cite them), to explore hybrid content-based and collaborative recommender systems, or even to explore graph visualization methods. Also considering the citations we could compute some bibliometric measures that could be incorporated to the recommendation model in order to enhance the performance.

6   Concluding Remarks

In this paper we have described the building process of a new test document collection and some preliminary results obtained using it to evaluate expert finding and document filtering methods, although we have also outlined other possible uses of the collection. Our collection starts from the MEDLINE/PubMed collection of scientific articles, but also uses Scopus/Elsevier data with two purposes: disambiguate author names (using ORCID and ScopusID) and adding information about citations to the PubMed records.

Our PMSC-UGR collection is relatively large: it contains 20,406 authors, 762,508 articles and 3,593,931 citations. Although it does not include external relevance judgements, we can use the authors and the citers of articles (with either the titles or the abstracts of these articles acting as queries) to establish implicit (but incomplete) relevance judgements. As reported in the literature, this incompleteness of the judgements is not a serious problem to compare the performance of competing systems given the large size of the collection (more than 150,000 queries if we use for example a 80%-20% partition in training and test articles).

For future work we plan to use the collection to evaluate different methods for expert finding and document filtering, also taking into account the citation graphs and the MeshTerms. We also plan to make PMSC-UGR available as a community resource.

Acknowledgment. This work has been funded by the Spanish "Ministerio de Economía y Competitividad" under project TIN2016-77902-C3-2-P, and the European Regional Development Fund (ERDF-FEDER).

References

1. Bailey, P., Craswell, N., Soboroff, I., de Vries, A.P.: The CSIRO enterprise search collection. In: SIGIR Forum, vol. 41, pp. 42–45 (2007)
2. Balog, K., Fang, Y., de Rijke, M., Serdyukov, P., Si, L.: Expertise retrieval. Found. Trends Inf. Retrieval **6**, 127–256 (2012)
3. Beel, J., Gipp, B., Langer, S., Breitinger, C.: Research-paper recommender systems: a literature survey. Int. J. Digit. Libr. **17**, 305–338 (2016)
4. Bobadilla, J., Hernando, A., Fernando, O., Gutiérrez, A.: Recommender systems survey. Knowl.-Based Syst. **46**, 109–132 (2013)
5. Carterette, B., Pavlu, V., Kanoulas, E., Aslam, J., Allan, J.: Evaluation over thousands of queries. In: Proceedings of the 31st ACM SIGIR Conference, pp. 651–658 (2008)
6. Carterette, B., Smucker, M.: Hypothesis testing with incomplete relevance judgments. In: Proceedings of the 16th ACM CIKM Conference, pp. 643–652 (2007)
7. Craswell, N., de Vries, A.P., Soboroff, I.: Overview of the TREC 2005 enterprise track. In: Proceedings of the 14th TREC Conference (2005)
8. de Campos, L.M., Fernández-Luna, J.M., Huete, J.F., Redondo-Expósito, L.: Comparing machine learning and information retrieval-based approaches for filtering documents in a parliamentary setting. In: Moral, S., Pivert, O., Sánchez, D., Marín, N. (eds.) SUM 2017. LNCS (LNAI), vol. 10564, pp. 64–77. Springer, Cham (2017). https://doi.org/10.1007/978-3-319-67582-4_5
9. Hanani, U., Shapira, B., Shoval, P.: Information filtering: overview of issues, research and systems. User Model. User-Adap. Inter. **11**, 203–259 (2001)
10. Mangaravite, V., Santos, R.L.T., Ribeiro, I.S., Gonçalves, M.A., Laender, A.H.F.: The LExR collection for expertise retrieval in academia. In: Proceedings of the 39th ACM SIGIR Conference, pp. 721–724 (2016)
11. Sanderson, M., Zobel, J.: Information retrieval system evaluation: effort, sensitivity, and reliability. In: Proceedings of the 28th ACM SIGIR Conference, pp. 162–169 (2005)
12. Tang, J., Zhang, J., Yao, L., Li, J., Zhang, L., Su, Z.: ArnetMiner: extraction and mining of academic social networks. In: Proceedings of the 14th ACM SIGKDD Conference, pp. 990–998 (2008)
13. Wang, D., Liang, Y., Xu, D., Feng, X., Guan, R.: A content-based recommender system for computer science publications. Knowl.-Based Syst. **157**, 1–9 (2018)

Bayesian Optimization of the PC Algorithm for Learning Gaussian Bayesian Networks

Irene Córdoba[1]([✉]), Eduardo C. Garrido-Merchán[2], Daniel Hernández-Lobato[2], Concha Bielza[1], and Pedro Larrañaga[1]

[1] Departamento de Inteligencia Artificial, Universidad Politécnica de Madrid, Madrid, Spain
irene.cordoba@upm.es

[2] Departamento de Ingeniería Informática, Universidad Autónoma de Madrid, Madrid, Spain

Abstract. The PC algorithm is a popular method for learning the structure of Gaussian Bayesian networks. It carries out statistical tests to determine absent edges in the network. It is hence governed by two parameters: (i) The type of test, and (ii) its significance level. These parameters are usually set to values recommended by an expert. Nevertheless, such an approach can suffer from human bias, leading to suboptimal reconstruction results. In this paper we consider a more principled approach for choosing these parameters in an automatic way. For this we optimize a reconstruction score evaluated on a set of different Gaussian Bayesian networks. This objective is expensive to evaluate and lacks a closed-form expression, which means that Bayesian optimization (BO) is a natural choice. BO methods use a model to guide the search and are hence able to exploit smoothness properties of the objective surface. We show that the parameters found by a BO method outperform those found by a random search strategy and the expert recommendation. Importantly, we have found that an often overlooked statistical test provides the best over-all reconstruction results.

1 Introduction

Graphical models serve as a compact representation of the relationships between variables in a domain. An important subclass is the Bayesian network, where conditional independences are encoded by missing edges in a directed graph with no cycles. By exploiting these independences, Bayesian networks yield a modular factorization of the joint probability distribution underlying the data. Of particular interest are Gaussian Bayesian networks for modelling variables in a continuous domain, which have been widely applied in real scenarios such as gene network discovery [11] and neuroscience [1].

When learning graphical models from data, two main tasks are usually differentiated: structure and parameter learning. The former consists in recovering the

© Springer Nature Switzerland AG 2018
F. Herrera et al. (Eds.): CAEPIA 2018, LNAI 11160, pp. 44–54, 2018.
https://doi.org/10.1007/978-3-030-00374-6_5

graph structure, and the latter amounts to fitting the numerical quantities in the model. In Gaussian Bayesian networks, parameter learning involves using standard linear regression theory, whereas structure learning is not an easy task in general, given the combinatorial search space of acyclic digraphs. There are two main approaches one can find in the literature for structure discovery in Bayesian networks: score-and-search heuristics, where the search space is explored looking for the network which optimizes a given score function, and constraint-based approaches, where statistical tests are performed in order to include or exclude dependencies between variables.

A popular constraint-based method with consistency guarantees is the PC algorithm [6]. In this method, a backward stepwise testing procedure is performed for determining absent edges in the resulting graph. Thus, of critical importance are the choice of the statistical test to be performed, and the significance level at which the potential edges are going to be tested. However, both are usually fixed after a grid search or directly set by expert knowledge [2,6]. In the literature on Bayesian network structure learning some empirical studies explore exact structure recovery [9], the behavior of score-and-search algorithms [8], and the impact of the significance level in the PC algorithm for high dimensional sparse scenarios [2,6]. We are not aware, however, of any research work using elaborated methods for hyper-parameters selection in this context.

In this paper we show that Bayesian optimization (BO) can be used as an alternative methodology for choosing the significance level and the statistical test in the PC algorithm. BO has been recently applied successfully in different optimization problems [14,15]. We consider here a structure learning scenario in moderately sparse settings that is representative of those considered in [6]. We show that BO outperforms, in terms of structure recovery error, in a relatively small number of iterations, both a baseline approach based on a grid search and specific values set by expert knowledge obtained from previous results on this problem [6]. Furthermore, we also analyze what values for the statistical test and the significance level are recommended by the BO approach, and compare them with those often used by the relevant literature on the subject.

This article is organized as follows. In Sect. 2, we introduce the main concepts relative to Gaussian Bayesian networks that will be used throughout the rest of the paper. Then, in Sect. 3, we describe the PC algorithm, emphasizing its hyper-parameters and how they may affect its performance. Black box BO is outlined in Sect. 4, with emphasis on the particular characteristics of our problem. The experimental setting as well as the results we have obtained are described in Sect. 5. Finally, we conclude the paper in Sect. 6, where we also point out the main planned lines of future work.

2 Preliminaries on Gaussian Bayesian Networks

Throughout the remainder of the paper, $X_1, \ldots, X_p$ will denote p random variables, and $\boldsymbol{X}$ the random vector they form. For a subset of indices $I \subseteq \{1, \ldots, p\}$, $\boldsymbol{X}_I$ will denote the random vector corresponding only to the variables indexed by

I. We will use $\boldsymbol{X}_I \perp\!\!\!\perp \boldsymbol{X}_J \mid \boldsymbol{X}_K$ for denoting that $\boldsymbol{X}_I$ is conditionally independent of $\boldsymbol{X}_J$ given the values of $\boldsymbol{X}_K$, being I, K, J disjoint subsets of $\{1, \ldots, p\}$. Let $G = (V, E)$ be an acyclic digraph, where $V = \{1, \ldots, p\}$ is the vertex set and $E \subseteq V \times V$ is the edge set. When G is part of a graphical model, its vertex set V can be thought of as indexing a random vector $\boldsymbol{X} = (X_1, \ldots, X_p)$. In a Bayesian network, the graph G is constrained to be acyclic directed and with no multiple edges.

A common interpretation of edges in a Bayesian network is the ordered Markov property, although many more exist, which can be shown to be equivalent [7]. This property is stated as follows. For a vertex $i \in V$, the set of parents of i is defined as $\mathrm{pa}(i) := \{j : (j, i) \in E\}$. In every acyclic digraph, an ancestral order $\prec$ can be found between the nodes where it is satisfied that if $j \in \mathrm{pa}(i)$, then $j \prec i$, that is, the parents of a vertex come before it in the ancestral order. For notational simplicity, in the remainder we will assume that the vertex set $V = \{1, \ldots, p\}$ is already ancestrally ordered. The ordered Markov property of Bayesian networks can be written in this context as

$$X_i \perp\!\!\!\perp \boldsymbol{X}_{\{1, \ldots, i-1\} \setminus \mathrm{pa}(i)} \mid \boldsymbol{X}_{\mathrm{pa}(i)}$$

for all $i \in V$.

The above conditional independences, together with the properties of the multivariate Gaussian distribution, allow to express a Gaussian Bayesian network as a system of recursive linear regressions. Indeed, if for each $i \in V = \{1, \ldots, p\}$, we consider the regression of X_i on its predecessors in the ancestral order, $X_1, \ldots, X_{i-1}$, then from the results regarding conditioning on multivariate Gaussian random variables we obtain

$$X_i = \sum_{j=1}^{i-1} \beta^{ji|1, \ldots, i-1} X_j + \epsilon_i, \tag{1}$$

where the regression coefficient $\beta^{ji|1, \ldots, i-1} = 0$ when $j \notin \mathrm{pa}(i)$, and ϵ_i are independent Gaussian random variables with zero mean and variance equal to the conditional variance of X_i on $X_1, \ldots, X_{i-1}$. Therefore, both the structure and parameters of a Gaussian Bayesian network can be directly read off from the system of linear regressions in Eq. (1).

3   Structure Learning with the PC Algorithm

The PC algorithm for learning Gaussian Bayesian networks proceeds by first estimating the skeleton, that is, the underlying undirected graph, of the acyclic digraph, and then orienting it. That is, for each vertex $i \in V = \{1, \ldots, p\}$, it looks through the set of its neighbors, which we will denote as $\mathrm{ne}(i)$, and selects a node $j \in \mathrm{ne}(i)$ and subset $C \subseteq \mathrm{ne}(i) \setminus \{j\}$. Then, the conditional independence $X_i \perp\!\!\!\perp X_j \mid \boldsymbol{X}_C$ is tested on the available data. It is a backward stepwise elimination method, in the sense that it starts with the complete undirected graph, and then proceeds by testing conditional independences in order to remove edges, doing so incrementally in the size of the neighbor subset C.

The PC main phase pseudocode can be found in Algorithm 1. The output of Algorithm 1 is the skeleton, or undirected version, of the estimated Gaussian Bayesian network, which is later oriented. Algorithm 1 is typically called the *population* version of the PC algorithm [6], since it assumes that perfect information is available about the conditional independence relationships present in the data. This is useful for illustrating the behavior and main properties of the algorithm; however, in real scenarios this is unrealistic, and statistical tests must be performed on the data in order to determine which variable pairs, with respect to different node subsets, are conditionally independent.

Algorithm 1. The PC algorithm in its population version

Input: Conditional independence information about $\boldsymbol{X} = (X_1, \ldots, X_p)$
Output: Skeleton of the Gaussian Bayesian network
1: $G \leftarrow$ complete undirected graph on $\{1, \ldots, p\}$
2: $l \leftarrow -1$
3: **repeat**
4:    $l \leftarrow l + 1$
5:   **repeat**
6:      Select i such that $(i, j) \in E$ and $|\mathrm{ne}(i) \setminus \{j\}| \geq l$
7:      **repeat**
8:        Choose new $C \subseteq \mathrm{ne}(i) \setminus \{j\}$ with $|C| = l$
9:        if $X_i \perp\!\!\!\perp X_j \mid \boldsymbol{X}_C$ **then**
10:         $E \leftarrow E \setminus \{(i, j), (j, i)\}$
11:        **end if**
12:      **until** (i, j) has been deleted or all neighbor subsets of size l have been tested
13:   **until** All $(i, j) \in E$ such that $|\mathrm{ne}(i) \setminus \{j\}| \geq l$ have been tested
14: **until** $|\mathrm{ne}(i) \setminus \{j\}| < l$ for all $(i, j) \in E$

3.1 Significance Level and Statistical Test

The criteria for removing edges is related to the ordered Markov property and Eq. (1). In particular, from multivariate Gaussian analysis we know that for $i \in V$ and $j < i$,

$$\beta^{ji|1,\ldots,i-1} = 0 \iff \rho^{ji|1,\ldots,i-1} = 0,$$

where $\rho^{ji|1,\ldots,i-1}$ denotes the partial correlation coefficient between X_i and X_j with respect to $X_1, \ldots, X_{i-1}$. In the PC algorithm, at iteration l, the null hypothesis $H_0 : \rho^{ji|C} = 0$ is tested against the alternative hypothesis $H_1 : \rho^{ji|C} \neq 0$, where C is a subset of the neighbors of i (excluding j) in the current estimator of the skeleton such that $|C| = l$.

The significance level at which H_0 will be tested, which we will denote in the remainder as α, is typically smaller or equal than 0.05, and serves to control the type I error. The other parameter of importance is the statistical test itself. The usual choice for this is a Gaussian test based on the Fisher's Z transform of the partial correlation coefficient [2,6], which is asymptotically normal. However, there are other choices available in the literature that could be considered and can be found in standard implementations of the algorithm. For example, the `bnlearn` R package [13] provides the standard Student's t test for the untransformed partial correlation coeficient, and the χ^2 test and a test based on the shrinkage James-Stein estimator, for the mutual information [4].

3.2   Evaluating the Quality of the Learned Structure

When performing structure discovery in graphical models, there are several ways of evaluating the results obtained by an algorithm. As a starting point, one could use standard error rates, such as the true positive and false positive rates. These rates simply take into account the original acyclic digraph $G = (V, E)$, and the estimated one $\hat{G}$, with edge set $\hat{E}$. Then, E with $\hat{E}$ are compared element-wise. This is a common approach in Bayesian networks.

We have preferred however to use the Structural Hamming Distance (SHD) [16]. This measure is motivated as follows. In Bayesian networks, there is not a unique correspondence between the model and the acyclic digraph that represents it. That is, if we denote as $\mathcal{M}(G)$ the set of multivariate Gaussian distributions whose conditional independence model is compatible (in the sense of the pairwise Markov property and Eq. (1)) with the acyclic digraph G, then we may have two distinct acyclic digraphs G_1 and G_2 such that $\mathcal{M}(G_1) = \mathcal{M}(G_2)$. In such case, G_1 and G_2 are said to be Markov equivalent.

The SHD measure between two acyclic digraph structures G_1 and G_2 takes into account this issue of non unique correspondence. In particular, it counts the number of operations that have to be performed in order to transform the Markov equivalence class of one graph into the other. Thus, given two acyclic digraphs that are distinct but Markov equivalent, their true positive and false positive rates could be nonzero, while their SHD is guaranteed to be zero.

4   Black-Box Bayesian Optimization

Denote the SHD objective function as $f(\boldsymbol{\theta})$, which depends on the parameters in the PC algorithm, $\boldsymbol{\theta} = (\alpha, T)$, that are going to be optimized, α, the significance level, and T, the independence test. We can view $f(\boldsymbol{\theta})$ as a black-box objective function with noisy evaluations $y_i = f(\boldsymbol{\theta}) + \epsilon_i$, with ϵ_i being a, typically, Gaussian noise term. With BO the number of evaluations of f needed to solve the optimization problem are drastically reduced. Let the observed data until step $t - 1$ of the algorithm be $\mathcal{D}_{t-1} = \{(\boldsymbol{\theta}_i, y_i)\}_{i=1}^{t-1}$. At iteration t of BO, a probabilistic model $p(f(\boldsymbol{\theta}) \mid \mathcal{D}_{t-1})$, typically a Gaussian process (GP) [12], is fitted to the data collected so far. The uncertainty about f provided by the probabilistic model is then used to generate an acquisition function $a_t(\boldsymbol{\theta})$, whose value at each input location indicates the expected utility of evaluating f there. Therefore, at iteration t, $\boldsymbol{\theta}_t$ is chosen as the one that maximizes the acquisition function. The described process is repeated until enough data about the objective has been collected. When this is the case, the GP predictive mean for $f(\cdot)$ can be optimized to find the solution of the optimization problem, or we can provide as a solution the best observation made so far.

The key for BO success is that evaluating the acquisition function is very cheap compared to the evaluation of the actual objective, because it only depends on the GP predictive distribution for the objective at any candidate point. The GP predictive distribution for $f(\boldsymbol{\theta}_t)$, the candidate location for next iteration, is given by a Gaussian distribution characterized by a mean μ and a variance σ^2 with values

$$
\begin{aligned}
\mu &= \boldsymbol{k}_*^T (\mathbf{K} + \sigma_n^2 \mathbf{I})^{-1} \boldsymbol{y}, \\
\sigma^2 &= k(\boldsymbol{\theta}_t, \boldsymbol{\theta}_t) - \boldsymbol{k}_*^T (\mathbf{K} + \sigma_n^2 \mathbf{I})^{-1} \boldsymbol{k}_* .
\end{aligned}
\tag{2}
$$

where $\boldsymbol{y} = (y_1, \ldots, y_{t-1})^t$ is a vector with the objective values observed so far; σ_n^2 is the variance of the additive Gaussian noise; $\boldsymbol{k}_*$ is a vector with the prior covariances between $f(\boldsymbol{\theta}_t)$ and each y_i; $\mathbf{K}$ is a matrix with the prior covariances among each y_i; and $k(\boldsymbol{\theta}_t, \boldsymbol{\theta}_t)$ is the prior variance at the candidate location $\boldsymbol{\theta}_t$. The covariance function $k(\cdot, \cdot)$ is pre-specified; for further details about GPs and example of covariance functions we refer the reader to [12]. Three steps of the BO process are illustrated graphically in Fig. 1 for a toy minimization problem.

In BO methods the acquisition function balances between exploration and exploitation in an automatic way. A typical choice for this function is the information-theoretic method Predictive Entropy Search (PES) [5]. In PES, we are interested in maximizing information about the location of the optimum value, $\boldsymbol{\theta}^*$, whose posterior distribution is $p(\boldsymbol{\theta}^* | \mathcal{D}_{t-1})$. This can be done through the negative differential entropy measure of $p(\boldsymbol{\theta}^* | \mathcal{D}_{t-1})$. Through several operations, an approximation to PES is given by

$$
a(\boldsymbol{\theta}) = H[p(y | \mathcal{D}_{t-1}, \boldsymbol{\theta})] - \mathbb{E}_{p(\boldsymbol{\theta}^* | \mathcal{D}_{t-1})}[H[p(y | \mathcal{D}_{t-1}, \boldsymbol{\theta}, \boldsymbol{\theta}^*)]],
$$

where $p(y | \mathcal{D}_{t-1}, \boldsymbol{\theta}, \boldsymbol{\theta}^*)$ is the posterior predictive distribution of y given $\mathcal{D}_{t-1}$ and the minimizer $\boldsymbol{\theta}^*$ of f, and $H[\cdot]$ is the differential entropy. The first term of the previous equation can be analytically solved as it is the entropy of the predictive distribution and the second term is approximated by Expectation Propagation [10]. We can see an example of the PES acquisition function in Fig. 1.

5   Numerical Experiments

Since we will consider networks of different node size p, we will use in our experimental setting as the validation measure a normalized version of SHD with respect to the maximum edge number $p(p-1)/2$. The significance level α will range from 10^{-5} to 0.1, and BO will search through its corresponding decimal logarithmic space $[-5, -1]$. The statistical test will be represented using a categorical variable whose value indicates one of the above mentioned four tests. Namely, two test based on the partial correlation coefficient: a Gaussian

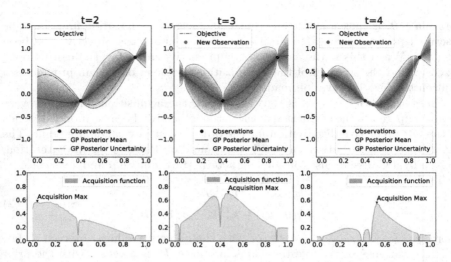

Fig. 1. An example of BO on a toy 1D noiseless problem, where a function (shown in red, dotted) sampled from a GP prior is optimized. The top figures show the GP (mean and standard deviation) estimation of the objective $f(\cdot)$ in blue. The acquisition function PES is shown in the bottom figures in yellow. The acquisition is high where the GP predicts a low objective and where the uncertainty is high. From left to right, iterations (t) of BO are executed and the points (black) suggested by the acquisition function are evaluated in the objective, we show in red the new point for every iteration. In a small number of iterations, the GP is able to almost exactly approximate the objective. (Color figure online)

test based on the Fisher's Z transform and the Student's T test; and two test based on the mutual information: the χ^2 test, and a test based on the shrinkage James-Stein estimator. As outlined before, this problem is specially suitable for BO, since we do not have access to gradients, the objective evaluations may be expensive and they may be contaminated with noise.

We have employed *Spearmint* (https://github.com/HIPS/Spearmint) for BO and the *pc.stable* function from the *bnlearn* R package [13] for the PC algorithm execution. We have run BO with the PES acquisition function over a set of Gaussian Bayesian networks generated following the simulation methodology of [6]. That is, the absent edges in the acyclic digraph G are sampled by using independent Bernoulli random variables with probability of success $d = n/(p-1)$, where p is the vertex number of G and n is the average neighbor size. The probability d can be thought of as an indicator of the density of the network: smaller d values mean sparser networks. The node size p is obtained from a grid of values $\{25, 50, 75, 100\}$, while the average neighbor size is $n \in \{2, 8\}$. Finally, we consider different sample sizes $N \in \{10, 50, 100, 500\}$. Therefore, we have a total of 32 different network learning scenarios, which are moderately high-dimensional and sparse, and representative of those that can be found in [6]. We create 40 different replicas of the experiment and report average results across them, in order to provide more robust results. In each of these replicas, the structure is randomly generated as described above, and its nonzero regression coefficients (Eq. (1)) are sampled from a uniform distribution on $[0.1, 1]$.

For BO, we have used the PES acquisition function and 10 Monte Carlo iterations for sampling the parameters of the GP. The acquisition function is averaged across these 10 samples. We have used the Mátern covariance function for $k(\cdot, \cdot)$ (Eq. (2)) and the transformation described in [3] so that the GP can deal with the categorical variable (the test type). We compare BO with a random search (RS) strategy of the average normalized SHD error surface and with the expert criterion (EC), taken from [6]. These authors recommend a value of $\alpha = 0.01$ and use the Fisher's Z partial correlation test. At each iteration, BO provides a candidate solution which corresponds to the best observation made so far. We stop the search in BO and RS after 30 evaluations of the objective, where we start to appreciate stability, specially for BO.

The average normalized SHD results obtained are shown in Fig. 2. We show the relative difference in log-scale with respect to the best observed result, since after normalizing the SHD differences become small because of the combinatorial network space. Therefore, the lower the values obtained, the better. We show the mean and standard deviation of this measure along the 40 replicas of the experiment, for each of the three methods compared (BO, RS and EC). We can see that EC is easily improved after only 10 iterations of BO and RS. Furthermore, BO outperforms RS providing significantly better results as more evaluations are performed. Importantly, the standard deviation of the results of BO are fairly small in the last iterations. This means that BO is very robust to the different replicas of the experiments.

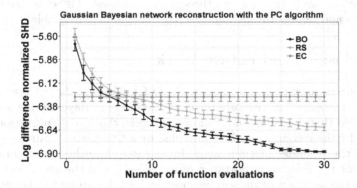

Fig. 2. Logarithmic difference with respect to the best observed average normalized SHD obtained in 40 replicas of the 32 considered Gaussian Bayesian networks.

Since the expert criterion is outperformed, we are interested in the parameter suggestions delivered by BO. In order to explore these results, we have generated two histograms that summarize the suggested parameters by BO in the last

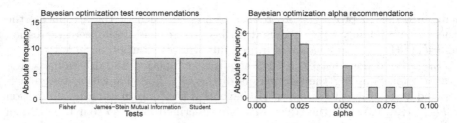

Fig. 3. Histograms with the recommended parameters by BO in the last iteration.

iteration, shown in Fig. 3. We observe that the most frequently recommended test is the James-Stein shrinkage estimator of the mutual information [4], while the most frequent recommendation for the significance level is concentrated at values lower than 0.025.

These results are very interesting from the viewpoint of graphical models learning. The first observation is that the optimal value obtained for the significance level is fairly close to the one suggested in [6]. However, the SHD results are arguably better for the BO than for the human expert. This may be explained by the second interesting result we have obtained. Namely, the shrinkage James-Stein estimator of the mutual information is suggested more times than the extended Fisher's Z partial correlation test. Therefore, in the context of sparse, high-dimensional networks, where we may have $p > N$ (such as in our experimental set-up and the one in [6]), it may be better to focus on the selection of the statistical test, rather than on carefully adjusting the significance level. In the literature, however, it is often done the other-way-around, and more effort is put on carefully adjusting the significance level. The code used for the experiments and figures throughout the paper is publicly available[1].

6   Conclusions and Future Work

In this paper we have proposed the use of BO for selecting the optimal parameters of PC algorithm for structure recovery in Gaussian Bayesian networks. We have observed that, in a small number of iterations, the expert suggestion is outperformed by the recommendations provided by a BO method. Furthermore, an analysis of the recommendations made by the BO algorithm shows interesting results about the relative importance of the selection of the statistical test, as opposed to the selection of the significance level. In the literature, however, it is often that the selection of the significance level receives more attention.

For future work, we plan to explore other objective measures that do not rely on knowing the true graph structure, such as network scores. We would also like to compare the performance of other acquisition functions as well as alternative hyper-parameter optimization methods such as genetic algorithms. Finally, we

---

[1] https://github.com/irenecrsn/bopc.

plan to extend this methodology to consider multi-objective optimization scenarios and also several constraints, since current BO methods are able to handle these problems too.

Acknowledgements. We acknowledge the use of the facilities of Centro de Computación Científica (CCC) at Universidad Autónoma de Madrid, and financial support from Comunidad de Madrid, grant S2013/ICE-2845; from the Spanish *Ministerio de Economía, Industria y Competitividad*, grants TIN2016-79684-P, TIN2016-76406-P, TEC2016-81900-REDT; from the Cajal Blue Brain project (C080020-09, the Spanish partner of the EPFL Blue Brain initiative); and from Fundación BBVA (Scientific Research Teams in Big Data 2016). Irene Córdoba is supported by grant FPU15/03797 from the Spanish *Ministerio de Educación, Cultura y Deporte*.

References

1. Bielza, C., Larrañaga, P.: Bayesian networks in neuroscience: a survey. Front. Comput. Neurosci. **8**, 131 (2014)
2. Colombo, D., Maathuis, M.H.: Order-independent constraint-based causal structure learning. J. Mach. Learn. Res. **15**(1), 3741–3782 (2014)
3. Garrido-Merchán, E.C., Hernández-Lobato, D.: Dealing with categorical and integer-valued variables in Bayesian optimization with Gaussian processes (2018). arXiv:1805.03463
4. Hausser, J., Strimmer, K.: Entropy inference and the James-Stein estimator, with application to nonlinear gene association networks. J. Mach. Learn. Res. **10**, 1469–1484 (2009)
5. Hernández-Lobato, J.M., Hoffman, M.W., Ghahramani, Z.: Predictive entropy search for efficient global optimization of black-box functions. In: Advances in Neural Information Processing Systems, pp. 918–926 (2014)
6. Kalisch, M., Bühlmann, P.: Estimating high-dimensional directed acyclic graphs with the PC-algorithm. J. Mach. Learn. Res. **8**, 613–636 (2007)
7. Lauritzen, S.L., Dawid, A.P., Larsen, B.N., Leimer, H.G.: Independence properties of directed Markov fields. Networks **20**(5), 491–505 (1990)
8. Malone, B., Järvisalo, M., Myllymäki, P.: Impact of learning strategies on the quality of Bayesian networks: an empirical evaluation. In: Proceedings of the Thirty-First the Conference on Uncertainty in Artificial Intelligence, pp. 562–571 (2015)
9. Malone, B., Kangas, K., Järvisalo, M., Koivisto, M., Myllymäki, P.: Empirical hardness of finding optimal Bayesian network structures: algorithm selection and runtime prediction. Mach. Learn. **107**(1), 247–283 (2018)
10. Minka, T.P.: Expectation propagation for approximate Bayesian inference. In: Proceedings of the Seventeenth Conference on Uncertainty in Artificial Intelligence, pp. 362–369 (2001)
11. Ness, R.O., Sachs, K., Vitek, O.: From correlation to causality: statistical approaches to learning regulatory relationships in large-scale biomolecular investigations. J. Proteome Res. **15**(3), 683–690 (2016)
12. Rasmussen, C.E.: Gaussian processes in machine learning. In: Bousquet, O., von Luxburg, U., Rätsch, G. (eds.) ML -2003. LNCS (LNAI), vol. 3176, pp. 63–71. Springer, Heidelberg (2004). https://doi.org/10.1007/978-3-540-28650-9_4
13. Scutari, M.: Learning Bayesian networks with the bnlearn R package. J. Stat. Softw. **35**(3), 1–22 (2010)

14. Shahriari, B., Swersky, K., Wang, Z., Adams, R.P., de Freitas, N.: Taking the human out of the loop: a review of Bayesian optimization. Proc. IEEE **104**(1), 148–175 (2016)
15. Snoek, J., Larochelle, H., Adams, R.P.: Practical Bayesian optimization of machine learning algorithms. In: Advances in Neural Information Processing Systems, pp. 2951–2959 (2012)
16. Tsamardinos, I., Brown, L.E., Aliferis, C.F.: The max-min hill-climbing Bayesian network structure learning algorithm. Mach. Learn. **65**(1), 31–78 (2006)

Identifying the Machine Learning Family
from Black-Box Models

Raül Fabra-Boluda[✉], Cèsar Ferri, José Hernández-Orallo,
Fernando Martínez-Plumed, and María José Ramírez-Quintana

DSIC, Universitat Politècnica de València, València, Spain
{rafabbo,cferri,jorallo,fmartinez,mramirez}@dsic.upv.es

Abstract. We address the novel question of determining which *kind* of machine learning model is behind the predictions when we interact with a black-box model. This may allow us to identify families of techniques whose models exhibit similar vulnerabilities and strengths. In our method, we first consider how an adversary can systematically query a given black-box model (oracle) to label an artificially-generated dataset. This labelled dataset is then used for training different surrogate models (each one trying to imitate the oracle's behaviour). The method has two different approaches. First, we assume that the family of the surrogate model that achieves the maximum Kappa metric against the oracle labels corresponds to the family of the oracle model. The other approach, based on machine learning, consists in learning a meta-model that is able to predict the model family of a new black-box model. We compare these two approaches experimentally, giving us insight about how explanatory and predictable our concept of family is.

Keywords: Machine learning families · Black-box model
Dissimilarity measures · Adversarial machine learning

1 Introduction

Machine Learning (ML) is being increasingly used in confidential and security-sensitive applications deployed with publicly-accessible query interfaces, e.g., FICO or credit score models, health, car or life insurance application models, IoT Systems Security, medical diagnoses, facial recognition systems, etc. However, because of these public interfaces, an attacker can query the model with special chosen inputs, get the results and learn how the model works from these input-output pairs –using ML techniques. This corresponds to the typical adversarial machine learning problem [5,11,15]. In an attack scenario, the attacker can take advantage of the knowledge of the type of learning technique (the ML family) the attacked model was derived from (and, in some cases, the true data distribution used to induce it) in order to explore intrinsic flaws and vulnerabilities. In this regard, several previous works have introduced specific strategies for attacking, extracting and stealing ML models of particular families such as *Support Vector*

© Springer Nature Switzerland AG 2018
F. Herrera et al. (Eds.): CAEPIA 2018, LNAI 11160, pp. 55–65, 2018.
https://doi.org/10.1007/978-3-030-00374-6_6

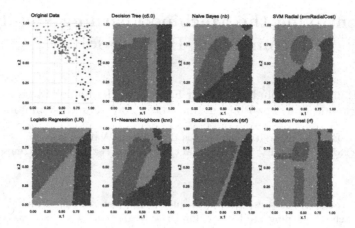

Fig. 1. Behaviour of different models learned over the same set of data (shown on the top-left plot). From left to right and from top to bottom: a decision tree, Naïve Bayes, a SVM, logistic regression, 11-Nearest Neighbour, Neural Network, and Random Forest.

Machines [3], *(deep) Neural Networks* [18,19], *Naive Bayes* [12], or even several online prediction APIs [22].

One of the main reasons for not having *general* techniques for exploiting black-box models may be due the intrinsic differences between ML techniques: different models constructed using different ML techniques might disagree not only on decision boundaries but also on how they extrapolate on areas with little or no training examples. Figure 1 illustrates this, where, on the top-left plot we see the original train data of a bivariate dataset that we use to learn several ML models (using different techniques). What we observe is that all the models behave similarly on dense zones (where the training examples are originally located), but their behaviour on sparse areas (without training examples) is unpredictable and depends on the learning technique used. We may say that these less dense zones are those more likely to contain vulnerabilities. Hence, for many applications, some characteristics of the model are more relevant than the model itself (e.g., what topologies the decision boundaries have). In other words, for many attacks it is more important to know what the model looks like than its full semantics.

In this paper we address the problem of determining which *kind* of ML technique (family) has been used to construct a model that behaves as a black box, given a relevant subset of queries. This could be seen as an initial step for an adversarial learning procedure where, once we have obtained some knowledge about the ML family of the model to be attacked, it is possible to apply specific successful techniques, such as those mentioned above. Regarding the method proposed, and given that, other than its behaviour, we do not have access to any information about the black-box model (original data distribution nor the learning algorithm used to train it). We just use the black-box model as an *oracle* for labelling a synthetic dataset, generated by following a specific query strategy. This is then used to learn different models (using different ML techniques belonging to different learning families) trying to imitate the oracle behaviour. We denote these new models as *surrogate models*. We analyse the performance of these surrogate models in order to predict the original family of the target black-box model. Furthermore, a more elaborated technique is also proposed employing meta classifiers and dissimilarity measures between surrogate models.

The paper is organised as follows. Section 2 briefly outlines the most relevant related work. Section 3 addresses the problem of predicting the ML family given a black-box model. The experimental evaluation is included in Sect. 4. Finally, Sect. 5 closes with the conclusions and future work.

2   Related Work

Although there is an extensive literature on the topic of model extraction related to learning theory (such as the probably approximately correct (PAC) model of learning [23] and its query-based variants [1,2]), as well as in the larger field of adversarial machine learning [5,11,15], in the vast majority of these approaches, the type of the model is assumed to be known [3,12,17–19,22].

However, we also find different approaches focused on the replication of the functionality of models whose type may be unknown. A simple way to capture the semantics of a black-box ML model consists of mimicking it to obtain an equivalent one. This can be done by considering the model as an oracle, and querying it with new synthetic input examples (queries) that are then labelled by the oracle and used for learning a new declarative model (the *mimetic model*) that imitates the behaviour of the original one. Domingos et al. [7] addressed this problem by creating a comprehensible mimetic model (decision trees) from an ensemble method. Similar posterior proposals focused on generating comprehensible mimetic models that also exhibit a good performance. In this regard, Blanco-Vega et al. analysed the effect of the size of the artificial dataset in the quality of the replica and the effect of pruning the mimetic model (a decision tree) on its comprehensibility [4], developing also an MML-based strategy [24] to minimise the number of queries. Papernot et al. [17] also applied a mimicking strategy aiming at crafting examples that game a black-box model in order to obtain the desired output: the crafted examples that are able to cheat the replica are likely to cheat the original model, by the property of transferability that they studied in this same work.

Unlike the previous approaches, in this paper we consider that an attacker's goal is not to replicate or extract the ML model, but to obtain key actual model characteristics, such as the ML family. This would be, in some cases, more relevant than the model itself as it can be considered as a crucial first step before applying those more specific techniques or approaches aforementioned. Regarding our approach, such as in the mimetic approach, we also consider the black-box model as an oracle that is used to label a set of artificially-generated input examples (we also assume that the original data is not available). But instead of learning one mimetic model, we use the labelled data for learning several surrogates models using different learning techniques that will be prove successful for the task at hand.

3   Model Family Identification

One way of determining the *ML family* of a model that is to be attacked could be to mimick it and analyse their particular decision boundaries layouts. Therefore, given the extracted topologies the boundaries of a black-box model have, we will be able to identify the ML family that usually has that kind of boundary. In this section we describe two approaches. Both just interact with the black-box model by means of queries.

3.1   Learning Surrogate Models

As we already mentioned in the introduction, in order to generate surrogate datasets, we can query an oracle O (the black-box model to be attacked) with artificial examples so that O labels them (Fig. 2 illustrates this). This allows us to build an artificial dataset labelled by O (what we call the *surrogate dataset SD*). As this dataset (the surrogate dataset SD) contains the output labels of O, it tends to capture the boundary patterns of O.

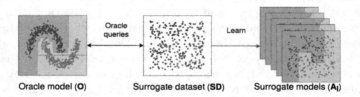

Oracle model (**O**)   Surrogate dataset (**SD**)   Surrogate models (**A₁**)

Fig. 2. Black-box models (oracles), trained over an unknown original dataset, are used to label synthetic surrogate datasets (generated following specific query strategies), which are then used to train surrogate models.

We can follow different strategies to query O. A basic (and effective) strategy consists in generating the artificial examples following a uniform distribution (SD, the surrogate dataset), which provides a good coverage of the feature space. We expect that such coverage includes non-dense areas where the behaviour of the induced black-box models is likely to be unexpected, possibly containing intrinsic vulnerabilities for specific techniques. Since SD is likely to capture the decision boundaries from O, we can thus learn other models from SD that mimic the behaviour of O. These surrogate models, denoted by A_i, $1 \leq i \leq N$, with N is the number of model families we consider in this work. In other words, we learn from SD a surrogate model A_i per model family $i \in N$. Each surrogate model A_i belongs to a different model family, then each A_i might provide a different characterisation of SD and, indirectly, a characterisation of O.

The most straightforward procedure to analyse how a model A_i behaves with respect to a dataset SD consists on evaluating this SD with A_i by cross-validation. As we want to use the set of surrogates models to identify the oracle's family, it makes sense to use the *Cohen's kappa* coefficient (κ) [14] as a dissimilarity metric to estimate the degree of inter-rate agreement for qualitative items. Unlike other evaluation measures, the kappa statistic is used not only to evaluate a single classifier, but also to evaluate classifiers between themselves (corrected by chance), and it is generally thought to be a more robust measure than a simple percentage of the agreement (see [10]). Actually, with kappa we can also take into account how different classifiers disagree on boundaries caused by extrapolation noise, as we have previously seen in Fig. 1. In our case, we measure the agreement between the oracle (that has been used to label SD) and the surrogate model A_i when explaining the same dataset SD. We can evaluate the surrogate dataset SD (e.g., by using a train-test split or cross-validation) with the different surrogate models A_i, so that we would obtain a κ_i for each A_i. Figure 3 illustrates this procedure.

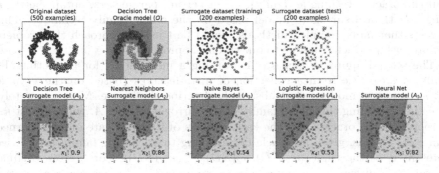

Fig. 3. Synthetic example to show the κ measure for different surrogate models A_i when compared to an specific oracle model O using a decision tree technique.

In the top of Fig. 3 we see (from left to right) the original dataset, the oracle model O, the surrogate training set (SD) and the test set (these two sets labelled by O). The training set SD is used to learn all the surrogate models A_i shown in the bottom row. The test set is used to evaluate each of the A_i (i.e., to obtain each κ_i score). It is easy to see that the surrogate model with greatest resemblance to O is the decision tree (bottom-left plot), as confirmed by its kappa value ($\kappa_1 = 0.90$), although some other techniques are also doing a good job (e.g., nearest neighbours). The surrogate decision tree shares more commonalities with the oracle (both models have high expressive power and their boundaries are always parallel to the axis) than any other since the techniques used to induce them, although different, belong to the same family. Regarding the nearest neighbours-based model, although it also has a high expressive power, the boundaries are clearly not parallel to the axes (they present a sawtooth pattern which would be particularly difficult to achieve using decision trees). This is reflected by a lower κ value. Other surrogate models such as Logistic Regression or Naïve Bayes (with less expressive power) obtain much worse results, as we can see either visually or on the basis of the κ measure.

3.2 Model Family Identification

From the previous example (Fig. 3) we saw that the family of the oracle model O was the one among the surrogate models A_i that achieved the highest κ. These suggests a straightforward procedure for family identification: we may expect that the family of O corresponds to the family of the A_i which provides the highest κ_i, since it implies the A_i achieves a high degree of agreement with O when explaining SD. Therefore, one first method we introduce consists in, given O, evaluating (using Cohen's Kappa) the generated SD with different surrogate models A_i (each A_i belonging to a different model family $i \in N$), and assigning the family of the A_i with highest κ value. However, this simple method might have its issues: it could be the case that other learning algorithms (different from the oracle's technique family), under certain parameters, obtain better κ using SD than a surrogate model based on the oracle family technique. This suggests that more complex methods should be tried to approach the problem more robustly, which takes us to our second approach.

The second approach consists in learning a meta-model for predicting the family of an oracle from a collection of meta-features (based on the κ value of the surrogate models) that abstractly describe the oracle. More concretely, instead of returning the model family of the surrogate model A_i with highest κ_i, we represent the oracle O by the kappa values of the surrogate models learned from SD. That is, given an oracle O belonging to a learning family $y \in N$, we represent it as the tuple $\langle \kappa_1(SD), \kappa_2(SD), \ldots, \kappa_N(SD) \rangle$. For a given original dataset, we can learn as many oracles as families $y \in N$, each one represented by the tuple of κ values of its corresponding surrogate models, as follows:

$$\langle \kappa_1(SD_1), \kappa_2(SD_1), \ldots, \kappa_N(SD_1) \rangle$$
$$\vdots$$
$$\langle \kappa_1(SD_N), \kappa_2(SD_N), \ldots, \kappa_N(SD_N) \rangle$$

If we apply this procedure for a set of D original datasets, we can construct a dataset collecting the tuples with the κ-based representation of the $D \times N$ oracles generated and adding the oracle family y to each tuple. This dataset can be used as the training set to learn a meta-model capable of predicting the model family y (the output) of a new incoming black-box model represented as a series of κ metrics (the input attributes of the meta-learning problem). This meta-model represents a similar approach to the top meta-model of Stacking [25], but in this case for the identification of families.

4   Evaluation

In this section we explain the experiments performed to validate our method for family identification[1]. All the experiments have been performed using R and, in particular, the package `Caret` to train the different ML models.

For the experiments, we have considered a number of ML techniques which have been widely used in practice, usually grouped in different families in terms of their learning strategy (see [8,9,16,21]). In particular, we have considered the following $N = 11$ model families: *Discriminant Analysis* (DA), represented by RDA technique; *Ensembles* (EN), represented by Random Forest; *Decision Trees* (DT), represented by C5.0 algorithm; *Support Vector Machines* (SVM) with Radial Basis Function Kernel; *Neural Networks* (NNET), represented by MLP, *Naïve Bayes* (NB), *Nearest Neighbours* (NN), *Generalized Linear Models* (GLM), represented by the *glmnet* technique; *Partial Least Squares and Regression* (PLSR), *Logistic and Multinomial Regression* (LMR) and *Multivariate Adaptive Regression Splines* (MARS). Model parameters were left to their default values.

In order to generate a dataset of meta-features that is large enough to evaluate our second approach, we employed $D = 25$ datasets (see Table 1).

Table 1. Characterisation of the 25 datasets from the UCI repository [6] used in the experiments: number of numerical (#*num*) and discrete (#*disc*) attributes, number of instances (#*inst*), and number of classes (#*class*).

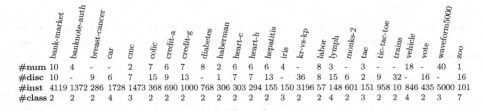

	bank-market	banknote-auth	breast-cancer	car	cmc	colic	credit-a	credit-g	diabetes	haberman	heart-c	heart-h	hepatitis	iris	kr-vs-kp	labor	lymph	monks-2	tae	tic-tac-toe	trains	vehicle	vote	waveform5000	zoo
#num	10	4	-	-	2	7	6	7	8	2	6	6	6	4	-	8	3	-	3	-	-	18	-	40	1
#disc	10	-	9	6	7	15	9	13	-	1	7	7	13	-	36	8	15	6	2	9	32	-	16	-	16
#inst	4119	1372	286	1728	1473	368	690	1000	768	306	303	294	155	150	3196	57	148	601	151	958	10	846	435	5000	101
#class	2	2	2	4	3	2	2	2	2	2	2	2	2	3	2	2	4	2	3	2	2	4	2	3	7

For each dataset, we trained $N = 11$ models (oracles) belonging to the different families introduced above (we learned $D \times N = 25 \times 11 = 275$ oracle models). For each of these oracles, we generated a surrogate dataset SD that we use to learn and evaluate the surrogate models A_i, belonging to the same N model families. Every example of SD was generated at random following the uniform distribution: for each numerical feature, we randomly generated a number between its minimum and maximum values. For each discrete feature, we randomly picked one of the possible values it may hold. The number of examples per SD was 100 per number of dimensions. The evaluation of A_i w.r.t. SD was performed using 5-fold stratified cross-validation to obtain each of the κ_i values

---

[1] For reproducibility and replicability purposes, all the experiments, code, data and plots can be found at https://github.com/rfabra/whos-behind.

(as previously described). At this point, we have a collection of 275 oracles represented as a series of κ_i measures, whose model family is known. This represents a dataset of meta-features, along with the family as label. Once the dataset of oracles is created, as explained in Sect. 3.2, we use a support vector machine algorithm to learn the meta-model, and we apply a leave-1-out evaluation, such that in each iteration the instances corresponding to one original dataset are used for test and the rest are used for training.

Table 2 (left) shows the confusion matrix for the experiments using the approach based solely on surrogate models and dissimilarity measures (kappa values). We observe that LMR, DT and MARS are the those with a higher positive identifications, with 22, 16 and 14 correct identifications respectively. Other cases perform very poorly, such as the DA, EN, SVM, NNET NB or PLSR families, with none or very few correct identifications. There are many cases in which a model is strongly confused with other models. For instance, the DA family tends to be confused with GLM (14 wrong identifications). SVM, NNET and GLM are often confused with LMR (12, 15 and 10 wrong identifications, respectively), NB gets confused with MARS (13 wrong identifications), and something similar happens with NN with SVM (11 wrong identifications). The overall accuracy has been 30%. All this suggests that the decision boundaries between families might not be so clear. In any case, as this is an 11-class classification problem, a random classification (baseline) would obtain an accuracy around 9%, so this straightforward approach improves the random baseline significantly and suggests that dissimilarity measures might be partially useful for the model family identification task.

Table 2. Confusion matrix (Real vs. Predicted Class) obtained when evaluating the approach based solely on surrogate models (Left, Acc: 30%) and the one based on meta-models (Right, Acc: 56%).

Family	DA	EN	DT	SVM	NNET	NB	NN	GLM	PLSR	LMR	MARS
DA	1	0	0	7	0	0	0	14	0	1	1
EN	0	7	2	4	0	0	1	1	0	6	4
DT	0	7	16	0	0	0	0	1	0	0	1
SVM	0	1	2	5	0	0	2	0	0	12	3
NNET	0	0	0	0	2	0	0	8	0	15	0
NB	2	1	1	0	2	1	0	4	0	1	13
NN	0	5	4	11	0	0	4	0	0	1	0
GLM	0	2	0	1	0	0	0	11	0	10	0
PLSR	1	0	0	0	2	0	0	18	0	3	1
LMR	0	0	0	0	0	0	0	2	0	22	1
MARS	0	2	3	1	0	0	0	0	0	4	14

Family	DA	EN	DT	SVM	NNET	NB	NN	GLM	PLSR	LMR	MARS
DA	9	1	1	0	1	2	0	0	10	0	1
EN	1	12	4	3	0	1	2	1	0	1	0
DT	0	1	21	0	0	1	0	0	0	0	2
SVM	1	5	1	10	0	0	4	1	0	3	0
NNET	0	0	0	0	19	0	0	5	1	0	0
NB	5	1	0	0	0	13	1	1	3	0	1
NN	1	2	1	1	0	0	19	0	0	0	1
GLM	0	1	0	0	10	1	0	9	1	1	1
PLSR	4	0	0	0	2	2	0	0	15	1	1
LMR	0	1	2	2	2	0	1	2	1	13	1
MARS	1	0	2	0	0	1	1	2	0	3	14

Table 2 (right) shows the confusion matrix obtained for the meta-model-based approach. In this case, we can observe a general improvement in the results with respect to the previous approach accuracy increases from 30% to 56%). Here, DT is the easiest family to identify, with 21 of 25 correct identifications. For other families, such as NNET or NN, the method also performs well, with 19 correct identifications. DA is mainly confused with PLSR, with 10 wrong identifications. Similarly, GLM is confused with NNET, also with 10 wrong identifications. From the results we can see GLM and DA are the hardest families to identify, since they are strongly confused with others. The rest of families (i.e., NB, LMR, PLSR or MARS) are correctly identified in most cases, showing no specific confusion patterns between them and other models.

In general terms, the results show that, although this is a particularly complex problem, the use of dissimilarity measures to differentiate ML families from one another seems an effective approach.

5   Conclusions and Future Work

In this work we addressed the problem of identifying the model family of a black-box learning model. We presented two approaches based on dissimilarity measures such as the Cohen's kappa coefficient. The first one consists in learning several surrogate models (from different learning families) from a set of artificial examples labelled by the black-box model (which acts as an oracle). Then we select the family of the surrogate model with the best kappa value as the family of the black-box model. The second approach consists in using the kappa values as meta-features for representing the black-box model, and learning a meta-model that, given a black box model, predicts its learning family.

The experiments show that the first proposed approach, although it performs poorly as a ML family identification method, it is able to improve significantly the accuracy we would obtain using a random baseline. The second approach based on the meta-model performs much more accurately than the first one, laying special emphasis on the potential of using meta-models trained with abstract meta-features (based on dissimilarities) characterising the oracles.

In order to improve the results of our approach based on the meta-model, we plan to investigate the use of other measures for model divergence and diversity [13, 20] as meta-features. Another aspect that could be of interest would be the use of other query strategies for generating the surrogate dataset. Finally, as our approaches for model identification rely on the kappa measure, we also intend to establish a new definition of model family by applying a hierarchical learning clustering algorithm using the kappa metric to group the learning techniques for conforming the different families.

Acknowledgements. This material is based upon work supported by the Air Force Office of Scientific Research under award number FA9550-17-1-0287, the EU (FEDER), and the Spanish MINECO under grant TIN 2015-69175-C4-1-R, the Generalitat Valenciana PROMETEOII/2015/013. F. Martínez-Plumed was also supported by INCIBE under grant INCIBEI-2015-27345 (Ayudas para la excelencia de los equipos de investigación avanzada en ciberseguridad). J. H-Orallo also received a Salvador de Madariaga grant (PRX17/00467) from the Spanish MECD for a research stay at the CFI, Cambridge, and a BEST grant (BEST/2017/045) from the GVA for another research stay at the CFI.

References

1. Angluin, D.: Queries and concept learning. Mach. Learn. **2**(4), 319–342 (1988)
2. Benedek, G.M., Itai, A.: Learnability with respect to fixed distributions. Theor. Comput. Sci. **86**(2), 377–389 (1991)
3. Biggio, B., et al.: Security Evaluation of support vector machines in adversarial environments. In: Ma, Y., Guo, G. (eds.) Support Vector Machines Applications, pp. 105–153. Springer, Cham (2014). https://doi.org/10.1007/978-3-319-02300-7_4
4. Blanco-Vega, R., Hernández-Orallo, J., Ramírez-Quintana, M.J.: Analysing the trade-off between comprehensibility and accuracy in mimetic models. In: Suzuki, E., Arikawa, S. (eds.) DS 2004. LNCS (LNAI), vol. 3245, pp. 338–346. Springer, Heidelberg (2004). https://doi.org/10.1007/978-3-540-30214-8_29
5. Dalvi, N., Domingos, P., Sanghai, S., Verma, D., et al.: Adversarial classification. In: Proceedings of the 10th ACM SIGKDD International Conference on Knowledge Discovery and Data Mining, pp. 99–108. ACM (2004)
6. Dheeru, D., Karra Taniskidou, E.: UCI machine learning repository (2017). http://archive.ics.uci.edu/ml
7. Domingos, P.: Knowledge discovery via multiple models. Intell. Data Anal. **2**(3), 187–202 (1998)
8. Duin, R.P.W., Loog, M., Pękalska, E., Tax, D.M.J.: Feature-based dissimilarity space classification. In: Ünay, D., Çataltepe, Z., Aksoy, S. (eds.) ICPR 2010. LNCS, vol. 6388, pp. 46–55. Springer, Heidelberg (2010). https://doi.org/10.1007/978-3-642-17711-8_5
9. Fernández-Delgado, M., Cernadas, E., Barro, S., Amorim, D.: Do we need hundreds of classifiers to solve real world classification problems. J. Mach. Learn. Res. **15**(1), 3133–3181 (2014)
10. Ferri, C., Hernández-Orallo, J., Modroiu, R.: An experimental comparison of performance measures for classification. Pattern Recognit. Lett. **30**(1), 27–38 (2009)
11. Giacinto, G., Perdisci, R., Del Rio, M., Roli, F.: Intrusion detection in computer networks by a modular ensemble of one-class classifiers. Inf. Fusion **9**(1), 69–82 (2008)
12. Huang, L., Joseph, A.D., Nelson, B., Rubinstein, B.I., Tygar, J.: Adversarial machine learning. In: Proceedings of the 4th ACM Workshop on Security and Artificial Intelligence, pp. 43–58 (2011)
13. Kuncheva, L.I., Whitaker, C.J.: Measures of diversity in classifier ensembles and their relationship with the ensemble accuracy. Mach. Learn. **51**(2), 181–207 (2003)
14. Landis, J.R., Koch, G.G.: An application of hierarchical kappa-type statistics in the assessment of majority agreement among multiple observers. Biometrics **33**, 363–374 (1977)
15. Lowd, D., Meek, C.: Adversarial learning. In: Proceedings of the 11th ACM SIGKDD International Conference on Knowledge Discovery in Data mining, pp. 641–647. ACM (2005)
16. Martınez-Plumed, F., Prudêncio, R.B., Martınez-Usó, A., Hernández-Orallo, J.: Making sense of item response theory in machine learning. In: Proceedings of 22nd European Conference on Artificial Intelligence (ECAI). Frontiers in Artificial Intelligence and Applications, vol. 285, pp. 1140–1148 (2016)
17. Papernot, N., McDaniel, P., Goodfellow, I.: Transferability in machine learning: from phenomena to black-box attacks using adversarial samples. arXiv preprint arXiv:1605.07277 (2016)

18. Papernot, N., McDaniel, P., Jha, S., Fredrikson, M., Celik, Z.B., Swami, A.: The limitations of deep learning in adversarial settings. In: 2016 IEEE European Symposium on Security and Privacy (EuroS&P), pp. 372–387. IEEE (2016)
19. Papernot, N., McDaniel, P., Wu, X., Jha, S., Swami, A.: Distillation as a defense to adversarial perturbations against deep neural networks. In: 2016 IEEE Symposium on Security and Privacy (SP), pp. 582–597. IEEE (2016)
20. Sesmero, M.P., Ledezma, A.I., Sanchis, A.: Generating ensembles of heterogeneous classifiers using stacked generalization. Wiley Interdiscip. Rev.: Data Min. Knowl. Discov. 5(1), 21–34 (2015)
21. Smith, M.R., Martinez, T., Giraud-Carrier, C.: An instance level analysis of data complexity. Mach. Learn. 95(2), 225–256 (2014)
22. Tramèr, F., Zhang, F., Juels, A., Reiter, M.K., Ristenpart, T.: Stealing machine learning models via prediction APIs. In: USENIX Security Symposium, pp. 601–618 (2016)
23. Valiant, L.G.: A theory of the learnable. Commun. ACM 27(11), 1134–1142 (1984)
24. Wallace, C.S., Boulton, D.M.: An information measure for classification. Comput. J. 11(2), 185–194 (1968)
25. Wolpert, D.H.: Stacked generalization. Neural Netw. 5(2), 241–259 (1992)

Participatory Design with On-line Focus Groups and Normative Systems

Marlon Cárdenas[1(✉)], Noelia García Castillo[1,2(✉)], Jorge Gómez-Sanz[1(✉)], and Juan Pavón[1(✉)]

[1] Universidad Complutense de Madrid, 28040 Madrid, Spain
{marlonca,ngarc02,jjgomez,jpavon}@ucm.es
[2] Conscious Management Institute (CMI), Madrid, Spain
http://grasia.fdi.ucm.es

Abstract. Participatory design is a generally accepted practice for the construction of Ambient Assisted Living (AAL) systems. The involvement of experts and users in the conception and design of assistive solutions can lead to better systems. A common technique to involve users is called focus-group, which is mainly a moderated group meeting. Despite its benefits, it cannot be neglected the implicit cost of preparing and performing such meetings, and ensuring, later on, that the resulting assistive solution meets the requirements. A disruptive way to change this situation is the application of ICT technologies. This work contributes with a proposal for partial automation of focus-group techniques that support on-line evaluation of assistive solutions during the conception stage. Also, the paper addresses the formalization of the evaluation feedback through the use of normative systems.

1 Introduction

The participation of end-users in the design of Ambient Assisted Living (AAL) solutions is traditionally made through social sciences techniques such as focus groups. This technique implies having representatives of selected target groups that meet together in a room to discuss questions that are raised by a moderator. The goal is to gather feedback and then analyze the outcome to infer key features the system must have as well as the impression of end-users about the interest of the technology.

There are many kinds of focus groups such as *full groups* (90 to 120 min of discussion among 8 to 10 people), *minigroups* (similar to a full group but with 4 to 6 people), and *telephone group* (similar to the previous ones, but conducted through a phone call) [1]. They imply a variety of costs such as recruitment expenses, participants and moderators fees, report elaboration plus transcriptions, external stimuli costs (to foster discussions), travel, preparation and facility occupation, to cite some [1,2]. The possibility to consider an on-line version is attractive to reduce such costs and to provide some more agile working method, as well as the possibility to reach a wider participation. There is also an additional issue about what is the outcome of the focus group and how to apply it

© Springer Nature Switzerland AG 2018
F. Herrera et al. (Eds.): CAEPIA 2018, LNAI 11160, pp. 66–75, 2018.
https://doi.org/10.1007/978-3-030-00374-6_7

afterwards. If it is a document, there is a risk that the document is not applied correctly or it is even forgotten.

This work presents a web-based tool that assists in the execution of on-line focus groups. This is based on a customization of a widespread content management system, Joomla[1], and the use of simulations, which are shown to users as videos representing scenarios of the problem and solutions, to support the discussion. An important contribution also is the approach for characterizing the results of the focus group into a set of formal norms that can be used to validate the simulation and accumulate knowledge about what features the simulated scenarios must have. The normative characterization of focus group results has as an antecedent in the context of admission procedures [3]. That work showed how to apply focus group techniques to gain knowledge on the admission process to a higher education institution, and then proceeded to formalize that knowledge.

The rest of the paper is organized as follows. Section 2 introduces basic concepts on on-line focus groups and how they can be supported by ICT tools. Section 3 addresses the second issue, how to express the main results of the analysis through an operationalization using normative systems. Section 4 exemplifies the approach with a sample case study. Related works are discussed along each section. Section 5 presents the conclusions.

2   On-line Focus Group Requirements

Henning [4] introduces the features of a focus group. A focus group does not pursue the agreement. It is made of people with some similar background or shared experiences that makes them suitable for the experience. The discussion is focused on a single or limited number of topics. Discussion is essential and it must be facilitated by the moderator, who will pose questions to promote the discussion. A non-threatening experience is essential.

Krueguer [5] established six requirements for a focus group interview: it involves people; it is conducted in a series; participants are homogeneous and unfamiliar with each other; it is a method of data collection; data are qualitative; and the discussion is focused. According to Turney and Pockne [6], on-line text-based focus groups can meet those six criteria. The on-line arrangement could be compatible with such features, but others are harder. For instance, the discussion itself in a focus group is recorded and anyone is free to participate. On-line methods could be made via video/audio-conference to give a similar feeling, but this would increase the costs (transcriptions costs to be precise).

On the other hand, the on-line focus group can offer greater anonymity than the telephone focus group approach in [1], but it may constraint the participation if users are not used to the Internet (which, in fact, is much less frequent as years go by). Despite whether the group is on-line or not, a focus group requires previous preparation. The participants have to be selected and the session be carefully planned: what questions and topics to raise, what to do if threatening

---

[1] url: https://www.joomla.org/.

situations arise, where to invest the discussion time the most. There may be needs to create additional stimuli [1] so that participants know better what to comment about. In the context of AAL, we assume that preliminary questions would be oriented either to understand better the context of the AAL system to be built; or to point out some desirable feature of the AAL system; or to get feedback on some available feature of the AAL system.

After consulting the literature, we decided to do a text-based on-line focus group to facilitate the access to participants. The reason for this approach is mainly to provide a free-text input that allows each participant to express whatever they think without constraints. This generates a posterior problem related with information management, which will be discussed later on. The principal disadvantage of text-based focus groups is inadequate data richness and quality. On-line participants often give fewer details and examples of their opinions or short answers [7]. Also, on-line communication is prone to misinterpretation and may drive to flaming or confrontations due to more extreme opinions [8].

Despite all these risks, scholars argue that, if executed properly, on-line text-based focus groups can be useful [7], especially in certain situations in which task orientation is more desirable than social interaction [9]. This methodology saves time and money because the complete transcription is available instantly, it also implies that participants can reflect on what has been written and refer to previous exchanges that are stored in easily accessible archives [10]. Furthermore, running costs are lower as there is no renting and no participants' travels.

The size and composition of participants for on-line focus groups are key elements for the success [4]. Instead of performing a pre-selection, it may be worth to think the other way around: to open participation and devise a filter to know which participants meet the group criteria. After the filter, which can be implemented with a survey, it would be possible to conduct people to different groups and let them discuss the topics. Researchers would be free to ignore less experienced groups or to center their attention in some specific groups.

The access for the on-line focus group would be through a web page to reduce technology requirements to a minimum. This web page should include a short survey to filter out the participants. After filling in the survey, each participant would be driven to the proper discussion group page. The page should start with a brief description of the topic of the discussion plus a motivating video about the topic. There is an initial text explaining the aim of the research and how the participants can contribute. Then, the on-line participation would follow, with input forms that permit to give free text feedback, scoring, quote, or reply to other users. In the meantime, moderators could intervene to reorient the discussion or to prevent threatening situations.

These elements can be summarized with the following required features:

1. Create a web page with basic contents that introduce the research, the purpose of the focus group, including some motivating material, like videos.
2. Characterize participants through an initial survey to know more about their suitability towards the study. The initial survey is critical for later actions, since it permits to weight later actions.

3. Recognize which input is providing each individual. The review of each individual feedback needs to be done taking into account the background of the user. For instance, if the person is a health expert, the opinion will be more relevant than if it is a student of the subject.

4. Allow individuals to score the input to signify agreement or disagreement about the statement. This permits users to value which feedback is more relevant according to their opinion. Since users are identified and their profile is known, the importance of the feedback can be weighted too. It would represent situations where people tend to hum in approval when someone says something. Such signs constitute non-verbal cues that the moderator has to pay attention to [4]. An alternative to scoring to facilitate this non-verbal communication would be the use of emoticons. If scoring is anonymous for other participants, it becomes a more direct way of approving/disapproving whatever other people say, which suits better the nature of a focus group.

5. The existence of a score can be used to gamify the participation. Though it may add some bias to the discussion, if opinions with highest scores are highlighted, it can motivate people to contribute more actively.

6. Allow to quote or reply to posts in the discussion. The focus group should not promote the creation of separated discussion groups within the group and all the feedback should keep within the initial topics. Nevertheless, when one user needs to address or precise of another, such activities should not be prevented.

7. Support the moderator with the following basic functions: highlight feedback from the moderator, allow to kick out and ban IPs, also to post text that attracts the attention of the group. These functions pursue that the moderator can actually moderate the group if necessary and act rigorously when needed.

From these features, only the scoring is more extraordinary with respect to the traditional concept of focus group.

2.1 Tools for Implementing On-line Focus Groups

A content management system, such as Joomla, provides a set of modules with support for implementing the required functionality. The basic tool provides identification features (feature 3) plus the capability of setting up web pages with arbitrary content (feature 1). Modules for comments allow to handle moderation (feature 7) and user interaction (features 6 and 4). There is also a dash panel that includes indicators such as most popular topics, most active posters, and similar, which is in line with feature 5. Surveys can be conducted too within Joomla, what would allow to satisfy feature 2.

Figure 1 shows an example of an introductory page for an on-line focus group. On the left there is a simple survey for people that logs in. This survey can be configured by the developer depending on the subject and purpose of the research. The center of the page includes an introduction to the discussion group, with a motivating video. At the right there is the start of the discussion. The discussion happens as a set of comments added after a video that illustrates the problem and a potential solution. The user can approve or disapprove other people's comments, or reply to them. The video is the cited stimuli in the previous section. It is produced using a software tool for capturing AAL scenarios, called AIDE [11]. This software translates a description of the system into a 3D simulation of the AAL context and the AAL solution.

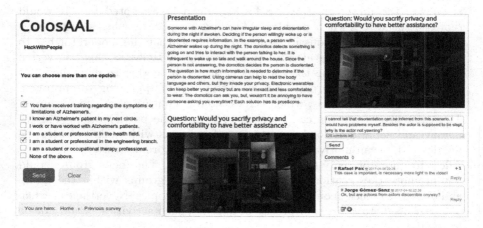

Fig. 1. The on-line discussion tool. Initial survey and the execution stage (two snapshots to the right side). The later is divided into two sections, the top part with the context (left), and the bottom part with the interaction (right)

The Joomla styles have been modified to fit into a 5 in. mobile phone screen, though it is still visible in a 4 in. screen. When the on-line focus group is finished, data is gathered to make a qualitative analysis. Joomla does not facilitate the later analysis of this information (and this is not in the scope of this paper).

3   Formalizing the Feedback from On-line Focus Groups

Feedback from the on-line discussion is formalized in two ways: as a visual specification of the video and as a deontic description of what is allowed and what is forbidden. The video alternative is not developed fully here. The stimuli video from Fig. 1 is generated with the AIDE tool [11]. Since the discussion topic and video do refer to the same topic, it is expected that feedback can be translated too to the video and generate a new stimuli for another, or the same, on-line focus group. The 3D simulation captures the expected behavior of the assistive solution.

The outcome of the discussion in the on-line focus group will be a set of norms to be applied to the AAL case study, which will return the number of violations that have occurred during the generation of a video such as the aforementioned.

This sets the development goal for the visual specification of the AAL system: to create a new specification that reduces the number of violations.

In this context, the representation of the knowledge is deontic because it bases on the use of permissions, obligations, and prohibitions. More specifically, the outcome of the discussion group could be expressed using deontic norms, in a language such as Drools, as it is done in [12], which is inspired in the work [13]. This formalism to express the norms is based on rule-based systems. In this kind of systems, there is an antecedent and a consequent, as the example in Fig. 2 shows. There evidences to be found may refer to information injected by the simulation or by other rules. The consequent creates new information, enacting a forward reasoning scheme.

A norm in this context is made of two elements: specific count-as rules [13] and definition of the norm state transitions, which will follow [12] representation. Mainly, the count-as rule will translate low level events into higher level evidences. In the example in Fig. 2, an *AgentPHATEvent* represents a low level simulation event. This event includes complete information about what is happening, in particular, what animation is the actor playing. The animation indicates what movements are being executed. In this case, the animation is translated as a separated information.

```
rule "countas: Identify animation being used"
  salience 100
        when // ANTECEDENT: EVIDENCES TO BE FOUND
        // An actor has performed some action and is performing some animation
            $ae: AgentPHATEvent(!animation.equals(""))
        // It is the last event referring to the same actor
        // and the animation is the last valid one
        not ( $ai: AgentPHATEvent($ae.getId().equals($ai.getId()),
                    !animation.equals($ae.getAnimation()),
                    $ae.getTime()<$ai.getTime()))
        then // CONSEQUENT CREATE NEW EVIDENCES TO TRIGGER ALL RULES
             // IF ANY ANTECEDENT DISAPPEARS OR CHANGES, THIS
             // EVIDENCE IS RETRACTED
        // A new animation is asserted
            insertLogical( new CurrentAnimation($ae.getId(),
                $ae.getAnimation(),$ae.getTime()) ) ;
end
```

Fig. 2. Example of a count-as rule to identify the animation performed by some actor

Time is important here. Drools implements temporal modal logic, but the version used for this development could not express time units or determine how soon, or late, something happens, only whether something happens after, before, or never. In order to cope with this issue, [12] proposed to include timestamps in the Drools facts and process them explicitly. A norm follows the life-cycle that is shown in Fig. 3. A norm is defined by a set of rules that define the transitions. Each norm specializes a generic transition rule.

By default, a norm will be in the *All OK* state. In general, the violation in the case of a prohibition happens because an evidence appears that should not exist. In the Fig. 4, the violation is asserted as soon as the evidence is found. In the

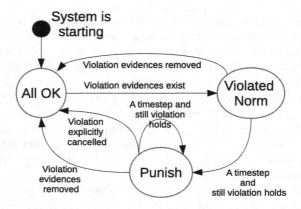

Fig. 3. Norm life-cycle as presented in [12]

```
rule "violation: template of prohibition violation"
  salience 100
        when
        // Some evidences appear that should not exist
              $ae: NameOfTheEvidence
        then
        // A new animation is asserted
              insertLogical(new NameOfTheViolatedNorm($ae.getId())) ;
end
rule "violation: template of obligation violation"
  salience 100
        when
        // Two evidences should happen together
              $ae: NameOfTheEvidence1
              not NameOfTheEvidence2(Math.abs(
              $ae.getTime()-getTime())<TIME_WINDOW)
        then
        // A new animation is asserted
              insertLogical(new NameOfTheViolatedNorm($ae.getId())) ;
end
```

Fig. 4. Template for prohibition/obligation transitions

case of an obligation, because an evidence that should exist, does not anymore. Both cases can be represented through the *existence* operator in the Drools rule system and activate the state *violated norm*. The absence of information in the case of the obligation (something is not happening) implies some difficulty. In the Fig. 4, the solution is to consider a time window sufficiently big referred to two evidences that should occur in time farther than *TIME_WINDOW*. For instance, a person receives assistance whenever the person falls can be expressed as the coexistence of "person is falling" and "person is being assisted" within a *TIME_WINDOW*.

The transition from *violated norm* to *punish* (see Fig. 3) is automatic as long as the evidences that made the norm violated still holds. When evidences do not hold anymore, the *All ok* state is restored. This is handled automatically by the truth maintenance of Drools and the *insertLogical* operator. *InsertLogical* retracts asserted facts when the antecedent of the rule that created them are no more valid, i.e., they were modified and the antecedent is not true anymore.

Violations need to be recorded, and this is done automatically with additional rules that are enabled when a *NameOfTheViolatedNorm* instance exists.

A norm will represent something the focus group determined that should occur or that should not occur. The difference between a prohibition and an obligation can be as simple as a negation of the existence of the evidence. The concrete application requires some practice, but, in general, it conforms a four step procedure:

1. Translate the events in the system as "countas" rules to generate more high level information.
2. Label the information with the appropriate simulation time stamp.
3. Combine these pieces of information to express what should/should not happen
   (a) Should happen. Then the user has to define two evidences that should happen within a *TIME_WINDOW*.
   (b) Should not happen. Then the user has to identify one evidence that should not exist.
4. Execute the simulation and compute the number of violations.

At the end of the translation of focus group evidences to norms, the result should be:

- A list of prohibition norms. Referring to events that should not happen.
- A list of obligation norms. Referring to events that should happen.

Such norms can be stored and associated to the simulation that is being evaluated.

4   Case Study

The discussion with some groups in the context of the ColoSAAL project (see the Acknowledgements section after the Conclusions), has some illustrative cases such as the one that is shown in Fig. 1 under the title 'Irregular sleep and disorientation'. This was supported by a simulation that depicts an scenario of a person waking up disoriented in the middle of the night. The simulation shows how the smart house assists the person.

Over the simulation, a text was included to explain the simulation and the starting question for participants' interaction: "Someone with Alzheimer has irregular sleep and disorientation during the night if awakened. Deciding if the person is disoriented requires some information. In the example, a person wakes up during the night. The sensors detect something is going on and the AAL system tries to interact with the person talking to her. It is infrequent to wake up so late and walk around the house. Since the person is not answering, the AAL system decides that the person is disoriented (how the system processes context-aware information is out of the scope of this paper but a discussion is made in [14] and [15]). The question is how much information is needed to determine if the person is disoriented. Using cameras can help to read the body language and others, but they invade the privacy. Electronic wearables can keep better the privacy but are more inexact and less comfortable to wear. The AAL system can ask the person, but it may be annoying to have someone asking everytime. Each solution has its pros and cons. *Would you sacrifice privacy and comfortability to have better assistance?"*

Then, the question to be discussed is how much privacy a person would give in to have better assistance. As developers, we know cameras are the most effective option, but also the most invading one. This initial feedback is translated into norms following the approach from Sect. 3. In this case, the natural language description is included for the sake of space.

- Obligation: There has to be more light than (set here the acceptable light threshold) to see. To check this, the simulation needs to return simulation parameters that permit to check luminosity level.
- Obligation: Actors' actions have to be visible. This obligation could be subsumed by the previous one.
- Obligation: The actor should yawn at some point to show the person is sleepy. For this to be checked, two evidences are needed: the existence of a yawn animation (first evidence) within the *TIME_WINDOW* while using the bathroom (second evidence).

Such norms would be expressed as Drools rules following the templates from Sect. 3 and, then, they would be applied to the simulation in order to produce a quantitative evaluation of how many times the simulation violated the norms. Then, the developer could reorient the simulation with the goal of reducing the violations.

5   Conclusions

This work has justified the need of cost-efficient techniques to implement focus-groups. Performing these on-line contributes to reduce the costs, and is an opportunity to facilitate a wider application of this technique. However, the lack of direct interaction of the group in a room at the same time requires some extra motivation or incentives for participation. These issues have been addressed in the design of a on-line focus group, which is based on a content management system (Joomla in this case), which has been properly configured. The main tool for motivation here is the use of simulations of scenarios of problems and potential solutions. This allows to translate gathered feedback into visual specifications and norms. The support of model-driven development [16] with the AIDE toolbox [11] facilitates an agile development cycle with several iterations, where the on-line focus group facilitates the involvement of end-users and domain experts.

Acknowledgements. We acknowledge support from the project "Collaborative Ambient Assisted Living Design (ColoSAAL)" (TIN2014-57028-R) funded by Spanish Ministry for Economy and Competitiveness; and MOSI-AGIL-CM (S2013/ICE-3019) co-funded by the Region of Madrid Government, EU Structural Funds FSE, and FEDER.

References

1. Greenbaum, T.L.: The Handbook for Focus Group Research. SAGE, Thousand Oaks (1998)
2. Morgan, D.L.: The Focus Group Guidebook, vol. 1. SAGE Publications, Thousand Oaks (1997)
3. Effah, J., Liu, K.: Virtual process modelling informed by organisational semiotics: a case of higher education admission. In: Liu, K., Gulliver, S.R., Li, W., Yu, C. (eds.) ICISO 2014. IAICT, vol. 426, pp. 42–51. Springer, Heidelberg (2014). https://doi.org/10.1007/978-3-642-55355-4_5
4. Hennink, M.M.: Focus Group Discussions. Series in Understanding Qualitative Research, 1st edn. Oxford University Press, Oxford (2014)
5. Krueger Richard, A., Anne, C.M.: Focus Groups. A Practical Guide for Applied Research. SAGE, Thousand Oaks (1994)
6. Turney, L., Pocknee, C.: Virtual focus groups: new frontiers in research. Int. J. Qual. Methods **4**(2), 32–43 (2005)
7. Vicsek, L.: Improving data quality and avoiding pitfalls of online text-based focus groups: a practical guide. Qual. Rep. **21**(7), 1232 (2016)
8. Oringderff, J.: "My way": piloting an online focus group. Int. J. Qual. Methods **3**(3), 69–75 (2004)
9. Walston, J.T., Lissitz, R.W.: Computer-mediated focus groups. Eval. Rev. **24**(5), 457–483 (2000)
10. Giles, D., Stommel, W., Paulus, T., Lester, J., Reed, D.: Microanalysis of online data: the methodological development of "digital ca". Discourse, Context Media **7**, 45–51 (2015)
11. Gómez-Sanz, J.J., Cardenas, M., Pax, R., Campillo, P.: Building Prototypes Through 3D Simulations. In: Demazeau, Y., Ito, T., Bajo, J., Escalona, M.J. (eds.) PAAMS 2016. LNAI, vol. 9662, pp. 299–301. Springer, Cham (2016)
12. Gomez-Sanz, J.J., Sánchez, P.C.: Domain independent regulative norms for evaluating performance of assistive solutions. Pervasive Mobile Comput. **34**, 79–90 (2017)
13. Álvarez-Napagao, S., Aldewereld, H., Vázquez-Salceda, J., Dignum, F.: Normative monitoring: semantics and implementation. In: Proceedings of The Multi-agent Logics, Languages, and Organisations Federated Workshops (MALLOW 2010), Lyon, France, 30 August–2 September 2010 (2010)
14. Fernández-De-Alba, J.M., Fuentes-Fernández, R., Pavón, J.: Architecture for management and fusion of context information. Inf. Fusion **21**, 100–113 (2015)
15. Alfonso-Cendón, J., Fernández-de Alba, J.M., Fuentes-Fernández, R., Pavón, J.: Implementation of context-aware workflows with multi-agent systems. Neurocomputing **176**, 91–97 (2016)
16. Gómez-Sanz, J.J., Pavón, J.: Meta-modelling in agent oriented software engineering. In: Garijo, F.J., Riquelme, J.C., Toro, M. (eds.) IBERAMIA 2002. LNCS (LNAI), vol. 2527, pp. 606–615. Springer, Heidelberg (2002). https://doi.org/10.1007/3-540-36131-6_62

Evaluation in Learning from Label Proportions: An Approximation to the Precision-Recall Curve

Jerónimo Hernández-González(✉)

Intelligent Systems Group, University of the Basque Country UPV/EHU,
Donostia, Spain
jeronimo.hernandez@ehu.eus

Abstract. In the last decade, the learning from label proportions problem has attracted the attention of the machine learning community. Many learning methodologies have been proposed, although the evaluation with real label proportions data has hardly been explored. This paper proposes an adaptation of the area under the precision-recall curve metric to the problem of learning from label proportions. The actual performance is bounded by minimum and maximum approximations. Additionally, an approximate estimation which takes advantage of low-uncertain bags is proposed. The benefits of this proposal are illustrated by means of an empirical study.

Keywords: Learning from label proportions
Weakly supervised classification · Evaluation · Precision-recall curve

1 Introduction

Supervised classification techniques try to infer the categorizing behavior of a problem of interest and build a classifier that reproduces it. To do so, a set of previous examples is gathered together with their certain category (class label). The objective is to accurately predict the category of new unlabeled examples.

During the last years, the basics of this traditional paradigm have been challenged. Whereas the collection of problem examples is becoming easier, the access to the respective (reliable) class labels remains difficult. Novel classification frameworks where the training dataset is not *certainly* labeled are receiving growing attention. Under the name of weakly supervised learning [9], specific methods are being proposed for learning from different types of weak labelings.

In learning from label proportions (LLP) [13,17], the information of supervision of the training data is aggregated by groups of examples. Individual examples are unlabeled but, for each group, the proportion of examples that belong to each category is known. In spite of this, the inferred classifier needs to be able to provide a label for individual examples. Musicant et al. [13] proposed the first adaptations of KNN, SVM and ANN. Quadrianto et al. described the

© Springer Nature Switzerland AG 2018
F. Herrera et al. (Eds.): CAEPIA 2018, LNAI 11160, pp. 76–86, 2018.
https://doi.org/10.1007/978-3-030-00374-6_8

learning guarantees of their proposal MeanMap [17]. Other techniques work on SVM [18,21], BNC [8], LDA [15], clustering-based solutions [2,19], ensembles [16], etc. Indeed, PAC-learning guarantees have been studied [5], showing that learning with this type of supervision is feasible. The explosion of this paradigm is motivated by an increasing number of real world problems that fit the LLP setting. Applications have been reported in poll prediction [11,20], physical processes [13], fraud detection [18], manufacturing [19], medicine [7], brain-computer interfaces [10], high energy physics [4], etc.

The main challenge when moving from a methodological proposal to dealing with real problems is the evaluation of the built classifiers. The vast majority of proposed methodologies are validated with traditional supervised classification datasets synthetically adapted to the LLP settings. Thus, the ground-truth labels are available for model evaluation. However, in real-world LLP problems the available data for both model learning and evaluation only involves groups of examples weakly supervised with label proportions (circumstantially, full labels might be available [7] and can ease the validation step). A bag-empirical loss, which measures how the label proportions of each bag are matched by a classifier, is commonly used. Patrini et al. studied its guarantees [14] and the PAC-learning study [5] is based on the same type of bag-level loss. Learning robust classifiers from LLP is guaranteed, although the use of this type of metric for model validation (and comparison) might not be sufficiently informative.

In this paper, we focus on the evaluation of classifiers when real-world data is provided and only label proportions are available as information of supervision. We approximate the precision-recall (PR) curve [3], which shows the performance of a probabilistic classifier as the decision threshold moves from 0 to 1, in the LLP setting. Absolute maximum and minimum possible values of precision and recall metrics are calculated. Furthermore, a reasonable minimum estimation of both metrics is considered and, based on this, a final approximation, guided by the information of the least uncertain bags, is proposed. Given per-bag precision and recall values, both micro and macro averaging is considered. A set of PR curves is displayed such that, on the whole, meaningful information about the performance of a classifier is provided and model comparison enabled.

The rest of the paper is organized as follows. Firstly, a formal definition of the problem is given, with particular attention on model evaluation in the LLP setting. Then, the proposal and its characteristics are presented. The paper finishes with discussion, conclusions and future work.

2   Learning from Label Proportions Problem

In this paper, we deal with the learning from label proportions (LLP) problem [17]. In this paradigm, only the way in which the information of supervision is provided for model training differs from the standard supervised classification framework [12]. It is described by a set of n predictive variables $(X_1, \ldots, X_n)$ and a single class variable C. Accordingly, $\mathcal{X}$ is defined as the instance space, that is, all the possible values that the random vector $(X_1, \ldots, X_n)$ can take;

and the label space $\mathcal{C}$ is the set of possible discrete values –class labels– that the class variable takes. This study focuses on binary classification problems, where the size of the label space is $|\mathcal{C}| = 2$. A problem example $(\boldsymbol{x}, c)$, with $\boldsymbol{x} \in \mathcal{X}$ and $c \in \mathcal{C}$, is a $(n+1)$-tuple that assigns a value to each variable. The objective is to learn a classifier, $\hat{h} : \mathcal{X} \to \mathcal{C}$, from a set of previous examples, which are supposed to be i.i.d. sampled from some underlying probability distribution, such that it is able to infer the label of new unseen examples.

As previously mentioned, the particularity of the LLP paradigm comes in the data. The examples of the training dataset D are grouped in b bags $D = \bigcup_{i=1}^{b} \boldsymbol{B}_i$, where $\boldsymbol{B}_i \cap \boldsymbol{B}_j = \emptyset, \forall i \neq j$. Each bag $\boldsymbol{B}_i = \{\boldsymbol{x}_1, \boldsymbol{x}_2, \ldots, \boldsymbol{x}_{m_i}\}$ groups m_i examples, where $\sum_{i=1}^{b} m_i = m$, and m_{ic} denotes the number of examples in $\boldsymbol{B}_i$ which have the label c. These m_{ic} values, called *counts* of the bag $\boldsymbol{B}_i$, sum up to m_i; i.e., $\sum_{c \in \mathcal{C}} m_{ic} = m_i$. Equivalently, this aggregated class information can be provided in terms of *proportions*, $p_{ic} = m_{ic}/m_i \in [0, 1]$, with $\sum_{c \in \mathcal{C}} p_{ic} = 1$.

Regarding the full labeling of the classical learning paradigm, the label proportions involve more loose information of supervision. However, as any bag has its own label proportions, each of them involves a particular degree of uncertainty [8]. It ranges from slight uncertainty, where a vast majority of the examples belong to the same class label ($m_{ic} >> m_{ic'}, c \neq c'$), to large uncertainty, when the examples in $\boldsymbol{B}_i$ are well-distributed (balanced) among class labels (i.e., the difference among m_{ic} values is minimized). The extreme scenario is found when all the examples of a single bag $\boldsymbol{B}_i$ belong to the same class label. In that case, the individual example might even be considered labeled.

Throughout this paper, the term *consistent completion* of a bag $\boldsymbol{B}_i$ is used to refer to any assignment of labels to all the examples of $\boldsymbol{B}_i$ that fulfills the label proportions. Note that the uncertainty of a bag can be calculated as the number of consistent completions.

2.1 Evaluation of the LLP Problem

For the last decade, the problem of learning from this type of data has been largely studied. Proposed techniques were repeatedly validated using fully supervised datasets synthetically adapted to the LLP settings. This allows researchers to learn from a LLP scenario and still use standard evaluation procedures.

In real applications, without the opportunity for such a workaround, only data supervised with label proportions can be used in the validation step. The main issue is that, given a bag and its label proportions, many possible completions are consistent and no clue is available about which one is the correct completion. Thus, a classifier $\hat{h}$ to be evaluated can be used to predict the class label of each example of a bag, $\hat{h}(\boldsymbol{x}), \forall \boldsymbol{x} \in \boldsymbol{B}_i$. As graphically shown in Fig. 1, a prediction over a bag with $p_{i+} = 4/7$ which produces the estimated proportions $\hat{p}_{i+} = 3/7$ may hide very different scenarios depending on the real (unknown) individual labeling. A common solution is a bag-empirical loss:

$$F(D, \hat{h}) = \frac{1}{m} \sum_{i=1}^{b} F(\boldsymbol{B}_i, \hat{h}) \tag{1}$$

with

$$F(\boldsymbol{B}_i, \hat{h}) = m_{i+} - \left(\sum_{j=1}^{m_i} \hat{h}(\boldsymbol{x}) \right)$$

where the positive label is represented as 1 and the negative class as 0. The underlying assumption of this optimistic metric is that, when the examples are ordered according to the assigned probability of being positive, the classes are separable (highest discriminative power) and the real positive examples receive the largest probability values (see the left representation of Fig. 1).

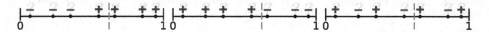

Fig. 1. Graphical examples of three consistent completions in relation to a probabilistic prediction for the 7 unlabeled examples of this toy bag (LP: $3 - / 4 +$). Each example is located on the horizontal axis according to the probability of belonging to the positive class assigned by the classifier. By fixing a threshold (dashed vertical line), the three right-most examples are classified as positive. Depending on the consistent completion considered, the performance of the classifier might range from a large accuracy (left figure) to minimum accuracy (middle figure). The right figure would represent a classifier without discriminative power (examples are randomly ordered). (Color figure online)

3   An Approximated PR Curve for the LLP Problem

The main contribution of this paper is a metric that covers different approximations to the area under the PR curve (AUC-PR) in the LLP setting. In the traditional supervised classification paradigm, the PR curve [3] draws a line among points with coordinates defined by the different paired recall (x-axis) and precision (y-axis) values. The different values are obtained by moving the decision threshold—used by a probabilistic classifier to decide the probability value from which examples are predicted as positive (below that value, they are predicted as negative). For each threshold value, recall and precision values are obtained (see Table 1 for their definition). In a pure LLP scenario, however, precision and recall values cannot be certainly calculated as the individual label of each validation example is unknown.

On the one hand, precision is defined as the probability of success when a positive example is predicted. As defined in Table 1, the denominator $(TP+FP)$ is the total number of examples predicted as positive by the classifier. In the LLP setting too, this number is known and does not need to be estimated. The real uncertainty is in the value of TP: the values of both TP and FP cannot be individually known. Therefore, the number of true positive examples (TP) needs to be approximated to calculate the precision value.

On the other hand, recall is defined as the probability of predicting as positive an actual positive example. According to Table 1, the denominator $(TP+FN)$ is the total number of real positive examples. This value is fixed for any threshold and can be certainly obtained in the LLP setting as the sum of positive counts over the bags, $(TP + FN) = \sum_{i=1}^{b} m_{i+}$. However, as in the previous case, the values of both TP and FN cannot be individually known. Again, TP needs to be approximated to calculate the recall value.

Table 1. Confusion matrix – counts of examples w.r.t. their real class label (by columns) and the class label predicted by a classifier $\hat{h}$ (by row)– (left table) and definition of the precision and recall metrics based on counts of a confusion matrix (right table).

<table>
<tr><td colspan="3"></td><td>real</td><td></td><td></td><td></td></tr>
<tr><td></td><td></td><td>+</td><td>-</td><td>Metric</td><td>Definition</td></tr>
<tr><td rowspan="2">predicted</td><td>+</td><td>TP</td><td>FP</td><td>Precision</td><td>$TP/(TP + FP)$</td></tr>
<tr><td>-</td><td>FN</td><td>TN</td><td>Recall</td><td>$TP/(TP + FN)$</td></tr>
</table>

In this study, up to 4 different estimations of TP are explored. First of all, an optimistic approximation is used. If an example is predicted as positive, it is considered as TP whenever the label proportions are fulfilled (the consistent completion represented by the left drawing of Fig. 1 is always assumed). This optimistic perspective is the same approach underlying the bag loss (Eq. 1). Secondly, the pessimistic approach considers TP examples in a bag only if the number of positive predicted examples exceeds the count of real negatives, m_{i-} (the middle consistent completion represented in Fig. 1 is always assumed). Note that this pessimistic approximation is barely realistic: A classifier $\hat{h}$ with the ability of providing such a separable ordering (although it is the other way around to the required one) cannot be considered the worst model. By simply using the negation of that classifier, $1 - \hat{h}$, the best-performing classifier could be reproduced. Based on this consideration, an approximation based on the random classifier (third estimation) is given. It marks how the solution of the random classifier should perform given the label proportions of a specific bag.

These three approximations are only based on the data and the classifier predictions without any other consideration. The fourth approximation, in contrast, uses information from the least uncertain bags to calculate an initial estimate for the TP value. Then, the initial estimate is updated with the information from more uncertain bags. The idea is that a promising estimation of TP—made with bags where the LLP setting supports the belief that the estimation might be accurate—is not degraded by the loose information of largely uncertain bags. The estimation of TP is adjusted based on the minimum recall estimate from bags with a large proportion of positive examples and on the minimum precision estimate from bags where the classifier predicted many examples as positive.

Formally, the optimistic approximation can be defined in a bag B_i as:

$$\overline{TP}_i = \max(0; \hat{m}_{i+} + m_{i+} - m_i)$$

where $\hat{m}_{i+}$ is the number of examples in B_i predicted as positive by the classifier. The pessimistic approximation is calculated by:

$$\underline{TP}_i = \min(m_{i+}; \hat{m}_{i+})$$

The approximation considering the random classifier is based on expected value of a Binomial distribution with parameters $\hat{m}_{i+}$ (number of predicted positive examples in bag B_i, as number of trials) and m_{i+}/m_i (actual proportion of positive examples in B_i, as probability of success):

$$\widetilde{TP}_i = \hat{m}_{i+} * m_{i+}/m_i$$

The fourth proposed approximation needs a few extra calculations. First of all, all the bags with m_{i+} over the average are identified, $\mathfrak{R}$. Similarly, all the bags with a number of predicted positives over the mean are identified, $\mathfrak{P}$. A weight that measures the certainty of the information provided by a bag is calculated for each B_i as: $W_i = 1 - (\overline{TP}_i - \underline{TP}_i)/(2 \cdot \max_{i'} \overline{TP}_{i'} - \underline{TP}_{i'})$. Making use of this information, the initial estimates of the recall and precision values are calculated:

$$\underline{Prec} = \left(\sum_{B_p \in \mathfrak{P}} W_p \cdot \underline{TP}_p \right) \Big/ \left(\sum_{B_p \in \mathfrak{P}} W_p \cdot \hat{m}_{p+} \right)$$

$$\underline{Rec} = \left(\sum_{B_r \in \mathfrak{R}} W_r \cdot \underline{TP}_r \right) \Big/ \left(\sum_{B_r \in \mathfrak{R}} W_r \cdot m_{r+} \right)$$

These values are used to calculate two different estimates of TP_i for each B_i. The largest estimate is selected and, in order to avoid ocasional nonsensical values, limited among $\widetilde{TP}_i$ and $\overline{TP}_i$:

$$ES_i = \min(\max(\widetilde{TP}_i; \underline{Prec} \cdot \hat{m}_{i+}; \underline{Rec} \cdot m_{i+}); \overline{TP}_i)$$

The final fourth approximated TP value is calculated as the mean value of this reasonable minimum TP estimate, ES_i, and the absolute maximum $\overline{TP}_i$:

$$TP_i^* = (\overline{TP}_i + ES_i)/2$$

Macro Versus Micro Estimations. The characteristic dataset of the LLP problem divided in groups of examples allows for two different interpretations of both precision and recall metrics: macro and micro. The *macro* aggregation of the values of these metrics over the bags of the dataset would ignore the existence of the bags and obtain the TP, FP and FN values in a first step to calculate a

Table 2. Micro/macro definitions of *precision* and *recall* given b subgroups of data.

	Micro	Macro
Precision	$\frac{1}{b}\sum_{i=1}^{b}(TP_i/(TP_i + FP_i))$	$(\sum_{i=1}^{b} TP_i)/(\sum_{i=1}^{b} TP_i + FP_i)$
Recall	$\frac{1}{b}\sum_{i=1}^{b}(TP_i/(TP_i + FN_i))$	$(\sum_{i=1}^{b} TP_i)/(\sum_{i=1}^{b} TP_i + FN_i)$

single precision and recall value posteriorly. The *micro* aggregation would involve the calculation of a precision and recall value per bag to, posteriorly, average their values and obtain the single final precision and recall values. The formal representation of both interpretations is displayed in Table 2.

PR Curve and Interpolation. By using a macro or micro aggregation, precision and recall values are obtained once a decision threshold is fixed. All the possible thresholds are considered: all the different probability values assigned by the probabilistic classifier to the examples of the validation set. Specifically, the probability of being a positive example is considered. A line is drawn by connecting the points with coordinates ($recall_{th}$, $precision_{th}$). As gaps between points might be considerably large, the interpolation proposed by Davis and Goadrich [3] is used. Simple integration techniques can be used to calculate the area under the curve. In our proposal, four lines are drawn in the same plot, one for each approximation. In this way, overall information is provided bounding the real (unknown) performance (Fig. 2 displays an example). An implementation in language R is publicly available in the webpage associated with this paper [1].

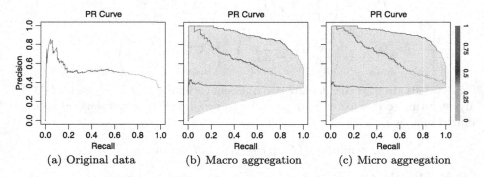

| (a) Original data | (b) Macro aggregation | (c) Micro aggregation |

Fig. 2. Examples of the PR curves of a naive Bayes classifier learned for the Pima dataset. The same classifier is evaluated in all the figures. In the first plot (standard), the validation data is fully labeled (original). In the other two plots (LLP), the same validation dataset with synthetically aggregated label proportions is used.

---

[1] http://www.sc.ehu.es/ccwbayes/members/jeronimo/aucpr_llp/.

4   Discussion

Weakly supervised learning is a relatively new subfield where the research community has mainly focused on methodological contributions to deal with the learning stage. In the case of LLP, to the extent of our knowledge, no proposal beyond bag-level loss (Eq. 1) has been presented to cope with the information provided by the label proportions for model evaluation. With the proposed evaluation metric, we aim to start filling that gap. Although the proper learning of classification models is guaranteed by the use of a level-bag loss [5,14], the availability of a broad collection of evaluation metrics may enhance the analysis of the performance of classifiers and enable model comparison and selection.

Table 3. Summary of the publicly available datasets [1,6] used in these experiments.

dataset	australian	bcw	pima	biodeg	titanic	splice	svmguide1
m	690	699	768	1055	2201	3175	7089
n	15	10	9	42	4	61	5
$p._-/p._+$	56/44	66/34	65/35	66/34	68/32	48/52	44/56

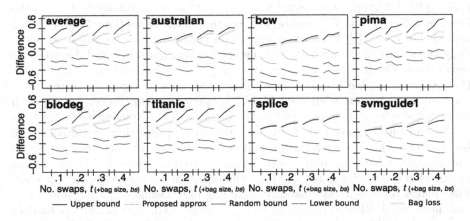

Fig. 3. Experimental results in different datasets and averaged over all of them (upper-left figure). Each figure shows the difference between the actual AUC-PR value and the different AUC-PR approximations using macro averaging. The difference between accuracy and bag loss is also displayed. To simulate different experimental conditions, an increasingly worse classifier is evaluated ($t \in \{10\%, \ldots, 40\%\}$; different lines) in LLP settings with bags of increasing size ($bs = \{5, 10, 25, 50\}$; different points in the lines).

This proposal is an overall technique that explores four different approximations to the precision-recall curve. Three of them use extreme estimations of

TP: An upper $(\overline{TP}_i)$ and lower $(\underline{TP}_i)$ bound, and a third estimation inspired by random guessing $(\widetilde{TP}_i)$. The fourth estimation (TP_i^*) takes advantage of the reasonable intuition that, in bags with large numbers of real or predicted positives, a more reliable first estimation might be obtained to delimit the estimations in the rest of bags. To support this discussion, a set of experiments has been carried out. Up to 7 publicly available fully supervised datasets (see Table 3) were used to learn naive Bayes classifiers. In the evaluation stage, validation data was transformed to different LLP scenarios. For the sake of simplicity, LLP scenarios were generated by aggregating the labels of groups of bs consecutive examples (for $bs \in \{5, 10, 25, 50\}$). Both macro and micro averaging was used to obtain the AUC-PR curve. Due to the lack of other performing metrics in the literature, only the bag loss (Eq. 1) has been considered for comparison. The final results were aggregated over 10 repetitions using different orderings for the validation data. As the real labels were in fact available, the real AUC-PR curve and accuracy values were also calculated. Finally, in order to observe the evolution of the performance drawn by the different metrics, the classifier was forced to perform worse. The predicted probabilities for a proportion t of examples of the validation set were swapped, $p(C = -|\boldsymbol{x}_j) \leftrightarrow p(C = +|\boldsymbol{x}_j)$, for $t \in \{10\%, 20\%, 30\%, 40\%\}$.

The results of these experiments are shown in Fig. 3 in terms of difference between AUC-PR values (LLP - standard). It can be easily observed that the proposed TP^* is the best approximation (it consistently shows difference values close to 0). Its behavior as larger bags are aggregated is similar to that of the lower bounds $(\underline{TP}$ and $\widetilde{TP})$ and opposite to that of the upper bound $(\overline{TP})$. Recall that the averaged $\underline{TP}$ value of low-uncertain bags is used as an initial estimate. Only the results of the macro averaging version are shown in Fig. 3 due to space restrictions. The results with micro averaging are similar, with slight differences (figures are available in the website associated with this paper). Note that a constant bag size has been used in LLP simulations. Differences among both averaging strategies might appear as the bag size of the different bags diverges.

As previously mentioned, $\overline{TP}$ follows the same optimistic perspective as the bag loss (Eq. 1). This appreciation is observed (black and gray lines) in Fig. 3. The datasets where this upper bound is closer to 0 are those where the performance of the classifier is better. The differences increase as bags are enlarged. That makes sense as the probability of finding in the same bag a FP and a FN that mask each other is higher as bags get larger. The rest of the approaches increase their difference negatively with the real AUC-PR value in large bags. In the case of large datasets (splice and svmguide1), the results are stable as the proportion of swaps, t, is increased (more mistakes are induced). The global upwards movement of all the lines as t increases is mainly related to the worse performance of the evaluated classifier. Thus, the evaluation given by using the proposed TP^* might be optimistic or pessimistic depending on the classifier itself (optimistic for poorly performing classifiers and pessimistic for highly accurate models). Thus, the specific results shown in this paper are strongly related to our selection of a naive Bayes classifier.

There is room for improvement thought novel approximations based on more accurate intuitions and/or evidences. Other metrics could also be approximated. For instance, an estimation of the TN value would straightforwardly lead to an AUC-ROC adaptation to the LLP setting. To complete the validation stage, it would be of interest to conceive a (cross) validation procedure that, to respect the integrity of bags, might possibly need to work with unstratified folds of different sizes. In this paper a training-validation division of the dataset is assumed.

To sum up, in this paper the area under the precision-recall curve metric is adapted for the learning from label proportions problem. By identifying an absolute minimum, a reasonable minimum and an absolute maximum value, the actual estimation is bounded. Finally, an approximate estimation which promotes the partial estimations gathered from low-uncertain bags is proposed. The usefulness of this metric has been shown throughout an empirical study.

Acknowledgments. This work has been partially supported by the Basque Government (IT609-13, Elkartek BID3A), and the Spanish Ministry of Economy and Competitiveness (TIN2016-78365-R).

References

1. Chang, C.C., Lin, C.J.: LIBSVM: a library for support vector machines (2011). http://www.csie.ntu.edu.tw/~cjlin/libsvmtools/datasets/
2. Chen, S., Liu, B., Qian, M., Zhang, C.: Kernel k-means based framework for aggregate outputs classification. In: IEEE ICDM Workshops, pp. 356–361 (2009)
3. Davis, J., Goadrich, M.: The relationship between precision-recall and ROC curves. In: Proceedings of 23rd ICML, pp. 233–240 (2006)
4. Dery, L.M., Nachman, B., Rubbo, F., Schwartzman, A.: Weakly supervised classification in high energy physics. J. High Energy Phys. **2017**(5), 145 (2017)
5. Fish, B., Reyzin, L.: On the complexity of learning from label proportions. In: Proceedings of 26th IJCAI, pp. 19–25 (2017)
6. Frank, A., Asuncion, A.: UCI machine learning repository (2010). http://archive.ics.uci.edu/ml
7. Hernández-González, J., Inza, I., Crisol-Ortiz, L., Guembe, M.A., Iñarra, M.J., Lozano, J.A.: Fitting the data from embryo implantation prediction: learning from label proportions. Stat. Methods Med. Res. **27**(4), 1056–1066 (2018)
8. Hernández-González, J., Inza, I., Lozano, J.A.: Learning Bayesian network classifiers from label proportions. Pattern Recognit. **46**(12), 3425–3440 (2013)
9. Hernández-González, J., Inza, I., Lozano, J.A.: Weak supervision and other nonstandard classification problems: a taxonomy. Pattern Recognit. Lett. **69**, 49–55 (2016)
10. Hübner, D., Verhoeven, T., Schmid, K., Müller, K.R., Tangermann, M., Kindermans, P.J.: Learning from label proportions in brain-computer interfaces: online unsupervised learning with guarantees. PLoS One **12**(4), e0175856 (2017)
11. Kück, H., de Freitas, N.: Learning about individuals from group statistics. In: Proceedings of 21st UAI, pp. 332–339 (2005)
12. Mitchell, T.: Machine Learning. McGraw Hill, New York City (1997)
13. Musicant, D.R., Christensen, J.M., Olson, J.F.: Supervised learning by training on aggregate outputs. In: Proceedings of 7th IEEE ICDM, pp. 252–261 (2007)

14. Patrini, G., Nock, R., Rivera, P., Caetano, T.: (Almost) no label no cry. In: Proceedings of NIPS 27, pp. 190–198 (2014)
15. Pérez-Ortiz, M., Gutiérrez, P.A., Carbonero-Ruz, M., Hervás-Martínez, C.: Adapting linear discriminant analysis to the paradigm of learning from label proportions. In: Proceedings of IEEE SSCI, pp. 1–7 (2016)
16. Qi, Z., Meng, F., Tian, Y., Niu, L., Shi, Y., Zhang, P.: Adaboost-LLP: a boosting method for learning with label proportions. IEEE Trans. Neural Netw. Learn. Syst., 1–12 (2017, early access)
17. Quadrianto, N., Smola, A.J., Caetano, T.S., Le, Q.V.: Estimating labels from label proportions. J. Mach. Learn. Res. **10**, 2349–2374 (2009)
18. Rüping, S.: SVM classifier estimation from group probabilities. In: Proceedings of 27th ICML, pp. 911–918 (2010)
19. Stolpe, M., Morik, K.: Learning from label proportions by optimizing cluster model selection. In: Gunopulos, D., Hofmann, T., Malerba, D., Vazirgiannis, M. (eds.) ECML PKDD 2011. LNCS (LNAI), vol. 6913, pp. 349–364. Springer, Heidelberg (2011). https://doi.org/10.1007/978-3-642-23808-6_23
20. Sun, T., Sheldon, D., O'Connor, B.: A probabilistic approach for learning with label proportions applied to the US presidential election. In: IEEE ICDM (2017)
21. Yu, F.X., Liu, D., Kumar, S., Jebara, T., Chang, S.: ∝SVM for learning with label proportions. In: Proceedings of 30th ICML, pp. 504–512 (2013)

Time Series Decomposition for Improving the Forecasting Performance of Convolutional Neural Networks

Iván Méndez-Jiménez and Miguel Cárdenas-Montes$^{(\boxtimes)}$ ⓘD

Centro de Investigaciones Energéticas Medioambientales y Tecnológicas,
Madrid, Spain
{ivan.mendez,miguel.cardenas}@ciemat.es

Abstract. Time Series forecasting is of high interest in the Big Data ecosystem. A larger data volume accessible in industry and science, and a higher profit from more accurate predictions have generated a growing application of Deep Learning techniques in the Time Series forecasting. In this work, the improvement of the forecasting capacity of Convolutional Neural Networks and Recurrent Neural Networks when using as input the trend, seasonal and remainder time series generated by the Seasonal and Trend decomposition using Loess, instead of the original time series observations, is evaluated. The benchmark used in this work is composed of eight seasonal time series with different lengths and origins. Besides, Convolutional Neural Networks and Recurrent Neural Networks, comparisons with Multilayer Perceptrons are also undertaken. As a consequence, an improvement in the forecasting capacity when replacing the original observations by their decomposition in Convolutional Neural Networks-based forecasting is stated.

Keywords: Time series analysis · Deep learning · Forecasting
Convolutional neural networks · Recurrent neural networks
Seasonal and trend decomposition using loess

1 Introduction

Nowadays all the activities related with the Big Data ecosystem are generating a growing interest in industry and science. Larger data volumes must be faster analysed with larger efficiencies for extracting valuable information. Time series modelling and forecasting, as part of the Big Data ecosystem, are forced to process larger data volumes offering accurate predictions for the future values of the observables. Among the deep architectures used for modelling and forecasting time series, Convolutional Neural Networks (CNN) [16,17] and Recurrent Neural Networks (RNN) [23] are frequently used.

CNNs are used in classification tasks with data with some structure, such as images and audio data. Besides, this architecture has been also applied for time series classification [27,28] and for time series prediction [15,22,25].

© Springer Nature Switzerland AG 2018
F. Herrera et al. (Eds.): CAEPIA 2018, LNAI 11160, pp. 87–97, 2018.
https://doi.org/10.1007/978-3-030-00374-6_9

RNN are a set of neural networks whose the main purpose is to process sequential data [13,23]. Whereas CNN aims at processing data with grid structure, such as images, the RNN are specialized networks for processing a sequence of data, including the application to time series analysis [5,12,24]. Similarly to CNN, the RNN can process sequences of different lengths without sequence-based specialization, the same as Multilayer Perceptrons (MLP).

One of the most identifiable features of CNN is parameter sharing. Parameter sharing allows extending and applying the model to examples of different form. Conversely, if sequence-based specialization is used, e.g. MLP, then separated parameters are generated for each value of the time index. This leads to the impossibility to generalize sequence lengths not seen during the training process, nor share statistical strength across different sequence lengths and across different positions in time.

With regards to parameter sharing, when CNN are used for analysing time series, strong similarities appear with RNN. However, in comparison with RNN, shared parameters in CNN are considered as shallow. In the convolution operation, a network shares parameters across the time among a reduced number of neighbouring members of the input. The concept of parameter sharing appears in the application of the same convolution kernel at each time step. In RNN, each member of the output is a function of the previous members of the output, and it is produced by using the same rule which has been applied to the previous outputs.

In this work eight seasonal time series are predicted with CNN and RNN by using as input the three series resulting from the Seasonal and Trend decomposition using Loess (STL), in contrast with the use as input of the original observations. In the CNN case, this approach implies to handle three CNNs for independent predictions of each of the three series from STL (namely CNN+STL). Then, the three predictions are added for evaluating the performance. Similarly for RNN (namely RNN+STL), three RNNs are created for independent predictions for the trend, the seasonal component and the remainder. In the comparison, MLP-based forecasting results are also included.

The time series composing the benchmark vary in their origin and length, ranging from 48 to 456 observations, and all of them exhibiting yearly seasonality (Sect. 2.1).

The tests undertaken demonstrate that the CNN performance improves when using as input the three components produced by the STL decomposition: trend, seasonal and remainder series instead the original observations. Conversely, in RNN a degradation of the performance is produced in the equivalent approach.

The rest of the paper is organized as follows: A brief description of the time series composing the benchmark, the techniques and algorithms used for analysing the data, and the statistics applied in this work, are presented in Sect. 2. The models comparison and the results obtained are presented and analysed in Sect. 3. Finally, Sect. 4 contains the conclusions of this work.

2   Methods and Materials

In this section, the algorithms selected for the performance comparison and the time series used as benchmark are briefly described.

2.1   Benchmark

The following time series have been used to evaluate the performance of the proposed approach.

– ^{222}Rn. This time series corresponds to the monthly medians of the ^{222}Rn level at LSC [1]. It has a length of 4 years (48 observations), being the shortest one of the time series used in this work.
– CO, NO_2 and O_3. These time series correspond to the CO, NO_2 and O_3 level pollution measurement station at Arturo Soria street in Madrid (Spain) [2]. They extend from January 2001 to December 2016 (192 observations). They are available at web site http://datos.madrid.es.
– Noise. This time series corresponds to the mean noise level at the station of Arturo Soria street in Madrid (Spain). It extends from January 1999 to December 2014 (192 observations). It is available at web site http://datos.madrid.es.
– BeerAU. This time series corresponds to the monthly beer production in Australia between January 1956 and December 1994 (456 observations). This information has been provided by Australian Bureau of Statistics and can be downloaded from **datamarket** web site [2].
– Births. This time series corresponds to the monthly New York City births for the period from January 1946 to December 1959 (168 observations). This information can be downloaded from **datamarket** web site [3].
– Passengers. This time series corresponds to the monthly international passengers (in thousands). It extends from January 1960 to December of 1977 (216 observations). This information can be downloaded from **datamarket** web site [21].

---

[1] Regarding the previous efforts in the analysis of the ^{222}Rn level at Canfranc Underground Laboratory (LSC), in [20] the initial efforts for modelling this time series using classical and deep-learning-based approaches are shown. In this paper, the times series is modelled using CNN, being the main focus on the monthly forecasting capacity for scheduling maintenance operations of the experiment hosted at LSC. The LSC is composed of diverse halls for hosting scientific experiments with requirements of very low-background. The two main halls, Hall A and Hall B —which are contiguous—, have instruments for measuring the level of ^{222}Rn.

[2] In [1] an assessment of the relationship between the SO_2 and the total suspended particulate levels and mortality in Madrid during the period 1986–1992 is presented. In this study the time series analysis is based on multivariate autoregressive integrated moving-average (ARIMA). Other publications relating Madrid acoustic or air pollution with population health can be found in [8,18].

For generating the training and the test sets, the rule of 75%-25% has been applied under the consideration of maintaining complete years in both sets. For all series and architectures, the sample size used is of 12 values.

2.2 Seasonal and Trend Decomposition Using Loess

The intuition behind the time series decomposition is that the time series is the composition of three more elementary series. On the one hand, a trend (T_t), which is responsible of long-term increase or decrease of data. It does not have to be linear. On the other hand, a seasonal pattern is the second component (S_t). It is influenced by seasonal factors, such as: the month, the day of the week, or the quarter of the year. Finally, the third component is the remainder or random component (R_t). If the decomposition is additive, then the values of the time series (Y_t) can be modelled as $Y_t = T_t + S_t + R_t$.

Diverse techniques for time series decomposition have been proposed. STL, *Seasonal and Trend decomposition using Loess* [7], was proposed taking into account the limitations of previous classical decomposition methods, for example X-12-ARIMA. In contrast with X-12-ARIMA, STL can handle any type of seasonality, not only monthly or quarterly; and the seasonal component can change over time, being the amount of the allowed change controlled by a parameter of the algorithm. Besides, the smoothness of the trend component can be also controlled by the algorithm.

In the current work, no-periodic STL decomposition has been used for decomposing the time series of the benchmark.

2.3 Multilayer Perceptrons

Multilayer Perceptrons (MLP) are biological-inspired composition of neurons, as fundamental elements, grouped in layers. Each layer is a non-linear combination of non-linear functions from the previous layer. Layers are ordered from the initial one, which receives the input data, to the last one, which produces the output, by passing from some intermediary ones or hidden, which map both data. The input of any neuron can be expressed as $f(w, \mathbf{x})$, where $\mathbf{x}$ is the vector input to the neuron, and w is the matrix of weights which is optimized through the training process. By increasing number of the hidden layers, more complicated relationships can be established. In the context of this analysis, MLP can be used to approximate non-linear relationships between dependent variables (the observations to be predicted) and independent variables (a set of previous and contiguous observations to this which must be predicted), in such a way they can be assimilated as extensions of generalized linear models [4].

In this work, Keras [6] has been used for implementing the MLP, CNN, and RNN architectures. Specifically, the MLP has been configured with two hidden layers composed of 24 and 12 neurons, with hyperbolic tangent as activation function, and trained with 100 epochs.

2.4 Convolutional Neural Networks

Convolutional Neural Networks (CNN) are specialized Neural Networks with special emphasis in image processing [13], although nowadays they are also employed in time series analysis [9, 26]. The CNN consists of a sequence of convolutional layers, the output of which is connected only to local regions in the input. These layers alternate convolutional, non-linear and pooling-based layers which allow extracting the relevant features of the class of objects, independently of their placement in the data example. The CNN allows the model to learn filters that are able to recognize specific patterns in the time series, and therefore it can capture richer information from the series. It also embodies three features which provide advantages over the MLP: sparse interactions, parameter sharing and equivariance to translation [13].

Although Convolutional Neural Networks are frequently associated to image or audio classification—2D grid examples—or video sequence—3D grid examples—, it can also be applied to time series analysis—1D grid examples—. When processing time series, instead of a set of images, the series has to be divided in overlapping contiguous time windows. These windows constitute the examples, where the CNN aims at finding patterns. At the same time, the application to time series modelling requires the application of 1D convolutional operators, whose weights are optimized during the training process.

The CNN employed are composed of two convolutional layers of 32 and 64 filters with `relu` as activation function, `MaxPooling1D`, and trained with 10 epochs.

2.5 Recurrent Neural Networks

Recurrent Neural Networks (RNN) are aimed at sequentially processing values $x^1, \ldots, x^\tau$ [19]. Similarly to CNN, they can scale over arbitrary sequence lengths, and benefit from parameters sharing through the model (Eq. 1).

Long Short-Term Memory (LSTM) introduces self-loops in the RNN schema, allowing these self-loops be conditioned on the context [14]. This architecture has the same inputs and outputs as an ordinary RNN, although it has more parameters and a system of gating units that controls the flow of information.

$$h = \sigma(W_{hh}\, h_{i-1} + W_{hx}\, x_i + b_h)$$
$$\hat{y}_i = W_{yh}\, h_i \tag{1}$$

where x_i is the input vector, y_i is the output vector, h_i is the hidden state, W_{hx} is input-to-hidden weights, W_{hh} is hidden-to-hidden weights, W_{yh} is hidden-to-output weights, and $\hat{y}$ is the predicted values. In the current work, a single hidden layer is used, with 24 LSTM elements, and trained with 100 epochs. Hyperbolic tangent is used as activation function.

2.6 Statistics

In order to ascertain if the proposed forecasting methods applied to the test set improve the prediction, two different types of tests can be applied: parametric and non-parametric. The difference between both relies on the assumption that data is normally distributed for parametric tests, whereas non explicit conditions are assumed in non-parametric tests. For this reason, the latter is recommended when the statistical model of data is unknown [10, 11]. Statistical inference is used in this work to infer which model produces better results, and if the differences are significant or not.

The Kruskal-Wallis test is a non-parametric test used to compare three or more groups of sample data. For this test, the null hypothesis assumes that the samples are from identical populations. The procedure when using multiple comparison to test whether the null hypothesis is rejected implies the use of a post-hoc test to determine which sample makes the difference. The most typical post-hoc test is the Wilcoxon signed-rank test with the Bonferroni or Holm correction.

The Wilcoxon signed-rank test belongs to the non-parametric category. For this test, the null hypothesis assumes that the samples are from identical populations, whereas the alternative hypothesis states that the samples come from different populations. It is a pairwise test that aims to detect significant differences between two sample means. If necessary, the Bonferroni or Holm correction can be applied to control the Family-Wise Error Rate (FWER). In our case, the Bonferroni correction is used for this purpose. FWER is the cumulative error when more than one pairwise comparison (e.g. more than one Wilcoxon signed-rank test) is performed. Therefore, when multiple pairwise comparisons are performed, Bonferroni correction allows maintaining the control over the FWER.

3   Experimental Results and Models Comparison

In Fig. 1, the MSE for 25 independent runs of algorithms and time series analysed are shown. As can be appreciated, the use as input of the series generated by STL tends to improve the performance when feeding CNN (CNN+STL implementation). For example, relevant improvements of the CNN+STL in relation to the rest of algorithms are obtained for the time series ^{222}Rn (Fig. 1(a)) and Births (Fig. 1(g)). For these time series, CNN+STL implementation obtains the overall best results.

Unfortunately, the most relevant effect of the proposed approach is the degradation of the performance of the RNN+STL in relation with RNN and the other approaches tested. The inspection of RNN+STL in comparison with RNN shows that RNN+STL is outperformed or performs similarly in all the cases except for the Noise time series (Fig. 1(e)), where is the overall best one.

For all the time series in the benchmark, CNN+STL outperforms CNN (Table 1), although it only outperforms MLP in the time series: ^{222}Rn, CO, NO_2, Births and Passengers.

Except for the Noise time series, the application of the Kruskal-Wallis test to the approaches: MLP, CNN and CNN+STL, indicates that the differences between the medians of the results are significant for a confidence level of 95% (p-value under 0.05), which means that the differences are unlikely to have occurred

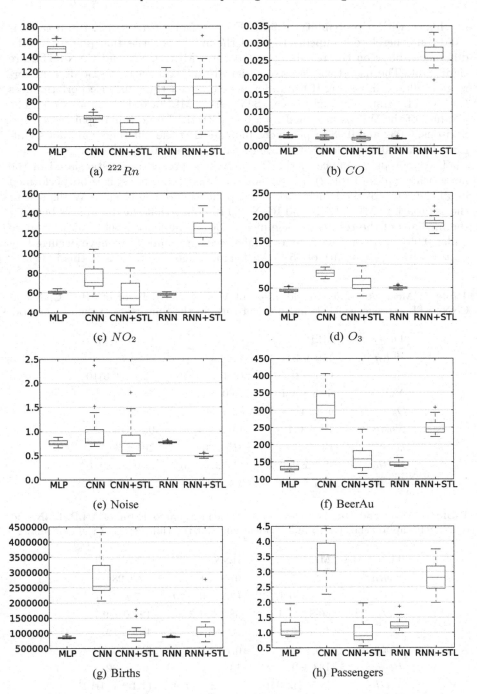

(a) ^{222}Rn

(b) CO

(c) NO_2

(d) O_3

(e) Noise

(f) BeerAu

(g) Births

(h) Passengers

Fig. 1. Mean Squared Error for 25 independent runs for each architecture and time series of the benchmark.

by chance with a probability of 95%. For the Noise time series, the p-value is 0.051, and therefore it impedes to reject the null hypothesis that points that the differences between the results of approaches: MLP, CNN and CNN+STL are significant. The use of the Wilcoxon signed-rank test with the Bonferroni correction allows discerning if the differences between the pair-wise performances CNN+STL versus CNN and CNN+STL versus MLP are significant or not. The p-values of the Wilcoxon signed-rank test with the Bonferroni of pair-wise comparisons, 0.032 and 0.98 respectively, indicate that the differences are not significant for a confidence level of 97.5% (p-value under 0.025).

The poor performance of the RNN+STL approach is substantiated in the numerical results (Table 2). Except for the Noise time series, it is outperformed in all the series of the benchmark. The application of the Kruskal-Wallis test to the approaches: MLP, RNN and RNN+STL, points that the differences between the medians of the results are significant for a confidence level of 95% (p-value under 0.05), which means that the differences are unlikely to have occurred by chance with a probability of 95%. Although, in most of the cases, this indicates

Table 1. Mean and standard deviation of Mean Squared Error of MLP, CNN and CNN+STL approaches for 25 independent runs for the time series of the benchmark.

Time series	MLP	CNN	CNN+STL
^{222}Rn	150 ± 7	59 ± 4	45 ± 7
CO	$(2.7 \pm 0.3)10^{-3}$	$(2.4 \pm 0.6)10^{-3}$	$(2.1 \pm 0.6)10^{-3}$
NO_2	60.8 ± 1.4	76.0 ± 12.4	58.7 ± 13.6
O_3	46 ± 4	82 ± 7	60 ± 18
$Noise$	1.0 ± 0.4	1.0 ± 0.4	0.8 ± 0.3
$Beer$	133 ± 9	315 ± 46	164 ± 35
$Births$	$(8.5 \pm 0.3)10^5$	$(2.8 \pm 0.6)10^6$	$(1.0 \pm 0.2)10^6$
$Passengers$	1.17 ± 0.30	3.4 ± 0.6	1.1 ± 0.4

Table 2. Mean and standard deviation of Mean Squared Error of MLP, RNN and RNN+STL approaches for 25 independent runs for the time series of the benchmark.

Time series	MLP	RNN	RNN+STL
^{222}Rn	150 ± 7	97 ± 10	92 ± 28
CO	$(2.7 \pm 0.3)10^{-3}$	$(2.1 \pm 0.2)10^{-3}$	$(27 \pm 3)10^{-3}$
NO_2	60.8 ± 1.4	58.1 ± 1.3	124.0 ± 9.7
O_3	46 ± 4	50 ± 3	187 ± 12
$Noise$	1.0 ± 0.4	0.77 ± 0.02	0.49 ± 0.03
$Beer$	133 ± 9	145 ± 7	252 ± 22
$Births$	$(8.5 \pm 0.3)10^5$	$(8.7 \pm 0.2)10^5$	$(1.1 \pm 0.4)10^6$
$Passengers$	1.17 ± 0.30	1.6 ± 0.2	2.8 ± 0.5

that RNN+STL is significantly outperformed, except for the Noise time series points that the improvement is significant.

4 Conclusions

In this study, the improvement in the forecasting capacity of CNN and RNN when using as input the trend, seasonal and remainder series generated by the STL decomposition, instead of the original observations, has been evaluated. For this purpose, eight seasonal time series have been used as benchmarks. In most of the cases, 7 of 8, a significant improvement in the prediction is achieved by the proposed CNN+STL approach in comparison with CNN; and in 5 of 8 cases when comparing with MLP. Unfortunately, for RNN implementations, the proposed approach (RNN+STL) produces a strong degradation of the performance.

As future work, an alternative and more efficient use of the information from the STL decomposition for generating a significant improvement of the performance of CNN, and specially of RNN, is proposed.

Acknowledgment. The research leading to these results has received funding by the Spanish Ministry of Economy and Competitiveness (MINECO) for funding support through the grant FPA2016-80994-C2-1-R, and "Unidad de Excelencia María de Maeztu": CIEMAT - FÍSICA DE PARTÍCULAS through the grant MDM-2015-0509.

IMJ is co-funded in a 91.89 percent by the European Social Fund within the Youth Employment Operating Program, for the programming period 2014–2020, as well as Youth Employment Initiative (IEJ). IMJ is also co-funded through the Grants for the Promotion of Youth Employment and Implantation of Youth Guarantee in Research and Development and Innovation (I+D+i) from the MINECO.

References

1. Alberdi Odriozola, J.C., Díaz Jiménez, J., Montero Rubio, J.C., Mirón Pérez, I.J., Pajares Ortíz, M.S., Ribera Rodrigues, P.: Air pollution and mortality in Madrid, Spain: a time-series analysis. Int. Arch. Occup. Environ. Health **71**(8), 543–549 (1998). https://doi.org/10.1007/s004200050321
2. BeerAU Time Series. https://datamarket.com/data/set/22xr/monthly-beer-production-in-australia-megalitres-includes-ale-and-stout-does-not-include-beverages-with-alcohol-percentage-less-than-115-jan-1956-aug-1995
3. Births Time Series. https://datamarket.com/data/set/22nv/monthly-new-york-city-births-unknown-scale-jan-1946-dec-1959
4. Bishop, C.M.: Neural Networks for Pattern Recognition. Oxford University Press Inc., New York (1995)
5. Chniti, G., Bakir, H., Zaher, H.: E-commerce time series forecasting using LSTM neural network and support vector regression. In: Proceedings of the International Conference on Big Data and Internet of Thing, BDIOT 2017, pp. 80–84. ACM, New York (2017). https://doi.org/10.1145/3175684.3175695
6. Chollet, F., et al.: Keras (2015). https://github.com/fchollet/keras
7. Cleveland, R.B., Cleveland, W.S., McRae, J., Terpenning, I.: STL: a seasonal-trend decomposition procedure based on loess. J. Off. Stat. **6**(1), 3–33 (1990)

8. Díaz, J., García, R., Ribera, P., Alberdi, J.C., Hernández, E., Pajares, M.S., Otero, A.: Modeling of air pollution and its relationship with mortality and morbidity in Madrid, Spain. Int. Arch. Occup. Environ. Health **72**(6), 366–376 (1999). https://doi.org/10.1007/s004200050388

9. Gamboa, J.C.B.: Deep learning for time-series analysis. CoRR abs/1701.01887 (2017). http://arxiv.org/abs/1701.01887

10. García, S., Fernández, A., Luengo, J., Herrera, F.: A study of statistical techniques and performance measures for genetics-based machine learning: accuracy and interpretability. Soft Comput. **13**(10), 959–977 (2009)

11. García, S., Molina, D., Lozano, M., Herrera, F.: A study on the use of non-parametric tests for analyzing the evolutionary algorithms' behaviour: a case study on the CEC'2005 special session on real parameter optimization. J. Heuristics **15**(6), 617–644 (2009)

12. Garcia-Pedrero, A., Gomez-Gil, P.: Time series forecasting using recurrent neural networks and wavelet reconstructed signals. In: 2010 20th International Conference on Electronics Communications and Computers (CONIELECOMP), pp. 169–173, February 2010. https://doi.org/10.1109/CONIELECOMP.2010.5440775

13. Goodfellow, I., Bengio, Y., Courville, A.: Deep Learning. MIT Press (2016). http://www.deeplearningbook.org

14. Hochreiter, S., Schmidhuber, J.: Long short-term memory. Neural Comput. **9**(8), 1735–1780 (1997). https://doi.org/10.1162/neco.1997.9.8.1735

15. Lago, J., Ridder, F.D., Schutter, B.D.: Forecasting spot electricity prices: deep learning approaches and empirical comparison of traditional algorithms. Appl. Energy **221**, 386–405 (2018). https://doi.org/10.1016/j.apenergy.2018.02.069. http://www.sciencedirect.com/science/article/pii/S030626191830196X

16. LeCun, Y.: Generalization and network design strategies. University of Toronto, Technical report (1989)

17. Lecun, Y., Bottou, L., Bengio, Y., Haffner, P.: Gradient-based learning applied to document recognition. Proc. IEEE **86**(11), 2278–2324 (1998). https://doi.org/10.1109/5.726791

18. Linares, C., Díaz, J., Tobías, A., Miguel, J.M.D., Otero, A.: Impact of urban air pollutants and noise levels over daily hospital admissions in children in Madrid: a time series analysis. Int. Arch. Occup. Environ. Health **79**(2), 143–152 (2006). https://doi.org/10.1007/s00420-005-0032-0

19. Lipton, Z.C.: A critical review of recurrent neural networks for sequence learning. CoRR abs/1506.00019 (2015). http://arxiv.org/abs/1506.00019

20. Méndez-Jiménez, I., Cárdenas-Montes, M.: Modelling and forecasting of the ^{222}Rn radiation level time series at the Canfranc Underground Laboratory. In: de CosJuez, F. (ed.) HAIS 2018. LNCS, vol. 10870, pp. 158–170. Springer, Cham (2018). https://doi.org/10.1007/978-3-319-92639-1_14

21. Passengers Time Series: https://datamarket.com/data/set/22u3/international-airline-passengers-monthly-totals-in-thousands-jan-49-dec-60

22. Qiu, X., Zhang, L., Ren, Y., Suganthan, P.N., Amaratunga, G.A.J.: Ensemble deep learning for regression and time series forecasting. In: 2014 IEEE Symposium on Computational Intelligence in Ensemble Learning, CIEL 2014, Orlando, FL, USA, 9–12 December 2014, pp. 21–26 (2014). https://doi.org/10.1109/CIEL.2014.7015739

23. Rumelhart, D.E., Hinton, G.E., Williams, R.J.: Learning representations by back-propagating errors. Nature **323**(6088), 533–536 (1986). https://doi.org/10.1038/323533a0

24. Walid, Alamsyah: Recurrent neural network for forecasting time series with long memory pattern. J. Phys.: Conf. Ser. **824**(1), 012038 (2017). http://stacks.iop.org/1742-6596/824/i=1/a=012038
25. Wang, H.Z., Li, G.Q., Wang, G.B., Peng, J.C., Jiang, H., Liu, Y.T.: Deep learning based ensemble approach for probabilistic wind power forecasting. Appl. Energy **188**, 56–70 (2017). https://doi.org/10.1016/j.apenergy.2016.11.111. http://www.sciencedirect.com/science/article/pii/S0306261916317421
26. Wang, Z., Yan, W., Oates, T.: Time series classification from scratch with deep neural networks: a strong baseline. CoRR abs/1611.06455 (2016). http://arxiv.org/abs/1611.06455
27. Zheng, Y., Liu, Q., Chen, E., Ge, Y., Zhao, J.L.: Time series classification using multi-channels deep convolutional neural networks. In: Li, F., Li, G., Hwang, S., Yao, B., Zhang, Z. (eds.) WAIM 2014. LNCS, vol. 8485, pp. 298–310. Springer, Cham (2014). https://doi.org/10.1007/978-3-319-08010-9_33
28. Zheng, Y., Liu, Q., Chen, E., Ge, Y., Zhao, J.L.: Exploiting multi-channels deep convolutional neural networks for multivariate time series classification. Front. Comput. Sci. **10**(1), 96–112 (2016). https://doi.org/10.1007/s11704-015-4478-2

Asymmetric Hidden Markov Models with Continuous Variables

Carlos Puerto-Santana[(✉)], Concha Bielza, and Pedro Larrañaga

Technical University of Madrid, Madrid, Spain
ce.puerto@alumnos.upm.es, {mcbielza,pedro.larranaga}@fi.upm.es

Abstract. Hidden Markov models have been successfully applied to model signals and dynamic data. However, when dealing with many variables, traditional hidden Markov models do not take into account asymmetric dependencies, leading to models with overfitting and poor problem insight. To deal with the previous problem, asymmetric hidden Markov models were recently proposed, whose emission probabilities are modified to follow a state-dependent graphical model. However, only discrete models have been developed. In this paper we introduce asymmetric hidden Markov models with continuous variables using state-dependent linear Gaussian Bayesian networks. We propose a parameter and structure learning algorithm for this new model. We run experiments with real data from bearing vibration. Since vibrational data is continuous, with the proposed model we can avoid any variable discretization step and perform learning and inference in an asymmetric information frame.

Keywords: Hidden Markov models · Bayesian networks
Model selection · Structure learning · Time series
Information asymmetries · Linear Gaussian Bayesian network

1 Introduction

Hidden Markov Models (HMMs) have been used to predict and analyse dynamic and sequential data, e.g., in speech recognition or gene prediction. These models assume a hidden variable which explains an observed variable. However, they rely on the assumption of an equal emission probability function for every state (except for changes in parameters) of the hidden variable, which for the case of multiple observable variables may lead to a huge unnecessary amount of parameters to be learned and produce models with data overfitting and poor problem insights, specially when few data is available.

Many attempts have tried to capture asymmetries within probabilistic graphical models. For example, Bayesian multinets [5] describe different local graphical models depending on the values of certain observed variables; similarity networks [7] allow to build independent influence diagrams for subsets of a given domain. Context-specific independence in Bayesian networks [1] have tree structured conditional probability distributions with a d-separation-based algorithm

© Springer Nature Switzerland AG 2018
F. Herrera et al. (Eds.): CAEPIA 2018, LNAI 11160, pp. 98–107, 2018.
https://doi.org/10.1007/978-3-030-00374-6_10

to determine statistical dependencies between variables according to contexts. It has been shown that the use of these asymmetries within the model can improve the inference and learning procedures [2].

In HMMs, which can be seen as probabilistic graphical models, asymmetries could be emulated to be as those in Bayesian multinets, similarity networks or context-specific independence in Bayesian networks. A Chow-Liu tree or a conditional forest is used in [9] to model the observed variables given the hidden state. More recently asymmetric hidden Markov models (As-HMMs) are proposed in [2], where a local graphical model (not necessarily a tree or a forest) is associated to each state of the hidden variable. However, only models with discrete observable variables were discussed, leaving continuous observable variables forced to be discretized. The number of parameters depends upon the discretization method which can affect the inference and learning phases. In this paper we extend As-HMMs to deal with continuous variables, which permits avoiding discretization steps and the errors that may come from this process.

The structure of this document is the following. Section 2 recalls HMMs in a general way. Section 3 covers As-HMMs and introduce the asymmetric linear Gaussian hidden Markov models (AsLG-HMMs) which are capable of modelling As-HMMs with continuous variables. We also discuss the parameter and structure learning of AsLG-HMMs. Section 4 presents experiments with real vibrational data from bearings. The results obtained using AsLG-HMM are compared against the ones from using mixtures of Gaussian hidden Markov models (HMM-MoG). The paper is sounded off in Sect. 5 with conclusions and comments on possible future research lines.

2   Hidden Markov Models

Let $\mathbf{X}^T = (\mathbf{x}^0, ..., \mathbf{x}^T)$ be the observed variables $\mathbf{X} = \{X_1, ..., X_M\}$ over time, where $\mathbf{x}^t = (x_1^t, ..., x_M^t)$, $t = 0, 1, ..., T$. We assume that the observed variable $\mathbf{X}$ is influenced by a discrete variable Q which is hidden and has N possible values, i.e. $\mathrm{dom}(Q) = \{1, 2, ..., N\}$. A HMM is a double chain stochastic process, where the evolution of hidden states $\mathbf{Q}^T = (q^0, ..., q^T)$, of Q is assumed to fulfill the Markov property. Moreover $\mathbf{x}^t$ is assumed to be independent of itself over time given q^t.

Definition 1. *A HMM is a triplet $\lambda = (\boldsymbol{A}, \boldsymbol{B}, \boldsymbol{\pi})$ where $\boldsymbol{A} = [a_{i,j}]_{i,j=1}^N$ is a matrix representing the transition probabilities between the hidden states i, j over time t, i.e. $a_{i,j} = P(q^{t+1} = j | q^t = i)$; $\boldsymbol{B}$ is a vector representing the emission probability of the observations given the hidden state, $\boldsymbol{B} = [b_j(\mathbf{x}^t)]_{j=1}^N$, where $b_j(\mathbf{x}^t) = P(\boldsymbol{X} = \mathbf{x}^t | q^t = j)$ is a probability density function; and $\boldsymbol{\pi}$ is the initial distribution of the hidden states $\boldsymbol{\pi} = [\pi_j]_{j=1}^N$ where $\pi_j = P(q^0 = j)$.*

Given a model $\lambda = (\mathbf{A}, \mathbf{B}, \boldsymbol{\pi})$, it is possible to compute the probability of the complete information i.e., $P(\mathbf{Q}^T, \mathbf{X}^T | \lambda)$ as:

$$P(\mathbf{Q}^T, \mathbf{X}^T | \lambda) = \pi_{q^0} \prod_{t=0}^{T-1} a_{q^t, q^{t+1}} \prod_{t=0}^{T} b_{q^t}(\mathbf{x}^t). \tag{1}$$

Three main tasks can be performed in the context of HMMs: first, compute the likelihood of $\mathbf{X}^T$, i.e. $P(\mathbf{X}^T|\lambda)$, which can be done using the forward-backward algorithm [12]. Second, compute the most probable sequence of states i.e., find the value of $\delta_t(i) = \max_{Q^{t-1}} P(\mathbf{X}^t, \mathbf{Q}^{t-1}, q^t = i|\lambda)$, $t = 0, ...T$, $i = 1, ..., N$, which can be solved using the Viterbi algorithm [12]. Third, learn the parameters λ, which can be done with the expectation maximization (EM) algorithm [12].

To execute the forward-backward algorithm, the forward and backward variables are needed, which are defined respectively as $\alpha_t(i) = P(q^t = i, \mathbf{x}^0, ..., \mathbf{x}^t|\lambda)$, $i = 1, .., N$, $t = 0, ..., T$, and $\beta_t(i) = P(\mathbf{x}^{t+1}, ..., \mathbf{x}^T|q^t = i, \lambda)$, $i = 1, ..., N$, $t = 0, ..., T$. Observe in particular that the forward variable can help us to compute the likelihood of $\mathbf{X}^T$, since: $P(\mathbf{X}^T|\lambda) = \sum_{i=1}^N P(\mathbf{X}^T, q^T = i|\lambda) = \sum_{i=1}^N \alpha_T(i)$.

For the second problem, it is noticeable that the variable $\delta_t(i)$ can be seen as: $\delta_t(j) = \max_{i=1,..N}\{\delta_{t-1}(i)a_{i,j}\}b_j(\mathbf{x}^{t+1})$, $j = 1, .., N, t = 0, ..., T$, which can be solved recursively. For more details about this algorithm, see [12].

For the third problem, we will recall how to learn the parameters λ^* using the Baum-Welch method or equivalently, the EM algorithm [12]. The algorithm consists of two steps, the expectation (E) step and the maximization (M) step. We assume that a prior $\lambda^0 = (\mathbf{A}^0, \mathbf{B}^0, \boldsymbol{\pi}^0)$ is known. In the E step, we compute the distribution of the latent variable Q, i.e., $P(\mathbf{Q}^T|\mathbf{X}^T, \lambda^0)$. In the M step, we find λ^* as solution of the problem: $\lambda^* = \arg\max_\lambda H(\lambda|\lambda^0)$, where $H(\lambda|\lambda^0)$ is defined as

$$H(\lambda|\lambda^0) = \sum_{\mathcal{Q}} P(\mathbf{Q}^T|\mathbf{X}^T, \lambda^0) \ln P(\mathbf{Q}^T, \mathbf{X}^T|\lambda), \qquad (2)$$

where $\mathcal{Q} := \mathrm{dom}(Q^T)$. In [3] it is shown that $P(\mathbf{X}^T|\lambda^*) \geq P(\mathbf{X}^T|\lambda^0)$ with equality if and only if $H(\lambda^*|\lambda^0) = H(\lambda^0|\lambda^0)$ and $P(\mathbf{Q}^T|\mathbf{X}^T, \lambda^0) = P(\mathbf{Q}^T|\mathbf{X}^T, \lambda^*)$. Therefore, iterating the E and the M steps produce improvements in the model likelihood.

The E step for HMMs, for a prior λ^0, is done by calculating the quantities $\gamma_t(i) = P(q^t = i|\mathbf{X}^T, \lambda^0)$ and $\xi_t(i,j) = P(q^t = i, q^{t+1} = j|\mathbf{X}^T, \lambda^0)$, $i,j = 1, ..., N$, $t = 0, ..., T$ which can be computed using the forward and backward variables in the following way:

$$\gamma_t(i) = \frac{\alpha_t(i)\beta_t(i)}{\sum_{u=1}^N \alpha_t(u)\beta_t(u)}, \qquad (3)$$

$$\xi_t(i,j) = \frac{\alpha_t(i)a_{ij}b_j(\mathbf{x}^{t+1})\beta_{t+1}(j)}{\sum_{u=1}^N \sum_{v=1}^N \alpha_t(u)a_{uv}b_v(\mathbf{x}^{t+1})\beta_{t+1}(v)}. \qquad (4)$$

The M step for HMM requires to maximize the $H(\lambda|\lambda^0)$ function, which is done by using Eqs. (1) and (2):

$$H(\lambda|\lambda^0) = \sum_{\mathcal{Q}} P(\mathbf{Q}^T|\mathbf{X}^T, \lambda) \ln(\pi_{q^0}) + \sum_{\mathcal{Q}} \sum_{t=0}^{T-1} P(\mathbf{Q}^T|\mathbf{X}^T, \lambda) \ln(a_{q^t, q^{t+1}})$$

$$+ \sum_{\mathcal{Q}} \sum_{t=0}^{T} P(\mathbf{Q}^T|\mathbf{X}^T, \lambda) \ln(b_{q^t}(\mathbf{x}^t)), \qquad (5)$$

with restrictions $\sum_{i=1}^{N} \pi_i = 1$ and $\sum_{j=1}^{N} a_{i,j} = 1$, $i = 1, ..., N$. The updating formulas for parameters $\mathbf{A}$ and $\boldsymbol{\pi}$ can be computed using Lagrange multipliers. The resulting formulas are:

$$\pi_i^*(i) = \gamma_0(i), \quad i = 1, ..., N, \qquad a_{i,j}^* = \frac{\sum_{t=0}^{T-1} \xi_t(i,j)}{\sum_{t=0}^{T-1} \gamma_t(i)}, \quad i, j = 1, ..., N. \quad (6)$$

The updating formula for parameter $\mathbf{B}$ relies on the assumptions made over the observable variables and the emission probabilities. In the next section we develop the formulas to update parameter $\mathbf{B}$ in the context of an As-HMM with Gaussian variables.

3   Asymmetric Linear Gaussian Hidden Markov Models

In this section we recall definitions and relevant aspects from As-HMMs mentioned in [2]. Then we model the asymmetric emission probabilities using linear Gaussian Bayesian networks and deduce the update algorithm for parameter $\mathbf{B}$ and discuss its complexity. Finally, we describe the structure learning procedure.

3.1   Definitions

First of all, we recall the definition of linear Gaussian Bayesian network (LGBN), see [10]. This model will be used to describe the asymmetries in HMMs with continuous variables.

Definition 2. *Let $\boldsymbol{X} = (X_1, ..., X_M)$ be a continuous random variable. A linear Gaussian Bayesian network over $\boldsymbol{X}$ is a tuple $\mathcal{B}(R, G)$, where G is a directed acyclic graph (DAG). The parents of X_m are given by G and are represented by an ordered vector of length k_m denoted as $\boldsymbol{Pa}(X_m) = (U_{m,1}, ..., U_{m,k_m})$, $m = 1, ..., M$, with $U_{m,l} \in \boldsymbol{X}$, $l = 1, ..., k_m$. R is formed by the local distribution of each X_m conditioned on $\boldsymbol{Pa}(X_m)$. The joint density function satisfies:*

$$P(\boldsymbol{X}) = \prod_{m=1}^{M} \mathcal{N}(X_m | \beta_{m,0} + \beta_{m,1} U_{m,1} + \cdots + \beta_{m,k_m} U_{m,k_m}, \sigma_m^2), \quad (7)$$

where $\mathcal{N}$ denotes the one-dimensional Gaussian probability density function and β_i, k, $i = 1, ..., N$, $k = 0, ..., k_m$ are real numbers.

Now, state-specific Bayesian network, As-HMMs and AsLG-HMMs are defined. The idea is to give a distinct LGBN to every state of the hidden variable. As a consequence, in every state the parents of each variable are different. This representation captures asymmetries in the data.

Definition 3. *Let $\boldsymbol{X} = (X_1, ..., X_M)$ and Q be random variables. For each $q \in dom(Q)$, we associate a Bayesian network over $\boldsymbol{X}$ called state-specific Bayesian network for q, $\mathcal{B}_q(R_q, G_q)$. We define the following conditional distribution:*

$$P_q(\boldsymbol{X}) := P(\boldsymbol{X}|q) = \prod_{m=1}^{M} P_q(X_m | \boldsymbol{Pa}_q(X_m)). \quad (8)$$

Definition 4. *An asymmetric hidden Markov model over the random variables* $(\boldsymbol{X}, Q)$*, being* Q *the hidden variable, is a model* $\lambda = (\boldsymbol{A}, \boldsymbol{B}, \boldsymbol{\pi})$ *with initial distribution* $\boldsymbol{\pi} = [\pi_j]_{j=1}^{N}$ *where* $\pi_j = P(q^0 = j)$*, transition probabilities between the hidden states* $\boldsymbol{A} = [a_{i,j}]_{i,j=1}^{N}$ *where* $a_{i,j} = P(q^{t+1} = j | q^t = i)$ *and the emission density function vector* $\boldsymbol{B} = [b_j(\mathbf{x}^t)]_{j=1}^{N}$ *where* $b_j(\mathbf{x}^t) = P_j(\mathbf{x}^t)$ *i.e. a state-specific Bayesian network.*

Definition 5. *An asymmetric linear Gaussian hidden Markov model over* $\boldsymbol{X} = (X_1, ..., X_M)$ *the continuous random variables and* Q *the hidden discrete random variable, is an As-HMM* $\lambda = (\boldsymbol{A}, \boldsymbol{B}, \boldsymbol{\pi})$ *with the property that for each* $q \in dom(Q)$ *a state-specific linear Gaussian Bayesian network* $\mathcal{B}_q(R_q, G_q)$ *is associated. If the parents of variable* X_m *for the state* q *are an ordered column vector of length* k_m^q *denoted as* $\boldsymbol{Pa}_q(X_m) = (U_{m,1}^q, ..., U_{m,k_m^q}^q)$*,* $m = 1, ..., M$*, with* $U_{m,l}^q \in \boldsymbol{X}$*,* $l = 1, ..., k_m^q$*, then the emission probabilities* $\boldsymbol{B} = [b_j(\mathbf{x}^t)]_{j=1}^{N}$ *have the form:*

$$b_j(\mathbf{x}^t) = P_j(\mathbf{x}^t) = \prod_{m=1}^{M} \mathcal{N}\big(x_m^t | \beta_{m,0}^j + \beta_{m,1}^j U_{m,1}^j + \cdots + \beta_{m,k_m^j}^j U_{m,k_m^j}^j, (\sigma_m^j)^2\big), \quad (9)$$

Observe that each linear Gaussian model for each state $q \in dom(Q)$ is determined by the set of coefficients $\mathcal{T}_q := \bigcup_{m=1}^{M} \{\beta_{m,0}^q, ..., \beta_{m,k_m^q}^q\}$, since the mean of each variable is a function of these coefficients, see [10].

3.2 Learning Parameters

Now that we know how to represent the emission probabilities $\mathbf{B}$ for the case of AsLG-HMMs, we build the parameter update formulas. Assume the prior λ^0 is known and we execute the E step as in Sect. 2, therefore $\gamma_t(i)$, $i = 1, ..., N$, $t = 0, 1, ..., T$ quantities are defined. Let $D_q[F] := \sum_{t=0}^{T} f^t \gamma_t(q)$, with F any variable and $q \in dom(Q)$. First, we need the value of $\beta_{m,0}^q$ that maximizes the H function, hence we derive Eq. (5) with respect to $\beta_{m,0}^q$ and equate to zero. Observe that $\beta_{m,0}^q$ appears in the H function inside $b_{q^t}(\mathbf{x}^t) = P_{q^t}(\mathbf{x}^t)$, hence:

$$\frac{\partial H(\lambda|\lambda^0)}{\partial \beta_{m,0}^q} = \frac{\partial \sum_{t=0}^{T} \sum_{i=1}^{N} \gamma_t(i) \ln P_i(\mathbf{x}^t)}{\partial \beta_{m,0}^q} = \sum_{t=0}^{T} \gamma_t(q) \frac{\partial \ln P_q(\mathbf{x}^t)}{\partial \beta_{m,0}^q} = 0, \quad (10)$$

making the derivation assuming that $P_q(\mathbf{x}^t)$ is defined by an AsLG-HMMs and $u_{m,l}^{t,q}$ is the value of the l-parent of X_m for state q at time t, $l = 1, .., k_m^q$. Then we have:

$$\frac{\partial H(\lambda|\lambda^0)}{\partial \beta_{m,0}^q} = \sum_{t=0}^{T} -2 \frac{\gamma_t(q)}{(\sigma_m^q)^2} (\beta_{m,0}^q + \beta_{m,1}^q u_{m,1}^{t,q} + \cdots + \beta_{m,k_m^q}^q u_{m,k_m^q}^{t,q} - x_m^t) = 0. \quad (11)$$

This leads to the following expression:

$$D_q[X_m] = \beta_{m,0}^q D_q[1] + \beta_{m,1}^q D_q[U_{m,1}^q] + \cdots + \beta_{m,k_m^q}^q D_q[U_{m,k_m^q}^q]. \quad (12)$$

If Eq. (5) is derived with respect to the coefficients $\beta_{m,k}^n$ with $k = 1, ..., k_m^q$ as in Eq. (10) we obtain the following equations:

$$\begin{cases} D_q[X_m U_{m,1}^q] = \beta_{m,0}^q D_q[U_{m,1}^q] & + \cdots + \beta_{m,k_m^q}^q D_q[U_{m,k_m^q}^q U_{m,1}^q] \\ \quad\vdots & \qquad\qquad\vdots \\ D_q[X_m U_{m,k_m^q}^q] = \beta_{m,0}^q D_q[U_{m,k_m^q}^q] & + \cdots + \beta_{m,k_m^q}^q D_q[(U_{m,k_m^q}^q)^2]. \end{cases} \tag{13}$$

Equations (12) and (13) form a linear system of $k_m^q + 1$ unknowns with $k_m^q + 1$ equations. The solution of this system gives the coefficients $\{\beta_{m,0}^q, \beta_{m,1}^q, ..., \beta_{m,k_m^q}^q\}$, for each variable $m = 1, 2, ..., M$ and state $q \in \text{dom}(Q)$. Once, these coefficients are known, the mean $\mu_m^{t,q} = \beta_{m,0}^q + \beta_{m,1}^q u_{m,1}^{t,q} + \cdots + \beta_{m,k_m^q}^q u_{m,k_m^q}^{t,q}$ is estimated. To obtain the updating formula of $(\sigma_m^q)^2$, we must derive Eq. (5) with respect $(\sigma_m^q)^2$ and equate to zero:

$$\frac{\partial H(\lambda|\lambda^0)}{\partial(\sigma_m^q)^2} = \sum_{t=0}^{T} \gamma_t(q) \frac{\partial \ln P_q(\mathbf{x}^t)}{\partial(\sigma_m^q)^2} = 0. \tag{14}$$

Assuming that $P_q(\mathbf{x}^t)$ is defined by an AsLG-HMM, we have:

$$\frac{\partial H(\lambda|\lambda^0)}{\partial(\sigma_m^q)^2} = \sum_{t=0}^{T} \gamma_t(q) \left(\frac{(x_m^t - \mu_m^{t,q})^2}{(\sigma_m^q)^4} - \frac{1}{(\sigma_m^q)^2} \right) = 0, \tag{15}$$

which leads to the following expression:

$$((\sigma_m^q)^2)^* = \frac{\sum_{t=0}^{T} \gamma_t(q)(x_m^t - \mu_m^{t,q})^2}{\sum_{t=0}^{T} \gamma_t(q)}. \tag{16}$$

We discuss now the complexity of computing the $\mathcal{T} := \bigcup_{i=1}^{N} \mathcal{T}_i$ coefficients. Assume that we have N states, M variables and that the factorization for each state is the most complex i.e., every variable is dependent of the others. This implies that $|\mathcal{T}_i| = 1 + 2 + 3 + \cdots + M = \frac{M(M+1)}{2}$, $i = 1, ..., N$, therefore $|\bigcup_{i=1}^{N} \mathcal{T}_i| = \frac{NM(M+1)}{2}$. It is known that the complexity of solving a linear system of k variables is at most $O(k^3)$ (using for example Gauss-Jordan algorithm). Hence for the worst case scenario the complexity of determining the coefficients for a single state is $O(1^3) + O(2^3) + \cdots + O(M^3) = O(\frac{M^2(M+1)^2}{4})$. Therefore to compute the coefficients for every state, the complexity is $O(NM^2(M+1)^2)$. On the other hand, for the simplest factorization i.e., every variable is independent of the others given the state, the complexity of determining the coefficients is $O(NM)$, since we must solve M linear systems of one variable for N states.

3.3   Learning Structure

For the structure learning task the SEM algorithm, proposed in [4] is used. Assume the prior model $\lambda^{0,0} = (\mathbf{A}^{0,0}, \mathbf{B}^{0,0}, \boldsymbol{\pi}^{0,0})$ for the initial model M^0. One

SEM iteration consists of using the EM algorithm to get the parameters $\lambda^{0,*} = (\mathbf{A}^{0,*}, \mathbf{B}^{0,*}, \boldsymbol{\pi}^{0,*})$. Next, using the estimation $\lambda^{0,*}$ and $P(\mathbf{Q}^T | \mathbf{X}^T, \lambda^{0,*})$ got in the E step, we look for a model M^1 such that maximizes a given score function, usually the Bayesian information criterion (BIC). Once the model M^1 has been found, we set $\lambda^{1,0} := \lambda^{0,*}$ i.e., the found parameters are used as prior parameters for the next iteration of the SEM algorithm. As noticed in [2] the BIC score can be deduced and reduced from Eq. (5) as follows:

$$\text{Score} = \sum_{q=1}^{N} \sum_{t=0}^{T} \gamma_t(q) \ln P_q(\mathbf{x}^t) - \frac{1}{2} \ln(T) \#(\mathcal{B}_q(R_q, G_q)), \qquad (17)$$

where $\#(\mathcal{B}_q(R_q, G_q))$ is the number of parameters used for the state-specific linear Gaussian Bayesian network for state q. We must also mention that any algorithm can be used to optimize the score function. In particular, the simulated annealing introduced in [8] was used for this study. Recall that [6] proved the convergence of this method which gives us an ending guarantee of the optimization process.

4 Experiments

For the experiments we use a real dataset. The data comes from bearing vibrational information, see [11]. The data is filtered using spectral kurtosis algorithms and envelope techniques as in [14]; next, the bearing fundamental frequencies and its harmonics are extracted: ball pass frequency outer (BPFO) related to the bearings outer race, ball pass frequency inner (BPFI) related to the bearings inner race, ball spin frequency (BSF) related to the bearings rollers and the fundamental train frequency (FTF) related to the bearings cage. The mechanical set-up is shown in Fig. 1. In real life applications, bearings are fundamental components inside of tool machines. Is desirable to surveillance the bearing health state. However, the health state is a hidden variable; hence, HMMs can be applied to estimate the bearing health. In the literature, the health estimation is usually done with mixtures of Gaussian hidden Markov model (MoG-HMM) as in [13].

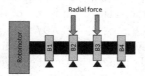

Fig. 1. Graphical representation of the mechanical set-up. A rotomotor spins a shaft at a rotational speed of 2000RPM coupled with four Rexnord ZA-2115 double row bearings with labels B1, B2, B3 and B4. A constant radial load of 2721.554 kg is applied to bearings 2 and 3. Vibrational data is recorded until one of the bearings fails. A signal record of 0.1s is taken every twenty minutes. The sampling rate is 20 kHz.

There is a training and a testing signal. The learning signal consist of 2156 records and the testing signal of 6324 records. We have information of the fundamental frequencies and three harmonics of each frequency, hence 16 variables are used in both signals. We will assume that there are four possible health states. In the training dataset, B3 fails due to its inner race and B4 due to its rollers. In the testing dataset B3 fails due to its outer race. The results of using AsLG-HMMs and MoG-HMMs with three mixtures are shown in Table 1. We show results for log-likelihood (LL), BIC score, and number of parameters (#). Notice that the number of parameters needed by a MoG-HMM is $NK((M^2 + M)/2 + 1)$, where K is the number of mixtures components.

Table 1. Likelihood and BIC results for test signal.

B	MoG-HMM			AsLG-HMM		
	#	LL	BIC	#	LL	BIC
1	1644.0	−170045.02	−177232.72	560.0	−162654.42	**−165028.46**
2	1644.0	−204349.46	−211537.16	560.0	−178040.19	**−180414.23**
3	1644.0	−349099.49	−356287.2	137.0	−270698.52	**−271226.57**
4	1644.0	−84479.32	−91667.03	133.0	−74495.96	**−75006.56**

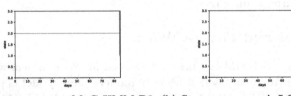

(a) State sequence MoG-HMM B3. (b) State sequence AsLG-HMM B3.

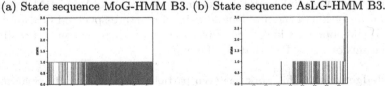

(c) State sequence MoG-HMM B4. (d) State sequence AsLG-HMM B4.

Fig. 2. Sequences of states predicted by Viterbi algorithm for B3 and B4 due to B3 is the failure bearing and B4 has the lowest BIC score. (a) and (c) are state sequences predicted with MoG-HMMS. (b) and (d) are state sequences predicted with AsLG-HMMs.

From the results obtained, it can be seen that the BIC scores from AsLG-HMMs are better than the ones obtained from MoG-HMMs. Also, if we observe Fig. 2, we see that the health evolution of the B3 and B4 predicted by the MoG-HMMs are not easy to read. In (a) at the end of the bearings life the sequence

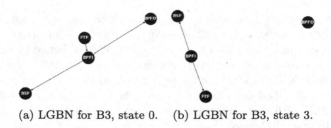

(a) LGBN for B3, state 0.   (b) LGBN for B3, state 3.

Fig. 3. Different state specific Gaussian Bayesian network structures obtained for different states. Here, an illustrative model structure is built using only fundamental frequencies.

jumps between all the states and in (c) any relevant information is shown. On the other hand, the state sequence predicted by AsLG-HMMs reveals a change in the bearings health in its last days of life in (c) and in (d) shows an evolutionary sequence.

On the other hand, to illustrate the better problem insight that As-HMMs provide, we train an AsLG-HMMs with only the fundamental frequencies (four variables) and observe the obtained state-specific LGBN for B3, see Fig. 3. As we see for state 0 (healthy state), FTF frequency determines the others, this was expected since FTF is close to the shaft frequency. Meanwhile in state 3 (failure state), the BPFI determine the BSF and FTF frequencies, which may indicate a failure in the bearings inner race.

5   Conclusions and Future Work

In this paper the AsLG-HMM has been introduced in order to deal with continuous variables in asymmetric hidden Markov models. This model is proposed to overcome overfitting models and discretization steps. Also AsLG-HMM provides useful interpretation of the problem domain, since state-specific LGBN are used. Also As-HMMs open a wide range of research lines, there are many possibilities and variations of As-HMMs that can be explored.

Acknowledgements. This work has been partially supported by the Spanish Ministry of Economy and Competitiveness through TIN2016-79684-P projects, by the Regional Government of Madrid through the S2013/ICE-2845-CASI-CAM-CM project, and by Fundación BBVA grants to Scientific Research Teams in Big Data 2016. Additionally, we give thanks to Etxe-Tar S.A. to provide the filtered datasets to perform the corresponding experiments.

References

1. Boutilier, C., Friedman, N., Goldszmidt, M., Koller, D.: Context-specific independence in Bayesian networks. In: Proceedings of the Twelfth International Conference on Uncertainty in Artificial Intelligence, pp. 115–123. Morgan Kaufmann Publishers Inc., Burlington (1996)

2. Bueno, M.L., Hommersom, A., Lucas, P.J., Linard, A.: Asymmetric hidden Markov models. Int. J. Approx. Reason. **88**, 169–191 (2017)
3. Dempster, A.P., Laird, N.M., Rubin, D.B.: Maximum likelihood from incomplete data via the EM algorithm. J. R. Stat. Soc. Ser. B (Methodological) **39**(1), 1–38 (1977)
4. Friedman, N.: The Bayesian structural EM algorithm. In: Proceedings of the Fourteenth Conference on Uncertainty in Artificial Intelligence, pp. 129–138. Morgan Kaufmann Publishers Inc., San Francisco (1998)
5. Geiger, D., Heckerman, D.: Knowledge representation and inference in similarity networks and Bayesian multinets. Artif. Intell. **82**(1), 45–74 (1996)
6. Granville, V., Krivanek, M., Rasson, J.P.: Simulated annealing: a proof of convergence. IEEE Trans. Pattern Anal. Mach. Intell. **16**(6), 652–656 (1994)
7. Heckerman, D.: Probabilistic similarity networks. Networks **20**(5), 607–636 (1990)
8. Kirkpatrick, S., Gelatt, C.D., Vecchi, M.P.: Optimization by simulated annealing. Science **220**(4598), 671–680 (1983)
9. Kirshner, S., Padhraic, S., Andrew, R.: Conditional Chow-Liu tree structures for modeling discrete-valued vector time series. In: Proceedings of the 20th Conference on Uncertainty in Artificial Intelligence, pp. 317–324. AUAI Press (2004)
10. Koller, D., Friedman, N.: Probabilistic Graphical Models: Principles and Techniques. The MIT Press, Cambridge (2009)
11. Qian, Y., Yan, R., Gao, R.X.: A multi-time scale approach to remaining useful life prediction in rolling bearing. Mech. Syst. Signal Process. **83**, 549–567 (2017)
12. Rabiner, L.R.: A tutorial on hidden Markov models and selected applications in speech recognition. In: Readings in Speech Recognition, pp. 267–296. Morgan Kaufmann, San Francisco (1990)
13. Tobon, D., Medjaher, K., Zerhouni, N., Tripot, G.: A data-driven failure prognostics method based on mixture of Gaussians hidden Markov models. IEEE Trans. Reliab. **61**(2), 491–503 (2012)
14. Wang, Y., Liang, M.: An adaptive SK technique and its application for fault detection of rolling element bearings. Mech. Syst. Signal Process. **25**, 1750–1764 (2010)

Measuring Diversity and Accuracy in ANN Ensembles

M. Paz Sesmero[1(✉)], Juan Manuel Alonso-Weber[1],
Alessandro Giuliani[2], Giuliano Armano[2], and Araceli Sanchis[1]

[1] Universidad Carlos III de Madrid,
Avd. Universidad 30, Leganés, Madrid, Spain
{msesmero,masm}@inf.uc3m.es, jmaw@ia.uc3m.es
[2] Università di Cagliari, Via Is Maglias 198, Cagliari, Italy
{alessandro.giuliani,armano}@diee.unica.it

Abstract. Performance of classifier ensembles depends on the precision and on the diversity of the members of the ensemble. In this paper we present an experimental study in which the relationship between the accuracy of the ensemble and both the diversity and the accuracy of base learners is analyzed. We conduct experiments on 8 different ANN ensembles and on 5 multiclass data sets. Experimental results show that a high diversity degree among the base learners does not always imply a high accuracy in the ensemble.

Keywords: Ensemble of classifiers · Diversity · Accuracy · ANN

1 Introduction

An ensemble of classifiers is a set of classifiers, also called base learners, whose individual decisions are combined to achieve a system that outperforms every one of its members [1].

As well as other ideas that have been applied in the field of Artificial Intelligence, the ensembles of classifiers respond to an attempt to emulate human behavior. In particular, the ensembles of classifiers try to replicate the performance of a human being when it must make a decision about an important matter. For example, it is common to seek a second or a third opinion before having a surgery, buying a product or taking an important financial decision. That is, a decision is considered more reliable if it is taken based on the opinion of different experts. Extrapolation of this proposition to Machine Learning leads to the development of ensembles of classifiers.

Many theoretical studies have demonstrated that the success of any ensemble of classifiers is related to the *accuracy* and *diversity* of the members of the ensemble. In other words, an ensemble of classifiers could improve the accuracy of any of its individual members if they have a low error rate (are accurate) and their errors are not coincident (are diverse). However, obtaining base learners which satisfy both requirements simultaneously is not an easy task because the lower the number of errors, the higher its correlation is.

© Springer Nature Switzerland AG 2018
F. Herrera et al. (Eds.): CAEPIA 2018, LNAI 11160, pp. 108–117, 2018.
https://doi.org/10.1007/978-3-030-00374-6_11

In this paper, we present an experimental study in which we examine the relationship between the accuracy of different ANN ensembles and both the accuracy and the diversity of their members.

The remainder of this paper is organized as follows. First, some background about the generation of the ensemble members is given. Then, we present a summary of the measures used to quantify the diversity and provide a brief review of some studies where diversity is used as a key factor in the design of the ensembles. Section 4, describes the experimental evaluation and the results obtained. Finally, Sect. 5 contains concluding remarks and future work.

2 Generating Base Learners

Different theoretical works have proven that the diversity among base learners is a necessary condition for obtaining an accurate ensemble. The techniques used to generate a pool of diverse classifiers are based on the idea that the behavior of a classifier depends on both, the dataset and the learning algorithm used to build it.

The techniques based on varying the training set can be subdivided in three groups [2]: manipulating the training examples, manipulating the input features and manipulating the output target. A summary of these techniques is presented bellow.

- Manipulating the training examples: According to this approach, each base learner is trained with a different subset of training data. This approach includes two of the most well-known ensemble methods: *Bagging* [3] and *Boosting* [4]. In *Bagging* each training dataset is generated by sampling with replacement. So, in the new datasets, some of the original examples may be repeated several times while others may be left out. In contrast, in *Boosting* each training dataset is generated by focusing on the instances that are incorrectly predicted by the previously trained classifiers.
- Manipulating the input features. Another general approach to obtain diverse classifiers is manipulating the set of features that is used to describe the examples. This group includes approaches where the new feature subsets are obtained by qualitative transformations of the original feature set [5] and approaches where the new feature subsets are obtained by random selection [6] or by applying different Feature Selection methods [7, 8].
- Manipulating the output target. A third approach for building diverse classifiers is recoding the class labels that are assigned to each training example. Among these techniques are the decomposition binary methods as OAO [9], OAA [10], ECOC [11] or OAHO [12] and systems as BCE [13] where the original multiclass problem is decomposed into several pairwise subproblems (a binary problem and a multiclass problem).

On the other hand, methods that vary the learning algorithm can be subdivided in two groups: Approaches that use different learning algorithms (called heterogeneous ensembles) and approaches that use different versions of the same learning algorithm. Among the methods that are included in this category are Randomization [14] and the proposed by Kolen and Pollack in [15]. In this last work authors proved that a pool of ANN that are trained on the same dataset, but with different initial weights derive on a set of diverse classifiers.

3 Diversity

Most works related to the generation of base classifiers have focused on obtaining a pool of diverse classifiers applying variations of the techniques that have been described in the previous section. Performance of these models have been tested based on the accuracy of the ensemble, but no reference about diversity among the base learners is given.

Given that diversity is a necessary condition for obtaining an accurate ensemble, in next section we summarize some of the more common measures to quantify diversity. Next, in Sect. 3.2. we provide a brief review of some studies where diversity is a key factor used in the design of the ensemble.

3.1 Diversity Measurements

As has been previously mentioned, diversity is a necessary condition for obtaining a good ensemble. However, there is not a unique definition of diversity and there are

Table 1. Summary of the main measures used to quantify ensemble diversity. Monotonically increasing/decreasing measures are identified with an ascending/descending arrow.

Name	Symbol	Definition	↑/↓	Pairwise
Q statistic	Q	$\dfrac{N^{11}N^{00}-N^{01}N^{10}}{N^{11}N^{00}+N^{01}N^{10}}$	↓	Y
Correlation coefficient	ρ	$\dfrac{N^{11}N^{00}-N^{01}N^{10}}{\sqrt{(N^{11}+N^{10})(N^{01}+N^{00})(N^{11}+N^{01})(N^{10}+N^{00})}}$	↓	Y
Disagreement measure	dis	$\dfrac{N^{01}+N^{10}}{N^{11}+N^{10}+N^{01}+N^{00}}$	↑	Y
Double-fault measure	DF	$\dfrac{N^{00}}{N^{11}+N^{10}+N^{01}+N^{00}}$	↓	Y
Kappa degree-of-agreement statistic	κ	$\dfrac{\sum\limits_{i=1}^{k}\frac{N_{ii}}{N}-\sum\limits_{i=1}^{k}\left(\frac{N_{i*}N_{*i}}{N\ N}\right)}{1-\sum\limits_{i=1}^{k}\left(\frac{N_{i*}N_{*i}}{N\ N}\right)}$	↓	Y
Plain disagreement measure	Div_plain	$\dfrac{1}{N}\sum\limits_{k=1}^{N}Is(C_i(x_k),=C_j(x_k))$	↑	Y
Ambiguity	Amb	$\dfrac{1}{LKN}\sum\limits_{l=1}^{L}\sum\limits_{n=1}^{N}\sum\limits_{k=1}^{K}\left(Is(C_l(x_k)=k)-\frac{N_k^n}{L}\right)$	↑	N
Entropy	E	$\dfrac{1}{N}\sum\limits_{n=1}^{N}\frac{1}{(L-\frac{L}{2})}\min\{l(x_k),L-l(x_k)\}$	↑	N

Where:

N is the cardinality of the test set;

K is the number of classes;

N^{ab} is the number of instances in the dataset, classified correctly (a = 1) or incorrectly (a = 0) by the classifier i, and correctly (b = 1) or incorrectly (b = 0) by the classifier j;

N_{ij} is the number of instances in the dataset, labeled as class i by the first classifier and as class j by the second classifier;

$C_i(x_k)$ is the class assigned by classifier i to instance k;

$Is()$ is a truth predicate; is the number of base classifiers that assign instance n to class k; and $l(x_k)$ is the number of classifiers that correctly classified instance k.

L is the number of base learners.

different statistical or mathematical measurements that are used in the literature to quantify this magnitude. Table 1 shows some of the more common measures used to quantify ensemble diversity [16, 17]. Measures where a high/low value means a high degree of diversity are identified with an ascending/descending arrow.

3.2   Applying Diversity

Some studies where diversity is used as a key factor in the design of the ensemble are presented below.

Zenobi and Cunningham [18] proposed a method for producing classifier ensembles that emphasizes diversity among the ensemble members. In this study, the base learner's choice is based on the diversity among the members of the ensemble. Experimental results proved that a choice based on diversity produces ensembles that are more accurate than those in which the choice is based on the error rate. Authors focused on ensembles of classifiers where diversity derives form using different feature subsets.

Gu and Jin [19] proposed an approach for generating ensembles where accuracy and diversity of the ensemble are simultaneously maximized using a Multi Objective Evolutionary Algorithm. Base learners were built manipulating the training examples (*Bagging*) or manipulating the input features. The diversity measures used in this work were Coincident Failure Diversity, Disagreement and Hamming Distance. According to the authors, experimental results shown that this methodology derives on multiple classifier ensembles that outperform single classifiers.

Löfström et al. [20] proposed an approach that uses genetic algorithms to search base learners that simultaneously optimize both an accuracy measure (ensemble accuracy or base classifier accuracy) and a diversity measure (double-fault or difficulty). The experiments shown that combinations of measures often resulting in better performance than when using any single measure.

Tsymbal et al. [17] analyzed the efficiency of five diversity measures (the plain disagreement, the fail/non fail disagreement, the Q statistic, the correlation coefficient and the kappa statistic) when they are applied to select the feature subsets that promote the greatest disagreement among the base classifiers. To evaluate each measure of diversity they calculated its correlation with the difference between the ensemble accuracy and the base classifier average accuracy. The best correlations where shown by the plain disagreement measure and the fail/non-fail disagreement measure.

Shipp and Kuncheva [21] analyzed the relationships between different methods of classifier combination and different measures of diversity. They concluded that whether identical accuracies are assumed to the base learners, they are less diverse than they could theoretically be. On the other hand, they shown a low correlation between the quantified measures of diversity and the studied combination methods. So, they conclude that the participation of diversity measures in designing classifier ensemble is an open question.

4   Experimental Setup

To evaluate the influence of the diversity on the accuracy of the ensemble, we have selected four different strategies for creating ANN classifier ensembles: Injecting randomness into the learning process, *Bagging*, *Boosting* and BCE. Also, to widen our study, these four classification models have been built using the full feature space and using the feature subsets obtained by applying *Best First* [22] and *Correlation Feature Selection* [23] as method of feature selection process[1]. This variation has a double objective: Increasing the diversity between the base learners and obtaining a further experimental validation.

The different datasets used in this experimentation are detailed in Table 2.

Table 2. Summary of the datasets used in the experimental evaluation.

Dataset	Number of instances	Number of features	Number of classes	N. instances maj/min class	Imbalance ratio	Source
OPTDIGITS	5620	64	10	572/554	1.03	[24]
LIBRAS	360	90	15	24/24	1.00	[24]
IMBALANCED SEMEION	1236	256	10	162/40	4.05	[24, 25]
SEMEION	1592	256	10	162/155	1.04	[24]
ASISTENTUR	1006	1024	9	478/22	21.73	[25]

4.1   Ensemble Accuracy Evaluation

In other to evaluate the accuracy of the different ensemble methods we have performed five replications of two-fold cross-validation process. Moreover, to control the variations that result from the ANN's inherent randomness, each classification model has been trained on each data set ten times and the extreme cases (better and worse performance) have been excluded. In our experiments, the base learners are one hidden layer ANN trained with the *Back-Propagation* algorithm. The number of hidden units is 20 for LIBRAS and SEMEION datasets and 30 for OPTDIGITS and ASISTENTUR datasets. The number of base learners for *InitRandom*, *Bagging* and *Boosting* has been fixed to 15. In BCE, the number of base learners is equal to the number of classes.

Table 3 shows the accuracy of the eight evaluated models on the five different datasets. Additionally, the accuracy of a single ANN (with and without feature selection) is included as a reference.

---

[1] Note than, by definition, BCE is built using a feature selection process. Nevertheless, in this work BCE, as the other classification models, is built using both, the full feature space (removing the feature selection step) and the feature subsets obtained by applying BF + CFS.

Table 3. Accuracy values (in %) for the evaluated models. The best values are in bold. Last row shows the average position of each ensemble when accuracy is ranked in ascending order.

	Full feature space					Feature subspace selected by CFS				
	ANN	RndInit	Bagging	Boosting	BCE	ANN	RndInit	Bagging	Boosting	BCE
OPTDIGITS	92.25	97.93	97.68	**97.99**	97.90	91.58	97.64	97.44	**97.86**	95.37
LIBRAS	72.99	77.81	77.08	**81.27**	77.65	63.06	64.53	75.66	70.65	**77.34**
IMBALANCED SEMEION	84.71	**91.06**	89.12	90.09	90.16	85.88	**90.35**	89.36	89.48	89.51
SEMEION	86.10	**91.39**	90.98	90.45	91.11	85.96	88.54	89.99	87.95	**90.67**
ASISTENTUR	93.72	**95.32**	94.36	94.69	95.03	93.34	93.74	85.07	93.98	**94.43**
RANKING	8.6	**1.4**	5.2	2.8	2.6	9.6	6.2	7.2	6.4	**5.0**

Experimental results presented in Table 3 indicate that, in spite of the manipulation of the input features is a diversity source, ensembles that are built using the full feature space are often more accurate than those that are built using a feature subset. Another interesting result is that the classification model that seems to offer the best accuracy is the ensemble where ANN are trained on the same dataset but with different initial weights (referred to as *RandomInit*). In addition, when the systems are built using a reduced number of features, BCE is the system that, in most cases, offers the best accuracy. Unsurprisingly, the implemented ensembles are always more accurate than single ANN.

4.2   Base Learner Accuracy

As has been mentioned in the introduction, an ensemble of classifiers is a pool of classifiers whose individual decisions are combined to achieve a system that outperforms every one of its members. To verify this assertion, we have computed the maximum, the minimum and the average accuracy of the base learners. These values are shown in Figs. 1a and 1b[2].

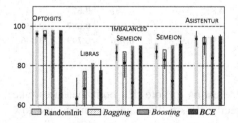 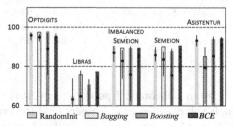

Fig. 1. Ensemble accuracy (bar), interval defined by the accuracy of the base learners (line) and the average accuracy of the base learners (point) when ensembles are built (a) using the Full Feature Space and (b) using the feature subspace selected by CFS.

---

[2] Enlarged figures are available at [26].

Values plotted in Figs. 1a and 1b shown that, some of the base learners are occasionally more accurate than the corresponding ensemble. Nevertheless, the ensemble always has a higher accuracy value than the average of its base learners. Therefore, we can conclude that, with some exceptions, the ensembles of classifiers outperform base learners.

4.3   Base Learner Diversity

To analyze the diversity degree of the base learners, we have quantified this magnitude using the measures mentioned in Sect. 3.1. Tables 4 and 5 show the average of two of these measures: The *Q statistic* (Q) and the *Kappa degree-of-agreement statistic* (κ). Values from each of the remaining measures are available at [26].

Table 4. Q statistic values for each ensemble. Bold face type indicates the best values in each category

Dataset	Full feature space				Feature subspace selected by CFS			
	RandomInit	Bagging	Boosting	BCE	RandomInit	Bagging	Boosting	BCE
OPTDIGITS	0.959	0.940	**0.637**	0.927	0.959	0.940	**0.640**	0.909
LIBRAS	0.962	0.768	**0.250**	0.932	0.984	0.734	**0.300**	0.864
IMBALANCED SEMEION	0.908	0.824	**0.481**	0.856	0.935	0.855	**0.621**	0.883
SEMEION	0.899	0.953	**0.487**	0.973	0.948	0.850	**0.662**	0.874
ASISTENTUR	0.971	0.926	**0.592**	0.893	0.993	0.882	**0.693**	0.884

Table 5. Kappa degree-of-agreement statistic for each ensemble. Bold face type indicates the best values in each category

Dataset	Full feature space				Feature subspace selected by CFS			
	RandomInit	Bagging	Boosting	BCE	RandomInit	Bagging	Boosting	BCE
OPTDIGITS	0.955	0.942	**0.829**	0.943	0.952	0.939	**0.824**	0.932
LIBRAS	0.846	0.663	**0.422**	0.802	0.882	0.638	**0.422**	0.716
IMBALANCED SEMEION	0.858	0.778	**0.592**	0.818	0.880	0.807	**0.677**	0.842
SEMEION	0.859	0.906	**0.605**	0.928	0.882	0.813	**0.687**	0.837
ASISTENTUR	0.945	0.906	**0.762**	0.901	0.968	**0.796**	0.799	0.896

From a theoretical point of view, it is interesting to note that both measures take high values when base learners tend to correctly classify the same instances and low values when errors are uncorrelated, that is, when the base learners are diverse. According to this claim, the experimental results show a low diversity between the base classifiers of all ensembles. As reported in [21], this could be an indication of the lack of any strong relationship between diversity measures and ensemble accuracy in real-life classification problems.

Another interesting result is that *Boosting* produces the most diverse base classifiers but, in most cases, it is not the most accurate ensemble (see Table 3). On the other hand, *RandomInit* is the more accurate ensemble but its base classifiers are not the more diverse.

To test the relationship between the diversity of the base learners and the accuracy of the ensemble, Tables 6 and 7 show the correlation between the diversity and the gain of the ensemble (difference between the ensemble accuracy and the average base learner accuracy). This correlation has been quantified using *Spearman's rank correlation coefficient* (RCC). Moreover, to evaluate the statistical significance of RCC we assume that, when N is the number of built ensembles, the statistic given by $t = RCC$ is distributed following Student's t distribution with N-2 degrees of freedom. So, if the null hypothesis is correct (the correlation between the two variables is null), the probability that this quantity is greater than 1.984 is less than 0.05. So, we may reject the null hypothesis in favor of the hypothesis that the two variables are correlated when t is greater than 1.984.

Table 6. Spearman's rank correlation for the ensemble diversity (Q statistic) and the gain of the ensemble. The best values are in bold. The shaded values represent the ensembles in which the *Q statistic* and the gain of the ensemble are uncorrelated.

Dataset	Full Feature Space				Feature Subspace Selected by CFS			
	RandomInit	Bagging	Boosting	BCE	RandomInit	Bagging	Boosting	BCE
OPTDIGITS	-0.51	-0.48	-0.12	**-0.55**	**-0.54**	-0.53	-0.16	-0.70
LIBRAS	-0.59	**-0.75**	-0.15	-0.55	**-0.53**	-0.52	0.33	-0.57
IMBALANCED SEMEION	-0.57	**-0.67**	-0.26	-0.54	-0.29	**-0.71**	-0.08	-0.54
SEMEION	-0.33	**-0.23**	-0.03	-0.40	-0.49	**-0.58**	-0.23	-0.16
ASISTENTUR	-0.62	-0.46	-0.18	**-0.63**	-0.04	**-0.47**	-0.35	-0.46

Table 7. Spearman's rank correlation for the ensemble diversity (Kappa degree-of-agreement statistic) and the gain of the ensemble. The best values are in bold. The shaded values represent the ensembles in which the *Kappa statistic* and the gain of the ensemble are uncorrelated.

Dataset	Full Feature Space				Feature Subspace Selected by CFS			
	RandomInit	Bagging	Boosting	BCE	RandomInit	Bagging	Boosting	BCE
OPTDIGITS	-0.73	0.46	0.05	**-0.78**	-0.66	-0.54	0.02	**-0.78**
LIBRAS	-0.64	**-0.75**	0.16	-0.58	-0.26	**-0.65**	0.09	-0.42
IMBALANCED SEMEION	-0.52	-0.50	-0.11	**-0.55**	-0.21	**-0.54**	0.07	-0.43
SEMEION	-0.03	**-0.07**	0.06	-0.18	**-0.29**	-0.29	0.10	-0.80
ASISTENTUR	-0.44	-0.36	0.24	**-0.52**	-0.08	**-0.36**	-0.19	-0.51

Experimental results reveal that in spite of *Boosting* has the most diverse base learners, there is no correlation between the diversity of the base learners and the gain of the ensemble. On the other hand, although the values of the Q and *kappa* statistics indicate that the base learners are not diverse, the RCC analysis shows that, except for *Boosting*, the gain of the ensemble is usually correlated with the diversity of the base learners.

5   Conclusions and Future Work

In this paper, we have presented an analysis of the relationship between the ensemble accuracy and both the accuracy and the diversity of the base learners. Experimental results show that obtaining a pool of base learners that are both accurate and diverse is a complex issue. Moreover, according to the experimental evaluation, we can conclude that base learners can be occasionally more accurate than the corresponding ensemble and that the influence of the diversity of base learners on the accuracy of the ensemble is difficult to predict.

It is important to remark that the studied ensembles have been built using different techniques for generating diverse base learners. Nevertheless, all members of the ensembles are based on a single learning algorithm. In particular, all base learners are ANN trained with the Back-Propagation algorithm.

As future work, we propose to analyze the influence of the diversity of base learners on the gain of the ensemble when the base learners are trained using different learning algorithm in both, heterogeneous and homogeneous ensembles.

Acknowledgments. This research was supported by the Spanish MINECO under projects TRA2016-78886-C3-1-R and TRA2015-63708-R, and by CAM under project S2013/MIT-3024.

References

1. Dietterich, T.G.: Ensemble methods in machine learning. In: Kittler, J., Roli, F. (eds.) MCS 2000. LNCS, vol. 1857, pp. 1–15. Springer, Heidelberg (2000). https://doi.org/10.1007/3-540-45014-9_1
2. Dietterich, T.G.: Machine-learning research. AI Mag. **18**, 97–137 (1997)
3. Breiman, L.: Bagging predictors. Mach. Learn. **24**, 123–140 (1996)
4. Schapire, R.E.: The strength of weak learnability. Mach. Learn. **5**, 197–227 (1990)
5. Sharkey, A.J.C., Sharkey, N.E.: Combining diverse neural nets. Knowl. Eng. Rev. **12**, 231–247 (1997)
6. Ho, T.K.: The random subspace method for constructing decision forests. IEEE Trans. Pattern Anal. Mach. Intell. **20**, 832–844 (1998)
7. Blum, A.L., Langley, P.: Selection of relevant features and examples in machine learning. Artif. Intell. **97**, 245–271 (1997)
8. Tsymbal, A., Pechenizkiy, M., Cunningham, P.: Diversity in ensemble feature selection (2003)

9. Anand, R., Mehrotra, K.G., Mohan, C.K., Ranka, S.: An improved algorithm for neural network classification of imbalanced training sets. IEEE Trans. Neural Netw. **4**, 962–969 (1993)
10. Hastie, T., Tibshirani, R.: Classification by pairwise coupling. Ann. Stat. **26**, 451–471 (1998)
11. Dietterich, T.G., Bakiri, G.: Solving multiclass learning problems via error-correcting output codes. J. Artif. Intell. Res. **2**, 263–286 (1995)
12. Murphey, Y.L., Wang, H., Ou, G.: OAHO: an effective algorithm for multi-class learning from imbalanced data. In: Proceedings of International Joint Conference on Neural Networks, pp. 406–411 (2007)
13. Sesmero, M.P., Alonso-Weber, J.M., Gutierrez, G., Ledezma, A., Sanchis, A.: An ensemble approach of dual base learners for multi-class classification problems. Inf. Fusion. **24**, 122–136 (2015)
14. Dietterich, T.G.: An experimental comparison of three methods for constructing ensembles of decision trees: bagging, boosting, and randomization. Mach. Learn. **40**, 139–157 (2000)
15. Kolen, J.F., Pollack, J.B.: Backpropagation is sensitive to initial conditions. Complex Syst. **4**, 269–280 (1990)
16. Kuncheva, L.I., Whitaker, C.J.: Measures of diversity in classifier ensembles and their relationship with the ensemble accuracy. Mach. Learn. **51**, 181–207 (2003)
17. Tsymbal, A., Pechenizkiy, M., Cunningham, P.: Diversity in search strategies for ensemble feature selection. Inf. Fusion **6**, 83–98 (2005)
18. Zenobi, G., Cunningham, P.: Using diversity in preparing ensembles of classifiers based on different feature subsets to minimize generalization error. In: De Raedt, L., Flach, P. (eds.) ECML 2001. LNCS (LNAI), vol. 2167, pp. 576–587. Springer, Heidelberg (2001). https://doi.org/10.1007/3-540-44795-4_49
19. Gu, S., Jin, Y.: Generating diverse and accurate classifier ensembles using multi-objective optimization. In: 2014 IEEE Symposium Series on Computational Intelligence in Multi-Criteria Decision-Making, Proceedings, Orlando, pp. 9–15 (2014)
20. Löfström, T., Johansson, U., Boström, H.: On the use of accuracy and diversity measures for evaluating and selecting ensembles of classifiers. In: Proceedings of the 7th International Conference on Machine Learning and Applications, ICMLA 2008, pp. 127–132 (2008)
21. Shipp, C.A., Kuncheva, L.I.: Relationships between combination methods and measures of diversity in combining classifiers. Inf. Fusion **3**, 135–148 (2002)
22. Xu, L., Yan, P., Chang, T.: Best first strategy for feature selection. In: 9th International Conference on Pattern Recognition, pp. 706–708 (1988)
23. Hall, M.A.: Correlation-based feature selection for machine learning (1999). http://www.cs.waikato.ac.nz/~mhall/thesis.pdf
24. Frank, A., Asuncion, A.: UCI Machine Learning Repository. http://archive.ics.uci.edu/ml/
25. Sesmero, M.P., Ledezma, A., Alonso-Weber, J.M., Gutierrez, G., Sanchis, A.: Control Learning and Systems Optimization Group. http://www.caos.inf.uc3m.es/datasets/
26. Sesmero, M.P.: Measures of Diversity and Accuracy. http://www.caos.inf.uc3m.es/diversity-and-accuracy-in-ann-ensembles/

Adapting Hierarchical Multiclass Classification to Changes in the Target Concept

Daniel Silva-Palacios[⊠][iD], Cesar Ferri[iD], and M. Jose Ramirez-Quintana[iD]

DSIC, Universitat Politècnica de València, Camí de Vera s/n, 46022 Valencia, Spain
dasilpa@posgrado.upv.es, {cferri,mramirez}@dsic.upv.es

Abstract. Machine learning models often need to be adapted to new contexts, for instance, to deal with situations where the target concept changes. In hierarchical classification, the modularity and flexibility of learning techniques allows us to deal directly with changes in the learning problem by readapting the structure of the model, instead of having to retrain the model from the scratch. In this work, we propose a method for adapting hierarchical models to changes in the target classes. We experimentally evaluate our method over different datasets. The results show that our novel approach improves the original model, and compared to the retraining approach, it performs quite competitive while it implies a significantly smaller computational cost.

Keywords: Hierarchical · Classification · Adaptation · Novelty

1 Introduction

The traditional setting in supervised learning assumes that the context at which the learning process takes place does not change from model training to deployment. This stationary conception of supervised problems contrasts with what happens in more realistic scenarios in which it is usual that the training context differs from the deployment context. Some illustrative cases where the learning context usually changes are object recognition and retailing. Thus, in practice, it is quite common that, for instance, an object recognizer that is trained using data easily collected (e.g., from the web), latter it is applied to images obtained from another device (e.g., a mobile); or consider a purchase profile learned from the selling data of a product *p1* but applied to predict the purchase profile for another product *p2*.

A general notion of context is presented in [9]. Here, the authors define the context as any information related to the data (data distribution, data representation, data quality, utility functions and task), and introduce a taxonomy of context changes that includes, among others, changes in the joint distribution of inputs and outputs from training to test (which is most commonly denoted as *dataset shift* [12,15]) and task changes.

© Springer Nature Switzerland AG 2018
F. Herrera et al. (Eds.): CAEPIA 2018, LNAI 11160, pp. 118–127, 2018.
https://doi.org/10.1007/978-3-030-00374-6_12

In this paper, we are interested in the particular case of changes in which the target concept varies by a novelty, i.e. a new class occurs in the test set that is unseen during the training stage. This is a challenging problem in areas such as computer vision [10,16–18], text categorization [11], or data stream [3,5,20].

The main consequence of a novelty is that it negatively affects the quality of the model by decreasing its accuracy, since the new instances from the novel class are misclassified. Hence, when a novelty is detected, the model needs to be adapted to the new context in order to maintain its performance. In incremental learning [2,6,8] we distinguish two main approaches to handle novelties: (1) to re-fit (retrain) the model using new and historical data (if available), which can also be combined with weights (to give more importance to the new data) or with a preprocessing to deal with the class imbalance; and (2) to modify the model (if possible) to also cover the new situation (usually these approaches are referred as reframing techniques).

The advantage of retraining is that any learning technique can be used which makes this approach easy to be applied. However, it could be computationally expensive (if the model has to be retrained many times) and it could be difficult to determine the suitable amount of old and new data to be used for the retraining. On the other hand, adaptive methods are more efficient, but depending on the learning technique, the adaptation of the model could be complicated. A general way of adapting a model is to determine which parts of the model should remain unchanged (i.e., how to reuse the old knowledge) and which parts should be updated/added (i.e., how to incorporate the new knowledge).

In this paper, we present an incremental method for adapting the hierarchical classification algorithm (HMC) presented in [19] to cope with new classes. The flexibility and modularity of hierarchical classification makes easier to update only the part of the hierarchy that is affected by the appearance of the new class. In order to do this, once a novelty is detected, our method consists of two phases: (1) the class hierarchy is modified to include a new internal node that has as children the new class n and its most similar class s in the training data; and (2) the hierarchy of classifiers is modified by replacing the leaf that corresponds to class s by a binary classifier that distinguish between classes s and n. In this way, we achieve a good balancing between the knowledge acquired during training and the knowledge emerging from the new instances. We experimentally evaluate our approach with a collection of datasets. The results show that our method is an alternative option to retraining the whole hierarchy of models.

This paper is organized as follow. In Sect. 2, we review some previous works. In sect. 3, we briefly present the HMC hierarchical classification method. Section 4 defines the adaptation approach we propose to deal with novelty problems. Section 5 presents the experiments that we conducted in order to evaluate our approach. Finally, Sect. 6 concludes the paper.

2   Related Work

In this section, we review some of the existing approaches proposed to solve the problem of predicting instances from new classes not seen during the training stage, which is generally denoted as *class-incremental learning* (CIL). CIL is generally considered a classification problem, thus many approaches propose the use of supervised and semi-supervised learning techniques for that purpose.

In [21], the authors introduced a CIL method to improve the performance of support vector machines (SVM) for text classification. For a problem with N classes, a set of $N - 1$ binary SVM classifiers is learned from the training instances. When a new class is detected a new binary SVM is learned by taking the instances from the new class as positive examples and the rest of instances as negative ones.

Most CIL methods are based on ensembles because they can be efficiently updated. *Learn++.NC* [14] is an incremental ensemble learning algorithm where each classifier is trained with a different subset of weighted instances. When instances from new classes are observed, a new classifier is learned and added to the ensemble. *Learn++.NC* defines a novel weighted voting mechanism that decides which classifiers from the ensemble are used to predict an instance as well as the weight for each individual prediction. Another ensemble-based CIL approach is presented in [5]. The method incorporates a forgetting mechanism for removing from the ensemble the old tree with lowest accuracy rate when the accuracy of a new tree is the highest one. In [13], an ensemble of completely random trees is created during training, later it is updated using the test instances detected as belonging to the new class by growing the trees until a stopping criterion is satisfied. An ensemble of models where each individual is learned by applying a K-medoid clustering algorithm is presented in [3]. The model is incrementally updated by adding the clusters obtained by a clustering algorithm with the examples of the emerging class. In [4], the authors propose the use of semi-supervised methods based on the idea that unlabeled data can provide helpful information during training to generate generalized models that perform better for seen and unseen classes. For that, they assume that the set of unlabeled examples contains sufficient instances of the new class.

As in our proposal, some of the approaches above mentioned assume that instances from a new class are identified by other existing mechanisms, and focus on how to update the models previously trained to incorporate this new class. Our model adaptation method can be seen as a multiclass CIL algorithm, that is not designed for any specific base learning technique (as usually happens in other previous approaches), and that can be applied even if the training sample is not available.

3   Hierarchical Multiclass Classification (HMC)

In this section we briefly outline the Hierarchical Multiclass Classification (HMC) method. For a further detailed description of HMC, we refer the reader to [19].

The underlying idea behind the HMC method is to decompose the original multiclass problem into a hierarchy of binary problems with the aim to improve the classification accuracy of a flat (non-hierarchical) multiclass classifier. Firstly, a class similarity measure is derived from the confusion matrix of a flat classifier following the reasoning that the classification errors in two classes i and j are an indicator of how similar both classes for the flat classifier are. Then, using this similarity metric, an agglomerative hierarchical clustering algorithm is applied to create the class hierarchy, whose leaves are the original classes.

Once the class hierarchy has been created, the second step in the HMC method is to build a hierarchy of classifiers by learning a binary classifier at every internal node (including the root) of the class hierarchy (using any base learning technique). In [19], the authors proposed two methods to apply the hierarchical model depending on whether the classifiers in the hierarchy predict class labels (top-down prediction) or class probabilities (top-down and bottom-up prediction). In the top-down prediction, to classify a new instance, the tree of classifiers is traversed in a top-down manner applying the classifiers in a path from the root until a leaf. In the top-down and bottom-up prediction, to classify a new instance the tree is first traversed in a top-down manner applying all the classifiers and estimating in each node the probability of belonging to each of its children. Then, the probability vectors at the leaves are propagated bottom-up until the root is reached. In this way, the components of the vector at the root represent the probability that the instance belongs to each one of the classes. For the experiments presented in Sect. 5, we will use the top-down and bottom-up prediction since it was shown in [19] that obtains the best performance.

4   A Method for Incrementally Adapting HMC Models

In this section we present our CIL algorithm to modify the HMC model in order to deal with a new class (the *target* class). For that purpose, the first step is to know which one of the original classes (the *source* classes) is the most similar to the target class. This will provide us the necessary information to know the location in the hierarchy where we have to insert the new node composed of the target class and its most similar source class.

4.1   Detecting the Similarity Between the Target and Source Classes

Let D be a training data, $C = \{c_1, \ldots, c_k\}$ the set of source classes (we will denote from now on as $C = \{1, \ldots, k\}$, for the sake of readability), F a flat multiclass model trained with D, H_C the class hierarchy generated using the classes in C and the predictions given by F, and M the hierarchical multiclass model generated by applying the HMC method. Let T be a set of unlabeled data (the deployment set), and $T_n \subseteq T$ the subset of instances in T from the target class n, $n \notin C$. Note that, as mentioned in Sect. 2, we assume that an existing mechanism detects the instances in T_n. Therefore, we can proceed considering that all instances in T_n are indeed labeled. Then, the M model is used to predict the class for the instances in T_n. Logically, M missclassifies all the instances in T_n since class n has not been used to train M. The following vector $P_n = \langle P_{n1}, \ldots, P_{nk} \rangle$ represents the performance of M over T_n, where $P_{ni}, 1 \leq i \leq k$, denotes the number of instances in T_n that are wrongly classified as being of class

i. Consequently, an easy way to discover which is the class in C most similar to t is to examine P_n looking for the component of highest value, since the higher the P_{ni} value is, the more similar classes t and i are for M. Hence, the most similar class s to a target class n is defined as $s = argmax_{i \in C}(P_{ni})$.

The next step consists of adapting the class hierarchy H_C to include the target class n. This is done by converting the leaf for the class s into an internal node with two children: the leaf for the class s and the leaf for the class n. We denote the adapted class hierarchy as H_C^*. As an example to illustrate the process, let us consider an invented dataset with five classes $\{a, b, c, d, e\}$ we use to train an HMC model M. Suppose that a new class f is detected in the deployment set. The result of applying M to the deployment instances from class f is $\{a : 1, b : 2, c : 0, d : 14, \mathbf{e:22}\}$. As observed, the source class most similar to f is class e. Figure 1a shows the class hierarchy inferred during the process of creating M. The shaded leaf indicates the place in the hierarchy where the update shall be done. Finally, the updated class hierarchy is depicted in Fig. 1b.

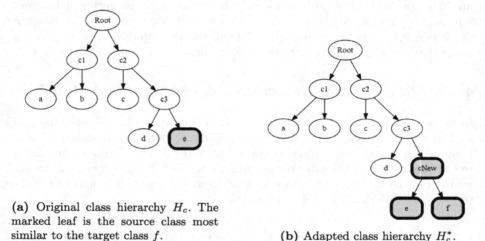

(a) Original class hierarchy H_c. The marked leaf is the source class most similar to the target class f.

(b) Adapted class hierarchy H_c^*.

Fig. 1. Example of the class hierarchy adaptation by the presence of a new class label.

4.2   Updating the HMC Classifier

Once the class hierarchy has been modified, we must update properly the hierarchical model M to make it consistent with H_c^*. This is done by learning a binary classifier corresponding to the last internal node created in H_c^*, and locally growing the tree M by replacing the leaf labeled with the class s (the most similar original class according to Sect. 4.1) by the recently created classifier. As the new binary classifier F_{ns} has to distinguish between classes n and s we must train it using instances of only these two classes. More concretely, F_{ns} is trained using $T_n \cup D_s$, where $D_s \in D$ is the subset of the training instances of class s. Note that, in our adaptation method, we reuse most of the previously acquired knowledge, since only a small part of the hierarchy of classifiers is changed to cover the novel class.

However, in most real applications, the original training data are no longer available for the deployment stage. In those cases, we are still able to adapt the HMC model by training the new binary classifier F_{ns} using the deployment data. More concretely, $T_n \cup T_{\hat{s}}$, where $T_{\hat{s}}$ is the set of all the deployment instances not belonging to the new class n, $e \in (T \setminus T_n)$, that are classified as being of class s by the M model. In Sect. 5, we will experimentally explore both scenarios.

5   Experiments

In this section we evaluate the performance of our proposal. For that reason we carried out a set of experiments following the method described in Sect. 4. The experiments were performed over 16 different datasets (Table 1) taken from the UCI[1], the LIBSVM[2] and KEEL [1] repositories. All the datasets are multivariate, multiclasss, and non-hierarchical a priori. The data sets were so that we included in our analysis problems with different numbers of classes.

Table 1. Information about the datasets used in the experiments: number of instances, attributes and classes.

Id	1	2	3	4	5	6	7	8
Dataset	Dermatology	Flare	Forest	Frogs	Glass	Optdigits	Pendigits	SatImage
NumInst	358	1066	523	7195	214	3823	7494	1795
NumAttr	34	19	27	22	9	64	16	36
NumClass	6	6	4	8	6	10	10	6
Id	9	10	11	12	13	14	15	16
Dataset	Segmentation	Sports	Texture	TrafficLight	Vehicle	Vertebral	Vowel	Zoo
NumInst	2310	8000	5500	300	846	310	990	101
NumAttr	18	13	40	10	18	6	13	16
NumClass	7	10	11	6	4	3	11	7

5.1   Experimental Setup

In the experiments we apply six classification techniques in an R^3 script by means of the libraries caret[4], rpart, e1071 and C50. Specifically, we use the following

---

[1] https://archive.ics.uci.edu/ml.
[2] https://www.csie.ntu.edu.tw/~cjlin/libsvmtools/datasets/.
[3] http://www.R-project.org/.
[4] http://CRAN.R-project.org/package=caret.

classification algorithms: a decision tree "C50", K-nearest neighbors "KNN", a recursive partitioning tree "RPART", a neural network "NNET", a parallel random forest "RF" and a support vector machine "SVM".

We evaluate our adaptive HMC method, referred here as the reframing approach (RA), against the alternative of completely retrain the HMC model, referred here as the retraining approach (TA).

In order to carry out a fair comparison between the retraining and the reframing approaches, we split the dataset into 50% for training (D), 25% for validation (V) and 25% for test (T). We used D for training the original HMC model, V for applying the retraining and reframing approaches, and T for the evaluation. Additionally, to be able to simulate the appearance of a new class, we removed one class from D to play the role of the new class. We repeated this process until all classes in the dataset were used as new ones. Additionally, the experiments were also conducted in two scenarios: the first setting assumed that the training set D was available for the reframing and retraining approaches, and the second setting (a more realistic one) considered that D was not available. Thus, in setting 1 we used $D \cup V$ to retrain and reframe the HMC model, whereas in setting 2 only V was used. Finally, we repeat the complete process 10 times, and we use accuracy as evaluation measure. The results show the average of all the executions.

5.2   Experimental Analysis

Table 2 shows the average results obtained by the reframing and retraining approaches for the two scenarios studied: the training data are available (Table 2, setting 1), and the training data are not available (Table 2, setting 2).

As observed, in setting 1 the retraining approach (TA) obtains better accuracy results for almost all classification tasks. Reframing is able to get better results in some datatsets in scenario 2 where training data is more limited.

Retraining the new models also implies a significant cost in resources. Table 3, shows the average of computational times in seconds spent by each classification technique to adapt the HCM classifier with the two approaches. We also compare both times dividing TA result by RA result. We can see that the reframing approach is much faster than retraining in both settings. Nevertheless, the difference in time depends on the classification technique used to create the local classifiers. In setting 1, RF got the lowest difference and SVM the highest and, in setting 2, the lowest is KNN and the highest is SVM. Setting 1 is in average slower than setting 2 since the first scenario considers more train data.

Following the recommendations of statistical analysis on multiple datasets of [7], we employ the adjusted p-values to find the statistical significance of the results obtained by our RA and TA methods. If we consider Shaffer and Bergmann methods, they produce the same p-values. For setting 1, there is a significant difference between means with a p-value of 0.0047. On the other hand in setting 2, the differences are no statistically significant (p-value is 0.3768). This shows that in the second scenario (a more realistic setting) reframing is able to

Table 2. Average of the accuracy and standard deviation per setting obtained by the different classification techniques applying the original hierarchical multiclass model (HMC), the reframing (RA) and the retraining (TA) approaches. The best result for every dataset is highlighted in bold. In setting 1, the training dataset D is available at the model adaptation time and in setting 2 it is not available.

DataSet	HMC	Setting 1		Setting 2	
		RA	TA	RA	TA
1	0.773	0.863 ± 0.061	$\mathbf{0.921 \pm 0.030}$	0.864 ± 0.062	$\mathbf{0.890 \pm 0.031}$
2	0.654	0.702 ± 0.064	$\mathbf{0.735 \pm 0.028}$	0.711 ± 0.071	$\mathbf{0.742 \pm 0.015}$
3	0.682	0.825 ± 0.056	$\mathbf{0.865 \pm 0.032}$	0.830 ± 0.059	$\mathbf{0.859 \pm 0.028}$
4	0.845	0.914 ± 0.067	$\mathbf{0.963 \pm 0.006}$	0.916 ± 0.064	$\mathbf{0.951 \pm 0.006}$
5	0.591	0.610 ± 0.078	$\mathbf{0.655 \pm 0.063}$	$\mathbf{0.629 \pm 0.070}$	0.616 ± 0.057
6	0.799	0.839 ± 0.029	$\mathbf{0.883 \pm 0.024}$	0.830 ± 0.029	$\mathbf{0.842 \pm 0.029}$
7	0.847	0.891 ± 0.032	$\mathbf{0.934 \pm 0.026}$	0.903 ± 0.029	$\mathbf{0.918 \pm 0.023}$
8	0.746	0.843 ± 0.028	$\mathbf{0.873 \pm 0.013}$	0.846 ± 0.027	$\mathbf{0.863 \pm 0.010}$
9	0.820	0.918 ± 0.028	$\mathbf{0.953 \pm 0.012}$	0.918 ± 0.028	$\mathbf{0.935 \pm 0.010}$
10	0.573	0.589 ± 0.013	$\mathbf{0.611 \pm 0.011}$	$\mathbf{0.595 \pm 0.013}$	0.594 ± 0.012
11	0.870	0.924 ± 0.018	$\mathbf{0.957 \pm 0.008}$	0.924 ± 0.019	$\mathbf{0.940 \pm 0.007}$
12	0.604	0.656 ± 0.060	$\mathbf{0.700 \pm 0.050}$	$\mathbf{0.651 \pm 0.061}$	0.639 ± 0.056
13	0.585	0.639 ± 0.056	$\mathbf{0.697 \pm 0.031}$	0.640 ± 0.056	$\mathbf{0.667 \pm 0.037}$
14	0.595	0.758 ± 0.063	$\mathbf{0.793 \pm 0.041}$	0.766 ± 0.072	$\mathbf{0.797 \pm 0.047}$
15	0.719	0.745 ± 0.033	$\mathbf{0.824 \pm 0.029}$	$\mathbf{0.738 \pm 0.032}$	0.654 ± 0.032
16	0.762	0.821 ± 0.097	$\mathbf{0.896 \pm 0.045}$	$\mathbf{0.827 \pm 0.074}$	0.805 ± 0.041

Table 3. Average time in seconds to adapt the model. Applying the reframing (RA) and the retraining (TA) approaches.

Techniques	Setting 1			Setting 2		
	RA	TA	TA/RA	RA	TA	TA/RA
C50	0.048	0.613	12.775	0.022	0.232	10.434
KNN	1.741	26.781	15.379	1.350	8.658	6.414
NNET	14.970	224.582	15.002	10.297	91.294	8.866
RF	0.247	2.770	11.216	0.111	0.872	7.884
RPART	0.026	0.380	14.432	0.015	0.166	11.350
SVM	0.074	3.311	47.045	0.026	0.552	21.366

obtain similar performance with respect to the retraining approach requiring less time for training the models.

6 Conclusions

In this paper, we propose an approach for adapting a hierarchical multiclass classifier when a new class appears in deployment. We have analysed our approach considering two scenarios: having access to the original training data or just only the deployment data. This second approach is more realistic since in many real cases the training data are not available due to limited computational resources or because the original data is obsolete.

Our approach employs the hierarchical structure of the HMC method and it is based on the notion of similarity between classes. This affinity allows us to easily determine which parts of the hierarchy must remain unchanged and which parts should be updated to deal with the novel class. In this way, we are able to maintain most of the original model in the original way and at the same time, we are able to incorporate the newly acquired knowledge from the new class.

The experiments conducted show that our reframing approach can obtain good performance with less computational time than the alternative of discarding the original model and retraining it from scratch.

As future work, we propose to analyse the effect of imbalance problems on the reframing approach. We are also interested in studying the possible application of our method to other dynamic scenarios where instances arrive in a continuous flow (e.g data streams).

Acknowledgements. This work was partially supported by the EU (FEDER) and the Spanish MINECO under grant TIN 2015-69175-C4-1-R, and by Generalitat Valenciana PROMETEOII2015/013. This work has been supported by the Secretary of Higher Education, Science and Technology (SENESCYT: Secretaría Nacional de Educación Superior, Ciencia y Tecnología), of the Republic of Ecuador.

References

1. Alcalá-Fdez, J., et al.: Keel data-mining software tool: data set repository, integration of algorithms and experimental analysis framework. J. Mult.-Valued Log. Soft Comput. **17**(2–3), 255–287 (2011)
2. Cauwenberghs, G., Poggio, T.: Incremental and decremental SVM learning. In: Advances in Neural Information Processing Systems, pp. 409–415 (2001)
3. Chandak, M.B.: Role of big-data in classification and novel class detection in data streams. J. Big Data **3**(1), 5 (2016)
4. Da, Q., Yu, Y., Zhou, Z.H.: Learning with augmented class by exploiting unlabeled data. In: Twenty-Eighth AAAI Conference on Artificial Intelligence (2014)
5. Farid, D.M., et al.: An adaptive ensemble classifier for mining concept drifting data streams. Expert. Syst. Appl. **40**(15), 5895–5906 (2013)

6. Ferri-Ramírez, C., Hernández-Orallo, J., Ramírez-Quintana, M.J.: Incremental learning of functional logic programs. In: Kuchen, H., Ueda, K. (eds.) FLOPS 2001. LNCS, vol. 2024, pp. 233–247. Springer, Heidelberg (2001). https://doi.org/10.1007/3-540-44716-4_15

7. García, S., Herrera, F.: An extension on statistical comparisons of classifiers over multiple data sets for all pairwise comparisons. J. Mach. Learn. Res. **9**, 2677–2694 (2008)

8. Giraud-Carrier, C.: A note on the utility of incremental learning. AI Commun. **13**(4), 215–223 (2000)

9. Hernández-Orallo, J., et al.: Reframing in context: a systematic approach for model reuse in machine learning. AI Commun. **29**(5), 551–566 (2016)

10. Jain, L.P., Scheirer, W.J., Boult, T.E.: Multi-class open set recognition using probability of inclusion. In: Fleet, D., Pajdla, T., Schiele, B., Tuytelaars, T. (eds.) ECCV 2014. LNCS, vol. 8691, pp. 393–409. Springer, Cham (2014). https://doi.org/10.1007/978-3-319-10578-9_26

11. Klinkenberg, R., Joachims, T.: Detecting concept drift with support vector machines. In: ICML, pp. 487–494 (2000)

12. Moreno-Torres, J.G., Raeder, T., Alaiz-RodríGuez, R.: A unifying view on dataset shift in classification. Pattern Recognit. **45**, 521–530 (2012)

13. Mu, X., Ting, K.M., Zhou, Z.H.: Classification under streaming emerging new classes: a solution using completely-random trees. IEEE Trans. Knowl. Data Eng. **29**(8), 1605–1618 (2017)

14. Muhlbaier, M.D., Topalis, A., Polikar, R.: Learn $^{++}$.NC: combining ensemble of classifiers with dynamically weighted consult-and-vote for efficient incremental learning of new classes. IEEE Trans. Neural Netw. **20**(1), 152–168 (2009)

15. Quionero-Candela, J., Sugiyama, M., Schwaighofer, A., Lawrence, N.D.: Dataset Shift in Machine Learning. The MIT Press, Cambridge (2009)

16. Ross, D.A., Lim, J., Lin, R.S., Yang, M.H.: Incremental learning for robust visual tracking. Int. J. Comput. Vis. **77**(1–3), 125–141 (2008)

17. Scheirer, W.J., Jain, L.P.: Probability models for open set recognition. IEEE Trans. Pattern Anal. Mach. Intell. **36**(11), 2317–2324 (2014)

18. Scheirer, W.J., de Rezende Rocha, A., Sapkota, A., Boult, T.E.: Toward open set recognition. IEEE Trans. Pattern Anal. Mach. Intell. **35**(7), 1757–1772 (2013)

19. Silva-Palacios, D., Ferri, C., Ramírez-Quintana, M.J.: Probabilistic class hierarchies for multiclass classification. J. Comput. Sci. **26**, 254–263 (2018)

20. ZareMoodi, P., Beigy, H., Siahroudi, S.K.: Novel class detection in data streams using local patterns and neighborhood graph. Neurocomputing **158**, 234–245 (2015)

21. Zhang, B.F., Su, J.S., Xu, X.: A class-incremental learning method for multi-class support vector machines in text classification, pp. 2581–2585. IEEE (2006)

Measuring the Quality of Machine Learning and Optimization Frameworks

Ignacio Villalobos$^{(\boxtimes)}$, Javier Ferrer, and Enrique Alba

Universidad de Málaga, Málaga, Spain
{nacho,ferrer,eat}@lcc.uma.es

Abstract. Software frameworks are daily and extensively used in research, both for fundamental studies and applications. Researchers usually trust in the quality of these frameworks without any evidence that they are correctly build, indeed they could contain some defects that potentially could affect to thousands of already published and future papers. Considering the important role of these frameworks in the current state-of-the-art in research, their quality should be quantified to show the weaknesses and strengths of each software package.

In this paper we study the main static quality properties, defined in the product quality model proposed by the ISO 25010 standard, of ten well-known frameworks. We provide a quality rating for each characteristic depending on the severity of the issues detected in the analysis. In addition, we propose an overall quality rating of 12 levels (ranging from A+ to D−) considering the ratings of all characteristics. As a result, we have data evidence to claim that the analysed frameworks are not in a good shape, because the best overall rating is just a C+ for Mahout framework, i.e., all packages need to go for a revision in the analysed features. Focusing on the characteristics individually, maintainability is by far the one which needs the biggest effort to fix the found defects. On the other hand, performance obtains the best average rating, a result which conforms to our expectations because frameworks' authors used to take care about how fast their software runs.

Keywords: Maintainability · Reliability · Performance · Security
Quality

1 Introduction

The international research community considers essential the replicability of the experimentation carried out in an article for progress in science. Actually, when the experimental artifacts are available, other researchers can validate the published work. With this requirement in mind, researchers began to make available their algorithms, which finally have become in large software frameworks [1,2]. These frameworks are collections of algorithms designed for solving complex problems that are freely available for the research and industrial communities.

© Springer Nature Switzerland AG 2018
F. Herrera et al. (Eds.): CAEPIA 2018, LNAI 11160, pp. 128–139, 2018.
https://doi.org/10.1007/978-3-030-00374-6_13

The main advantage for the users of these frameworks is that they can run lots of algorithms without the effort of designing and implementing them. The main problem, at the same time, is that they do not know on how algorithms are actually implemented, and even tend to trust in the quality without any data or evidence that they are at least good software pieces.

In the artificial intelligence research field, Machine Learning and Optimization Frameworks (MLOFs) are widely used by the community. The machine learning packages provide statistical techniques to progressively improve on a specific task from a huge dataset and extract knowledge. Besides, optimization frameworks are used for obtaining optimal solutions for complex continuous and discrete optimization problems. Artificial intelligence (AI) research community is particularly prolific in these fields due to the impact of the application of artificial intelligence techniques to the existing problems. However, finding experts in software engineering in AI is not that common, thus having the undesired result that most researchers build software without even applying the basics of software engineering that all graduate programs teach across the world.

Over the last few years, researchers have offered a wide variety of open source MLOFs that have been used in lots of articles [3]. Researchers use the frameworks' algorithms to look for evidence in favour of a hypothesis they try to support, refute, or validate. Actually, lots of works cite MLOFs because they use them in their experimentation. To illustrate this fact, we only have to enumerate the citations of some MLOFs: an article on the update of WEKA [2] has 16,653 citations[1], the framework Keel [1] has 1,102 citations (See footnote 1), and JMetal [4] has 712 citations (See footnote 1) only accounting for the seed articles introducing the tool, and we are just mentioning a few of them. Their popularity is then clear, because many researchers just want to focus on the application and not in the algorithm.

The main problem is that, generally speaking, the open source MLOFs are offered without any warranties of any kind concerning the safety, performance, bugs, inaccuracies, typographical errors, or other harmful characteristics of the software. The user let us alone with any problem or bug he/she can experience, if he/she is actually noticing such errors at all: sometimes the result of low quality is an unnoticed lose of time, or even a more complex situation where the user thinks to be running a given algorithm that actually is not performing the numerical steps as expected. In fact, we should take into account that some MLOFs users do not understand the provided implementation due to its complexity (not all users have programming skills) or they do not devote enough time to analyse it. In practice, users blindly trust the provided implementation and use it "as is". We want to highlight that a bad quality framework potentially may affect the already published results and the future ones of thousands of researchers. This is the reason why we analyse the quality of some open MLOFs in this paper.

Due to the lack of studies about the quality level the MLOFs have, our final goal is to offer a quantitative quality study of a subset of well-known MLOFs including four key aspects of the internal quality model of the ISO 25010 stan-

---

[1] At the moment of writing: 17, May 2018.

dard: maintainability, reliability, security, and performance efficiency. Our goal is to be positive and give constructive hints on how to solve the found problems, as well as to guide on the fastest ways (where/how) to do so. By no means this is a critic, much on the contrary we respect and endorse these frameworks and just want to contribute to their larger future success.

The main contributions of this work are the following ones: (1) To provide the first quality analysis for ten well-known MLOFs, (2) To evaluate four key attributes about the quality of the frameworks: maintainability, reliability, security, and performance efficiency, (3) To provide an estimation of the needed time to fix all the defects found in regards to each characteristic studied, and (4) To assign an overall mark for their quality so as to track their future improvements.

The rest of the paper is organized as follows. In Sect. 2, we define software quality and the different models proposed by the ISO 25010 standard. After that, we present the ten machine learning and optimization frameworks studied here, and we briefly describe how the analysis is performed. Additionally, we show the analysis results of the four studied quality aspects in four subsections. In Sect. 4, we discuss about the results obtained in the previous section and we propose an overall rating for the frameworks. Finally, in the last section we draw some conclusions and comment some interesting ideas for future work.

2   Forget on Opinions, Let's Go for Standards: ISO 25010

Informally speaking an average user wants a "good" software or service. This term could be very ambiguous and it is not quantifiable, specially when we talk about software. With the first of many definitions, from a professional point of view, quality has been described as the aptitude to accomplish to be used by a client (1970) or to conform with all the product requirements (1979). A few years later these definitions, the standard ISO 8402 (1986) introduced the now well-known definition for product and service quality. After that, with the standard ISO 9126 (1991), an standard finally add the term *Software product* to that definition. However, we want a quality development and a quality software product, but we also want to quantify the level of quality achieved.

Nowadays, the reference standard and the most used one to evaluate software quality is the standard ISO 25010. It defines a *quality in use* model and a *product quality* model. In this work we focus on the product quality model for analysing software products, in this case of the MLOFs, because of their key role in thousands of published articles. This model defines a taxonomy for the main characteristics to consider when you measure the software quality of a product. In the standard, the quality aspects are divided up in the following eight quality characteristics: maintainability, security, functionality, performance efficiency, reliability, usability, compatibility and portability. In this first paper we focus on four of them.

When you deal with a specific software project, we should keep in mind which are our priorities among the quality characteristics mentioned before. Depending on our defined target for the quality analysis, stakeholders or expected tasks of the software, we may be more interested in some quality aspects than in others. Moreover, we should update these quality requirements throughout the product life [5]. We would like to highlight that some requirements or constraints in certain characteristics could have a negative impact in others. Indeed, some aspects of the quality are opposed, e.g., if one try to get the maximum performance, the consequence could be a reduction on the security level.

3   A First Analysis of Static Features of MLOFs

Our goal in this paper is extracting insights from the source code of the MLOFs in order to quantify their overall quality. We study four software characteristics proposed in the product quality model to find possible issues. First, we analysed the maintainability due to the relevance of the capacity of the software to be modified (to extend or fix it). Second, we analysed the security aspect, owing to the relevance to ensure data integrity. After that, we analyse the performance efficiency due to the importance of saving computing resources and decreasing execution time. Finally, we analyse the *reliability* on account of the need to know whether all components perform correctly under specified conditions.

In order to perform all the analyses we have used the tool SonarQube in its version 7.0. We have chosen this tool because the wide acceptance from the developers community and the amount of extensions available. Moreover, SonarQube provides an easy way to be integrated with different tools through its API or plugins. For this study, we have created a quality profile with 295 rules to detect issues about maintainability, security, performance and reliability. Each rule has assigned a severity between the values: blocker, critical, major, minor. This rules' severity is assigned according to the defects that it detects.

But before detecting the weak points to improve in the feature, we need to know a bit the tools from a software perspective: before a deep study let us make some basic objective measurements to shape them better. As a case of study, we select ten MLOFs extracted from the literature, whose source code is developed in Java and are freely available. They have been chosen due to their extended use by the research community, their relevance and impact. Some of them are used for teaching, some for research, some for both. Some are widely known, some are humble code to guide researchers with no knowledge in computer science. We present their main characteristics in Table 1. In the first column we have the name of each software tool, after that we show the version analysed in this work. In the third column, we show the number of lines of code. In the next column we have the McAbee complexity per file, and finally, in the last column we have the number of classes. As we can see in the table, the biggest projects in number of lines of code and classes are Keel and Weka. Also they are the frameworks with the highest McAbee complexity per file. On contrary, the tiniest software packages are ssGA and Mahout but in this case, they are not the ones with the lowest Complexity per file, they are JCLEC and Watchmaker.

In the following subsections, we are presenting the results for the analysed quality characteristics. In each subsection we show a table with the rating

Table 1. Machine learning and optimization frameworks.

Project	Version	#LOC	Complexity per file	#Classes
ECJ [6]	25	53,771	18.49	620
JCLEC [7]	4.0	16,652	7.72	323
Jenes [8]	765	11,508	18.06	185
jMetal [4]	5.4	43,144	13.14	609
Keel [1]	3.0	585,337	36.89	3,808
Mahout [9]	0.13.1	1,255	11.14	30
moea-frame [10]	2.12	33,888	12.97	506
ssGA [11]	1.1	672	12.77	13
Watchmaker [12]	0.3.0	5,639	4.79	140
WEKA [2]	14812	353,923	34.81	2,383

obtained by each framework. This rating ranges from A to E where A is the best mark, and E is the worst. For each characteristic, it gets considering the following criterion. If the framework does not have any issue, its get the best mark, an A. Then, if it has only at least a minor issue, it will get a B qualification. If it has at least a major issue, it will get a C. If it has at least a critical issue, it will obtain a D and, finally, if it has one or more blocker issues, the framework will obtain an E. We also show the estimated effort needed (in minutes) to solve all the issues detected and the number of issues for each severity. This effort is calculated regarding to the number of implicated lines of code and an estimation of 18 min per line change needed.

3.1 Maintainability

In this subsection, we analyse the selected MLOFs from the point of view of the maintainability. A good source code maintainability can be measured (ISO 25010) as the degree of effectiveness and efficiency with which the software can be modified by developers, adding new functionalities or fixing existing errors. The most violated maintainability rules in this analysis are the following ones: (a) Local variables, parameters, and methods should comply the Java naming convention, (b) useless assignments for a local variable, and (c) empty statements must be removed. Note that these rules has small impact in the rating because their severity is minor. On contrary, some of the more severe maintainability issues found are the following ones: (a) clone implementation should not be overridden, and (b) Child class fields should not shadow parent class fields.

In Table 2 we show the results for each framework order by their rating and effort. None of them get the best qualification nor the second one: the majority of the frameworks obtain the two worst qualification (D and E). As we can see, the larger frameworks are more likely to have severe issues. In addition, sometimes there is a large difference in number of issues between two frameworks with the same rating. The reason is that the severity prevails over number in the found issues. Regarding the effort needed to fix all the issues, Keel needs the larger expected effort: it will take more than five years of work to solve them all, however, we must consider that Keel has more than half a million lines of code. In contrast, ssGA only has four blocker issues with a total effort of less than two workdays, although it has less than one thousand lines of code.

Table 2. Rating, effort and issues per severity for maintainability.

Project	Rating	Effort (min)	Issues			
			Blocker	Critical	Major	Minor
Mahout	C	1,057	0	0	25	118
Watchmaker	D	1,574	0	25	149	268
JCLEC	D	5,164	0	11	248	344
moea-frame	D	10,609	0	170	182	683
ssGA	E	728	4	0	12	133
Jenes	E	3,879	21	14	121	403
jMetal	E	21,111	10	73	642	1,790
ECJ	E	22,173	88	168	923	1,277
WEKA	E	117,360	48	996	5134	13,888
Keel	E	903,509	324	1,757	17,084	62,649

3.2 Security

The security characteristic measures the degree to which a software protects information and data, i.e., the probability that the software has a security risk. The most common issues found in this study regarding security are: (a) mutable fields should not be public static, (b) do not use deprecated code and (c) A method or attribute should be protected. Note again that these rules have a small impact in the rating because their severity is minor. On the other hand, some of the more severe violated rules founded are the following ones: (a) Credentials may be hard-coded, (b) do not call the Java garbage collector explicitly and let the virtual machine manage it, and (c) do not override object finalize method.

In the results showed in Table 3, we can observe that two frameworks, Weka and Keel, have blocker issues, and consequently they have the worst rating. They need a huge effort to fix all their issues related to security, specially Keel will take an expected year of work of one full time person. On the other hand, there are three frameworks (Mahout, ssGA and JCLEC) with a C rating because they do not have blocker and critical issues. Note that Watchmaker with only two issues has a D qualification due to the high severity of one of them (critical). As a consequence, Watchmaker could be improved in a short time, less than an hour, so it seems easy to improve its quality.

Table 3. Rating, effort and issues per severity for security.

Project	Rating	Effort (min)	Issues			
			Blocker	Critical	Major	Minor
Mahout	C	70	0	0	1	6
ssGA	C	280	0	0	27	1
JCLEC	C	660	0	0	21	43
Watchmaker	D	30	0	1	0	1
moea-frame	D	1,340	0	3	110	9
jMetal	D	2,185	0	2	64	122
Jenes	D	3,815	0	3	310	38
ECJ	D	14,245	0	5	274	1,093
WEKA	E	44,840	8	35	2,635	1,692
Keel	E	180,700	48	35	10,218	6,815

3.3 Performance

The performance efficiency characteristic measures the amount of computing resources (CPU, memory, I/O,...) used under some specific conditions. The most violated rules are minor issues that affect the performance are the following: (a) Use different methods for variable parsing, (b) method with a very high cognitive complexity and (c) constructor should not be use to instantiate primitive classes. On contrary, some of the more severe performance issues detected are the following ones: (a) Use of sleep instead wait when a lock is held, (b) constructors with a high number of parameters, and (c) do not use synchronized data structures when it is not needed.

In Table 4 we show the result of the analysis. At a first glance, we can observe that three frameworks need around one hour to solve all performance issues. On the other hand, the biggest projects need much more effort, specifically more than five months to solve all the performance issues. In addition, all the frameworks have a D qualification, except Watchmaker, which obtains the best qualification with a C because it has not blocker or critical issues in performance.

3.4 Reliability

Reliability measures the probability of failure-free software operation for a period of time. The most common violated rules found in our analysis in regards to reliability are the following ones: (a) do not throw generic exceptions, (b) do not write static fields from instance methods, and (c) cast operators before to perform maths operations. On the other hand, some of the more severe reliability issues are the following ones: (a) resources should be closed, (d) zero could be a possible denominator, and (c) do not use a threat instance as a monitor.

All the results for the reliability analysis are shown in Table 5. As we can see in the table, most of packages have a bad mark due to the existence of blocker

Table 4. Rating, effort and issues per severity for performance.

Project	Rating	Effort (min)	Issues			
			Blocker	Critical	Major	Minor
Watchmaker	C	40	0	0	3	0
ssGA	D	58	0	2	2	0
Mahout	D	64	0	5	0	0
JCLEC	D	398	0	20	12	2
Jenes	D	656	0	17	19	6
moea-frame	D	1,828	0	82	14	5
jMetal	D	2,783	0	59	66	42
ECJ	D	7,659	0	188	89	72
WEKA	D	83,507	0	1,152	2,344	510
Keel	D	230,037	0	2,864	6,275	3,914

issues. Despite that, three frameworks (Mahout, ssGA and Watchmaker) have a qualification of B, C and D, respectively. In the case of Mahout, it only has a minor issue and it could be resolved in only five minutes. Then, the ssGA has 18 major issues and they could be resolved in five hours. Finally, Watchmaker has only two issues but one of them is critical. From the point of view of the reliability, Keel and WEKA seem to need a deep improvement in reliability according to these evidences, because of the large number of issues and the long time needed to fix them.

Table 5. Rating, effort and issues per severity for reliability.

Project	Rating	Effort (min)	Issues			
			Blocker	Critical	Major	Minor
Mahout	B	5	0	0	0	1
ssGA	C	360	0	0	18	0
Watchmaker	D	10	0	1	0	1
Jenes	E	770	2	7	36	14
moea-frame	E	1,355	7	61	31	28
JCLEC	E	1,560	1	3	6	98
ECJ	E	1,868	9	21	99	58
jMetal	E	2,941	32	111	20	22
Keel	E	29,997	495	348	1,319	728
WEKA	E	48,238	144	300	2,051	110

4   Summary of Results and Global Discussion

In the previous sections we have presented the results obtained after analysing four key quality aspects. With the aim in mind of ranking the projects with a final qualification, we introduce an average rating summarizing the ratings of all analysed characteristics.

Given a rating $r_c \in \{A, B, C, D, E\}$ for a particular characteristic $c \in C$, we assign a value $r'_c \in \mathbb{N}$ in the range [1,5] such that the ratings $\{A, B, C, D, E\}$ corresponds to values $\{1, 2, 3, 4, 5\}$, respectively. After that, we average the r'_c to obtain an overall rating $r_o \in \mathbb{R}$ in the range [1.0, 5.0]. We compute r_o using the following equation:

$$r_o = \frac{\sum_{c \in C} r'_c}{|C|} \tag{1}$$

Finally, we translate r_o to an ordinal scale in the range A–D. We assign an overall rating A when $r_o \in [1.0, 2.0)$, an overall rating B when $r_o \in [2.0, 3.0)$, an overall rating C when $r_o \in [3.0, 4.0)$, and an overall rating D when $r_o \in [4.0, 5.0]$. In order to obtain a more precise overall rating we add a '+' symbol when r_o is in the first tertile of the range and a '−' symbol when the value is in the last 33%.

In Table 6 we show the ratings of each characteristic, the overall rating calculated as explained above, and the debt of the project. The debt is calculated as the percentage of *estimated* effort needed to fix all found issues divided by the total *estimated* effort to implement the project. When we compare the ratings obtained for all the characteristics, reliability obtains the worst results with 7 MLOFs rated with an E, followed by maintainability with 6 MLOFs rated with an E. In addition, in Fig. 1 we show the percentage of estimated effort to fix the found issues per characteristic. In the figure, it can be seen that maintainability requires almost 50% total effort in most frameworks.

None of the MLOFs have an E rating in performance, this indicates that the authors of the MLOFs has taken care about the performance efficiency of the algorithms. Actually, performance issues should directly affect the execution times reported in the articles, that is why we guess the performance aspect was carefully treated. In Fig. 1 we confirm our assumption being the performance and reliability the characteristics that require less remediation effort.

Overall, the best project in our comparison is Mahout with a rating of C+, followed by ssGA and Watchmaker rated with C-. It seems that MLOFs with a low complexity per file are less prone to issues. In order to confirm our expectation, we performed a Spearman's rank correlation test between the complexity per file and the overall rating for the MLOFs. We obtained a high coefficient $\rho = 0.908$, what means that the more the complexity in a file the more the probability to find an issue. In Table 6 we also showed the debt of each project. This measure removes the size component of a MLOF, so it shows the *relative* effort needed to fix the issues found. In this regard, Keel is the worst ranked framework with 7.66% and Watchmaker is the best with less than 1% of debt. Something

Table 6. Final qualification of each framework.

Project	Rating				Overall	
	Maintainability	Security	Reliability	Performance	Rating	Debt (%)
Mahout	C	C	B	D	C+	3.18
Watchmaker	D	D	D	C	C−	0.98
ssGA	E	C	C	D	C−	7.07
moea-frame	D	D	E	D	D+	1.49
JCLEC	D	C	E	D	D+	1.56
jMetal	E	D	E	D	D	2.24
Jenes	E	D	E	D	D	2.64
ECJ	E	D	E	D	D	2.85
WEKA	E	E	E	D	D−	2.77
Keel	E	E	E	D	D−	7.66

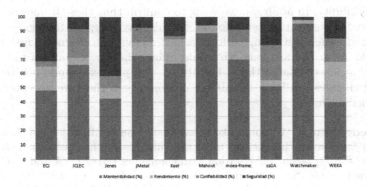

Fig. 1. Estimated effort (%) to fix the found issues per characteristic and framework.

that is not surprising, because Keel has the highest value in complexity per file, meanwhile Watchmaker has the lowest value.

5   Conclusions and Future Work

In this paper we have studied some important aspects of software products such as maintainability, reliability, performance efficiency and security characteristics of ten well-known machine learning and optimization frameworks. Note that these characteristics are part of the product quality model proposed by the ISO 25010 standard. After performing the analysis and classification of the defects detected, we are more concious about the current state of each development for the characteristics studied. The analysis revealed that, overall, none of the frameworks obtain the two best overall qualifications (A and B). This may be a concern for the researchers who had used these frameworks in their experiments.

Particularly, the best framework of our comparison is Mahout with a rating of C+, but if we take a look to the debt ratio, Watchmaker is the best in terms of effort needed to solve the found issues.

In conclusion, we can claim that maintainability is by far the most ignored aspect of the existing packages. A bad maintainability rating means that is more difficult for other developers to contribute or extend the frameworks. Regarding the reliability and the performance, they need a quite similar amount of time, between them, to improve their quality. From our point of view, they are more relevant when we focus on MLOFs because we prefer fast techniques that generate reliable solutions. Finally, about security we think that it is not as relevant as the rest of the characteristic because experiments are usually performed in a local and controlled environment. However, if researchers are going to use these packages in industrial real cases it could become a major issue.

There is a number of interesting findings we want to validate in a near future. We plan to study the rest of characteristics proposed in the ISO 25010 standard that we did not consider in this work such as functionality, portability, usability, and compatibility. In addition, we want to confirm that these frameworks could be improved if we fix the issues found. In this way, we plan to suggest fixes to the detected issues and then, perform a new analysis on the new version of the framework. After the analysis, we would know whether we improve all quality aspects or at least some of them. Finally, we will propose some *pull requests* to make our changes available for all the community if they are accepted by the frameworks' authors.

Acknowledgements. We would like to say thank you to all authors of these frameworks that make research easier for all of us. This research has been partially funded by CELTIC C2017/2-2 in collaboration with companies EMERGYA and SECMOTIC with contracts #8.06/5.47.4997 and #8.06/5.47.4996. It has also been funded by the Spanish Ministry of Science and Innovation and /Junta de Andalucía/FEDER under contracts TIN2014-57341-R and TIN2017-88213-R, the network of smart cities CI-RTI (TIN2016-81766-REDT) and the University of Malaga.

References

1. Alcalá-Fdez, J., et al.: KEEL data-mining software tool: data set repository, integration of algorithms and experimental analysis framework. J. Mult.-Valued Log. Soft Comput. **17**(2–3), 255–287 (2011)
2. Hall, M.A., Frank, E., Holmes, G., Pfahringer, B., Reutemann, P., Witten, I.H.: The WEKA data mining software: an update. SIGKDD Explor. Newsl. **11**, 10–18 (2009)
3. Parejo, J.A., Ruiz-Cortés, A., Lozano, S., Fernandez, P.: Metaheuristic optimization frameworks: a survey and benchmarking. Soft Comput. **16**, 527–561 (2012)
4. Durillo, J.J., Nebro, A.J.: jMetal: a Java framework for multi-objective optimization. Adv. Eng. Softw. **42**(10), 760–771 (2011)
5. Wagner, S.: Software Product Quality Control. Springer, Heidelberg (2013). https://doi.org/10.1007/978-3-642-38571-1

6. Luke, S.: ECJ evolutionary computation library (1998). http://cs.gmu.edu/~eclab/projects/ecj/
7. Ventura, S., Romero, C., Zafra, A., Delgado, J.A., Hervás, C.: JCLEC: a Java framework for evolutionary computation. Soft Comput. **12**(4), 381–392 (2008)
8. Troiano, L., De Pasquale, D., Marinaro, P.: Jenes genetic algorithms in java (2006). http://jenes.intelligentia.it
9. The Apache Software Foundation: Apache Mahout Project (2014). https://mahout.apache.org
10. Hadka, D., Reed, P.: Borg: an auto-adaptive many-objective evolutionary computing framework. Evol. Comput. **21**(2), 231–259 (2013)
11. Alba, E.: ssGA: Steady state GA (2000). http://neo.lcc.uma.es/software/ssga
12. Dyer, D.W.: Watchmaker framework for evolutionary computation (2006). https://watchmaker.uncommons.org/

Fuzzy Sets and Systems

Equivalence Relations on Fuzzy Subgroups

Carlos Bejines[1]([✉])[iD], María Jesús Chasco[1][iD], Jorge Elorza[1][iD],
and Susana Montes[2][iD]

[1] Universidad de Navarra, C/Irunlarrea, 1, 31008 Pamplona, Spain
cbejines@alumni.unav.es, {mjchasco,jelorza}@unav.es
[2] Universidad de Oviedo, C/San Francisco, 1, 33003 Oviedo, Spain
montes@uniovi.es

Abstract. We compare four equivalence relations defined in fuzzy subgroups: Isomorphism, fuzzy isomorphism and two equivalence relations defined using level subset notion. We study if the image of two equivalent fuzzy subgroups through aggregation functions is a fuzzy subgroup, when it belongs to the same class of equivalence and if the supreme property is preserved in the class of equivalence and through aggregation functions.

Keywords: Aggregation function · Fuzzy subgroup · Level subgroup
Isomorphism · Fuzzy isomorphism · Sup property

1   Introduction

Fuzzy subgroups were introduced by A. Rosenfeld in his paper entitled "Fuzzy Groups" [17]. A more general notion can be found in [1,18].

Equivalence relations give the possibility of classifying objects. On the set of fuzzy subgroups, different equivalence relations have defined during the last years. Given a group, Li, Chen, Gu and Wang, defined the notion of isomorphism between two fuzzy subgroups (see [11]). The relationship "being isomorphic to" is a equivalence relation. Two more natural equivalence relations were defined by Dixit (see [6]) and Murali and Makamba (see [15]). Both of them relate two fuzzy subgroups if they have the same level subgroups, but Murali and Makamba add the condition that the support of the fuzzy subgroups be the same. Therefore, the second one is stronger than the first one. Note that Murali and Makamba worked on finite groups (see [13–15]), but this paper is done without that restriction. Another, fuzzy isomorphism notion between fuzzy subgroups was first formulated by Ray (see [16]). It is an equivalence relation which is weaker than the isomorphism notion mentioned before.

A function $F : [0,1]^n \longrightarrow [0,1]$ is an aggregation function if it satisfies certain boundary conditions and an increasing condition.

Aggregation functions combine inputs that are usually interpreted as degrees of membership in fuzzy sets, preference, strength of evidence or support of a

© Springer Nature Switzerland AG 2018
F. Herrera et al. (Eds.): CAEPIA 2018, LNAI 11160, pp. 143–153, 2018.
https://doi.org/10.1007/978-3-030-00374-6_14

hypothesis. Many theoretical results have been obtained in this context (see [3,7,8]). As we will see, the domain of aggregation functions can be extended to the class of fuzzy subgroups pointwise.

The purpose of this work is to compare these equivalence relations on fuzzy subgroups and to study when some properties of equivalent fuzzy subgroups are preserved through an aggregation function.

We begin by recalling the basic notions which are needed along the paper. After, we introduce the four equivalence relations that we will analyse in Sect. 3. Section 4 focuses in properties of fuzzy subgroups which are preserved under aggregation functions.

2   Preliminaries

We review briefly some definitions and preliminaries results. The authors recommend the book [12] to deepen in the fuzzy groups theory. In the sequel, we will work with fuzzy subgroups of a group $(G, *)$. The group $(G, *)$ will be denoted by G and $x * y$ will be written as xy.

Definition 1 ([17]). *Let G be a group and μ a fuzzy subset of G. We say that μ is a fuzzy subgroup of G if:*

$(G1)$ $\mu(xy) \geq \min\{\mu(x), \mu(y)\}$ *for all $x, y \in G$.*
$(G2)$ $\mu(x) \geq \mu(x^{-1})$ *for all $x \in G$.*

From now on, the class of the fuzzy subgroups of G will be denoted by $\mathcal{F}(G)$.

For a fuzzy subset μ of an universal set X and $t \in [0, 1]$, we define the level subset μ_t and the strong level subset $\mu_t^>$ as follows:

$$\mu_t = \{x \in X \mid \mu(x) \geq t\} \qquad \text{and} \qquad \mu_t^> = \{x \in X \mid \mu(x) > t\}$$

If $\mu \in \mathcal{F}(G)$, its non-empty level subsets are subgroups of G called level subgroups of μ. A particular case is the support of μ, this is defined by $supp\ \mu = \mu_0^>$. The following characterization makes very convincing the use of level sets in the context of fuzzy subgroups.

Proposition 1 ([5]). *Let G be a group and μ a fuzzy subset of G, then μ is a fuzzy subgroup of G if and only if its non-empty level subsets (strong level subsets) are subgroups of G.*

Below, we describe the four equivalence relations which are studied in this paper. Given a group G, Li, Chen, Gu and Wang defined the isomorphism between two fuzzy subgroups using Zadeh's extension principle (see [11]), that is, given a fuzzy subset η of G and a function $f : G \longrightarrow G$, the fuzzy subset $f(\eta)$ is defined as

$$f(\eta)(y) = \begin{cases} \sup_{x \in G}\{\eta(x) \mid f(x) = y\} & \text{if } f^{-1}(y) \neq \emptyset \\ 0 & \text{if } f^{-1}(y) = \emptyset \end{cases}$$

Definition 2. *Given a group G and $\mu, \eta \in \mathcal{F}(G)$, we say that μ is isomorphic to η if there is an isomorphism $f : G \longrightarrow G$ fulfilling that $\mu = f(\eta)$.*

This isomorphism relation will be denoted by $\sim_I$. Note that if f is an isomorphism, for each $y \in G$, we have $f(\eta)(y) = \eta(f^{-1}(y))$. Therefore, the range of two isomorphic fuzzy subgroups is the same, that is, Im μ = Im η.

The following fuzzy isomorphism relation is defined by Ray in terms of level sets in [16].

Definition 3. *Given a group G and $\mu, \eta \in \mathcal{F}(G)$, we say that μ is fuzzy isomorphic to η if there is an isomorphism $f : supp\ \mu \longrightarrow supp\ \eta$ satisfying that*

$$\mu(x) < \mu(y) \Leftrightarrow \eta(f(x)) < \eta(f(y))$$

for all $x, y \in supp\ \mu$.

Fuzzy isomorphism relation will be denoted by $\sim_{FI}$. Note that if $\mu \sim_{FI} \eta$, then $|\text{Im }\mu| = |\text{Im }\eta|$, where $|\text{Im }\mu|$ denotes the cardinal of Im μ.

Given a group G, Dixit defined an equivalence relation between two fuzzy subgroups using also their level subgroups (see [6]). Moreover, Murali and Makamba defined another equivalence relation between fuzzy subgroups using their level subgroups and a particular condition on the support of them (see [15]).

Definition 4. *Given a group G and $\mu, \eta \in \mathcal{F}(G)$, we say that μ is D-equivalent to η if*

$$\mu(x) < \mu(y) \Leftrightarrow \eta(x) < \eta(y)$$

for all $x, y \in G$.

This equivalence relation has been called D-equivalence relation because it was introduced by Dixit and it will be denoted by $\sim_D$.

Definition 5. *Given a group G and $\mu, \eta \in \mathcal{F}(G)$, we say that μ is M-equivalent to η if $supp\ \mu = supp\ \eta$ and*

$$\mu(x) < \mu(y) \Leftrightarrow \eta(x) < \eta(y)$$

for all $x, y \in supp\ \mu$.

This equivalence relation has been called M-equivalence relation because it was introduced by Murali and Makamba and it will be denoted by $\sim_M$.

According to these definitions, it is worth pointing out the following clear fact: If $\mu \sim_M \eta$, then $\mu \sim_D \eta$ and $\mu \sim_{FI} \eta$.

3   The Connection Among the Equivalence Relations

In this section, we show the connection among the given equivalence relations. In [19], Zhang introduced the homologous notion between fuzzy subgroups. Two fuzzy subgroups μ and η are homologous if there is an isomorphism $f : G \longrightarrow G$

such that $\mu(x) = \eta(f(x))$ for all $x \in G$. Therefore, the homologous relation is exactly the relation defined in Definition 2.

In [10], Jain showed the following characterization of the equivalence relation $\sim_D$ using level subgroups.

Proposition 2 ([10]). *Let G be a group and $\mu, \eta \in \mathcal{F}(G)$. The following assertions are equivalent:*

1. $\mu \sim_D \eta$.
2. $\mu(x) \leq \mu(y)$ if and only if $\eta(x) \leq \eta(y)$ for all $x, y \in G$.
3. $\{\mu_t\}_{t \in \text{Im } \mu} = \{\eta_s\}_{s \in \text{Im } \eta}$
4. $\{\mu_t^>\}_{t \in \text{Im } \mu} = \{\eta_s^>\}_{s \in \text{Im } \eta}$

The following characterization for $\sim_M$ will be used in Proposition 7.

Lemma 1. *Let G be a group and $\mu, \eta \in \mathcal{F}(G)$. Then, $\mu \sim_M \eta$ if and only if the following conditions are satisfied:*

(i) $\mu_0^> = \eta_0^>$.
(ii) $\mu(x) \leq \mu(y)$ if and only if $\eta(x) \leq \eta(y)$ for all $x, y \in supp\ \mu$.

The following result shows the relationship between the isomorphism relation and the fuzzy isomorphism relation.

Proposition 3. *Let G be a group and $\mu, \eta \in \mathcal{F}(G)$. If $\mu \sim_I \eta$, then $\mu \sim_{FI} \eta$.*

Proof. Since $\mu \sim_I \eta$, there is an isomorphism $f : G \longrightarrow G$ satisfying $\mu(x) = \eta(f(x))$ for all $x \in G$. Consider the function $g : supp\ \mu \longrightarrow supp\ \eta$ defined by $g(x) = f(x)$ for all $x \in supp\ \mu$. First at all, we need to check that g is a well-defined function. Take $x \in supp\ \mu$ and suppose that $g(x) \notin supp\ \eta$, then $\mu(x) > 0$ and $\eta(g(x)) = 0$. Therefore $0 = \eta(g(x)) = \eta(f(x)) = \mu(x)$, a contradiction.

Let us see that g is an homomorphism. Since $supp\ \mu$ is a subgroup of G and $g = f|_{supp\ \mu}$, that is, the restriction of an homomorphism whose domain is a subgroup, we conclude that g is an homomorphism.

Let us see that g is injective, take $x, y \in supp\ \mu$. We have that

$$g(x) = g(y) \Rightarrow f(x) = f(y) \Rightarrow x = y$$

Let us see that g is surjective, take $y \in supp\ \eta$. Since f is an isomorphism, there is $x \in G$ such that $f(x) = y$. If $x \notin supp\ \mu$, then $\mu(x) = 0$, but $\mu(x) = \eta(f(x)) = \eta(y) = 0$, a contradiction. Hence, $x \in supp\ \mu$ with $g(x) = f(x) = y$.

Finally, since $\eta(g(x)) = \mu(x)$ for all $x \in supp\ \mu$, we conclude that

$$\mu(x) < \mu(y) \Leftrightarrow \eta(g(x)) < \eta(g(y))$$

for all $x, y \in supp\ \mu$. $\square$

The following examples show us that the other of implications are not true.

Proposition 4. *Let G a non-trivial group. Then,*

(1) *There are $\mu, \eta \in \mathcal{F}(G)$ satisfying $\mu \sim_{FI} \eta$ and $\mu \nsim_I \eta$.*
(2) *There are $\mu, \eta \in \mathcal{F}(G)$ satisfying $\mu \sim_D \eta$ and $\mu \nsim_{FI} \eta$.*
(3) *There are $\mu, \eta \in \mathcal{F}(G)$ satisfying $\mu \sim_M \eta$ and $\mu \nsim_I \eta$.*

For the Klein group G, that is, $G = \mathbb{Z}_2 \times \mathbb{Z}_2$. We have

(4) *There are $\mu, \eta \in \mathcal{F}(G)$ satisfying $\mu \sim_I \eta$ and $\mu \nsim_D \eta$.*

Proof. Consider the following fuzzy subgroups:

$$\mu(x) = \begin{cases} 1 & \text{if } x = e \\ 0.7 & \text{if } x \neq e \end{cases} \qquad \eta(x) = \begin{cases} 1 & \text{if } x = e \\ 0.4 & \text{if } x \neq e \end{cases} \qquad \nu(x) = \begin{cases} 1 & \text{if } x = e \\ 0 & \text{if } x \neq e \end{cases}$$

For (1), it is easy to check that $\mu \sim_{FI} \eta$, but $\mu \nsim_I \eta$ because $\operatorname{Im} \mu \neq \operatorname{Im} \eta$. For (2), we have that $\mu \sim_D \nu$ because they have the same levels, but $supp\,\mu \neq supp\,\nu$, hence $\mu \nsim_{FI} \nu$. For (3), we have that $\mu \sim_M \eta$ because they have the same levels and $supp\,\mu = supp\,\eta$, but $\operatorname{Im}\mu \neq \operatorname{Im}\eta$, hence $\mu \nsim_I \eta$. For (4), let G be the Klein group. Consider the following fuzzy subgroups:

$$\mu(x) = \begin{cases} 1 & \text{if } x = (0,0) \\ 0.7 & \text{if } x = (1,0) \\ 0.3 & \text{if } x \in \{(0,1),(1,1)\} \end{cases} \qquad \eta(x) = \begin{cases} 1 & \text{if } x = (0,0) \\ 0.7 & \text{if } x = (0,1) \\ 0.3 & \text{if } x \in \{(1,0),(1,1)\} \end{cases}$$

Note that μ and η are fuzzy subgroups by Proposition 1. Consider the isomorphism $f : \mathbb{Z}_2 \times \mathbb{Z}_2 \longrightarrow \mathbb{Z}_2 \times \mathbb{Z}_2$ defined by $f(0,0) = (0,0)$, $f(1,0) = (0,1)$, $f(0,1) = (1,0)$ and $f(1,1) = (1,1)$, since $\mu(x) = \eta(f(x))$ for all $x \in \mathbb{Z}_2 \times \mathbb{Z}_2$, we have that $\mu \sim_I \eta$, but $\mu \nsim_D \eta$ because they do not have the same levels. $\square$

The following picture shows the implications among the four equivalence relations above mentioned.

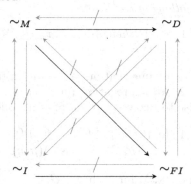

In [9], Iranmanesh and Naraghi prove that if the group G is cyclic and finite, then the M-equivalence relation and the fuzzy isomorphism relation induce the same classes of fuzzy subgroups. The following proposition extends this result to an arbitrary cyclic group. If the group is not cyclic, determining when they induce the same classes of fuzzy subgroups does not seem easy (open problem). We remember that a cyclic group is a group where there is an element a, which generates the group. That is, for each $b \in G$, $b = a^n$ for some $n \in \mathbb{Z}$.

Proposition 5. *Let G be a cyclic group and $\mu, \eta \in \mathcal{F}(G)$. Then $\mu \sim_M \eta$ if and only if $\mu \sim_{FI} \eta$.*

Proof. The direct implication is clear. Conversely, if G is a finite group, the proof can be found in [9]. If G is an infinite group, then G is isomorphic to $(\mathbb{Z}, +)$. The only isomorphisms in $\mathbb{Z}$ are the identity function $id(x) = x$ and the inverse function $f(x) = -x$. Given two fuzzy subgroups μ, η of $\mathbb{Z}$ with $\mu \sim_{FI} \eta$, we have that $\mu(x) < \mu(y) \Leftrightarrow \eta(id(x)) < \eta(id(y)) \Leftrightarrow \eta(x) < \eta(y)$ for all $x, y \in \mathbb{Z}$ or $\mu(x) < \mu(y) \Leftrightarrow \eta(f(x)) < \eta(f(y)) \Leftrightarrow \eta(-x) < \eta(-y) \Leftrightarrow \eta(x) < \eta(y)$ for all $x, y \in \mathbb{Z}$. Anyway, for all $x, y \in \mathbb{Z}$.

$$\mu(x) < \mu(y) \text{ if and only if } \eta(x) < \eta(y)$$

$\square$

The following picture shows the implications among the equivalence relations when G is a cyclic group.

$$\sim_I \xrightarrow{\hspace{2cm}} \sim_{FI} \xleftrightarrow{\hspace{2cm}} \sim_M \xrightarrow{\hspace{2cm}} \sim_D$$

The examples of Proposition 4 show that $\sim_{FI} \Rightarrow \sim_I$ and $\sim_D \Rightarrow \sim_M$ are not true.

4   Preserving Properties of Fuzzy Subgroups Through of Aggregation Functions

The definition of aggregation function and its properties can be found in [3].

Definition 6. *An n-ary operation $A : [0,1]^n \longrightarrow [0,1]$ is called an aggregation function if it fulfills the following conditions:*

1. *If $x_i \leq y_i$ for all $i \in \{1, ..., n\}$, then $A(x_1, ..., x_n) \leq A(y_1, ..., y_n)$.*
2. *$A(0, ..., 0) = 0$ and $A(1, ..., 1) = 1$.*

Unless otherwise stated we assume that our aggregation functions will be binary operations, that is, $n = 2$. An aggregation function $A : [0,1] \times [0,1] \longrightarrow [0,1]$ can be extended pointwise to fuzzy subgroup of G as follows:

$$A : [0,1]^G \times [0,1]^G \longrightarrow [0,1]^G$$
$$(\mu, \eta) \longmapsto A(\mu, \eta)$$

where $A(\mu, \eta)(x) = A(\mu(x), \eta(x))$ for each $x \in G$.

Definition 7. *Let G be a group and μ a fuzzy subgroup of G. We say that μ has the supreme property, sup property for short, if for each non-empty subset B of G, there exists $a \in B$ such that*

$$\sup_{x \in B} \{\mu(x)\} = \mu(a)$$

Given an aggregation function A and two fuzzy subgroups μ, η of a group G, in general, $A(\mu, \eta)$ is not a fuzzy subgroup. Depending on the equivalence relation used, if μ and η are equivalent, then $A(\mu, \eta)$ can be a fuzzy subgroup and, if one of them has the sup property, then the other one and $A(\mu, \eta)$ can have it too. We focus when it happens. Moreover, if $A(\mu, \eta)$ is a fuzzy subgroup, we study when it belongs to the same class of μ and η.

Now, suppose that μ and η are fuzzy subgroups with $\mu \sim_D \eta$. In [2], it is proved that $A(\mu, \eta)$ is a fuzzy subgroup for any aggregation function A. In addition, it is shown that if μ has the sup property, then η and $A(\mu, \eta)$ have the sup property. Therefore, also these results are true when the equivalence relation is $\sim_M$.

Unfortunately, when the group G is not cyclic, if μ is isomorphic to η or μ is fuzzy isomorphic to η, then $A(\mu, \eta)$ could not be a fuzzy subgroup as the following example shows us.

Example 1. Consider the Klein group $G = \mathbb{Z}_2 \times \mathbb{Z}_2$ and the fuzzy subgroups μ and η defined in (4) of Proposition 4. We have that $\mu \sim_I \eta$, and by Proposition 3, $\mu \sim_{FI} \eta$ too. However, taking the aggregation function $A : [0,1]^2 \longrightarrow [0,1]$ defined by $A(x, y) = \frac{x+y}{2}$ we obtain the following fuzzy set:

$$A(\mu, \eta)(x) = A(\mu(x), \eta(x)) = \begin{cases} 1 & \text{if } x = (0,0) \\ 0.5 & \text{if } x \in \{(1,0), (0,1)\} \\ 0.3 & \text{if } x = (1,1) \end{cases}$$

By Proposition 1, we conclude that $A(\mu, \eta)$ is not a fuzzy subgroup.

However, if μ is a fuzzy subgroup with the sup property and η is a fuzzy subgroup satisfying $\mu \sim_I \eta$ or $\mu \sim_{FI} \eta$, then η has the sup property.

Proposition 6. *Let G be a group and $\mu, \eta \in \mathcal{F}(G)$. Suppose that μ has the sup property, if $\mu \sim_{FI} \eta$ then η has the sup property.*

Proof. Since $\eta \sim_{FI} \mu$, there is an isomorphism $f : supp\ \eta \longrightarrow supp\ \mu$ satisfying $\mu(f(x)) \leq \mu(f(y)) \Leftrightarrow \eta(x) \leq \eta(y)$ for all $x, y \in supp\ \mu$. Note that $supp\ \mu = supp\ \eta$. Given $B \subseteq G$, we need to prove that there is $z \in B$ such that $\eta(z) = \sup_{x \in B}\{\eta(x)\}$. Suppose first that $B \subset supp\ \mu$ and consider the subset $f(B)$ of G defined by $f(B) = \{f(x) \mid x \in B\}$, since μ has the sup property, we have that there is $y \in f(B)$ such that $\sup_{x \in f(B)}\{\mu(x)\} = \mu(y)$. Note that $\sup_{x \in f(A)}\{\mu(x)\} = \sup_{x \in B}\{\mu(f(x))\}$. Moreover, $y = f(z)$, for some $z \in B$, hence $\mu(f(x)) \leq \mu(f(z))$ for all $x \in B$. This implies that $\eta(x) \leq \eta(z)$, then $\sup_{x \in B}\{\eta(x)\} \leq \eta(z)$. On the other hand, since $z \in B$, $\sup_{x \in B}\{\eta(x)\} \geq \eta(z)$. This implies that $\sup_{x \in B}\{\eta(x)\} = \eta(z)$. Suppose now that $B \not\subset supp\ \mu$ and consider $C = B \cap supp\ \mu$. If $C \neq \emptyset$ since $\sup_{x \in B}\{\eta(x)\} = \sup_{x \in C}\{\eta(x)\}$, the proof is analogous to the previous case. If $C = \emptyset$, then $B \subset supp\ \mu$, that is, $\mu(x) = 0$ for each $x \in B$. Since $supp\ \mu = supp\ \eta$, we have that $\eta(x) = 0$ for each $x \in B$ and we conclude η has the sup property. $\square$

Corollary 1. *Let G be a group and $\mu, \eta \in \mathcal{F}(G)$. Suppose that μ has the sup property, if $\mu \sim_I \eta$ then η has the sup property.*

Definition 8 ([4]). *Let $A : [0,1]^n \longrightarrow [0,1]$ be an aggregation function. We say that A is jointly strictly monotone if $x_i < y_i$ for all $i \in \{1, ..., n\}$, then*

$$A(x_1, ..., x_n) < A(y_1, ..., y_n)$$

It is known that A is jointly strictly monotone if and only if for all $\mu, \eta \in \mathcal{F}(G)$ with $\mu \sim_D \eta$, $A(\mu, \eta)$ belongs to the class of μ (see [2]). The following result shows what occurs in the M-equivalence case.

Proposition 7. *Let G be a group and $A : [0,1]^2 \longrightarrow [0,1]$ be an aggregation function, then A is jointly strictly monotone if and only if for all $\mu, \eta \in \mathcal{F}(G)$ with $\mu \sim_M \eta$, $A(\mu, \eta)$ belongs to the class of μ.*

Proof. First, we suppose A is jointly strictly monotone and we are going to prove that for every $\mu, \eta \in \mathcal{F}(G)$ satisfying $\mu \sim_M \eta$, we have

$$A(\mu, \eta) \in [\mu]_{\sim_M}$$

Considering Lemma 1, we begin by proving that $supp\ A(\mu, \eta) = supp\ \mu$. Take $x \notin supp\ \mu$, then $\mu(x) = 0$ and since $\mu \sim_M \eta$, we have $\eta(x) = 0$. Therefore, $A(\mu, \eta)(x) = A(\mu(x), \eta(x)) = A(0,0) = 0$. Now, take $x \in supp\ \mu$, then $\mu(x) > 0$ and since $\mu \sim_M \eta$, we have $\eta(x) > 0$. By reductio ad absurdum, suppose that $A(\mu, \eta)(x) = 0$. We have that $0 < \mu(x)$ and $0 < \eta(x)$ and $A(0,0) = 0 = A(\mu(x), \eta(x))$, which is a contradiction because A is jointly strictly monotone. Hence $supp\ A(\mu, \eta) = supp\ \mu$.

We continue by showing that $\mu(x) \geq \mu(y)$ if and only if $A(\mu, \eta)(x) \geq A(\mu, \eta)(y)$ for all $x, y \in supp\ \mu$. This part is proved as the proof described in [2] for the equivalence relation $\sim_D$. Given $x, y \in supp\ \mu$ such that $\mu(x) \geq \mu(y)$, since $\mu \sim_M \eta$, $\eta(x) \geq \eta(y)$. By the monotony of A, we have

$$A(\mu, \eta)(x) = A(\mu(x), \eta(x)) \geq A(\mu(y), \eta(y)) = A(\mu, \eta)(y)$$

For the other implication, given $x, y \in supp\ \mu$ with $A(\mu, \eta)(x) \geq A(\mu, \eta)(y)$, we need to prove $\mu(x) \geq \mu(y)$. By reductio ad absurdum, there exists $x, y \in supp\ \mu$ such that $A(\mu, \eta)(x) \geq A(\mu, \eta)(y)$ and $\mu(x) < \mu(y)$. From $\mu \sim_M \eta$, $\eta(x) < \eta(y)$. Since A is jointly strictly monotone, we have that $A(\mu(x), \eta(x)) < A(\mu(y), \eta(y))$, which is a contradiction.

Conversely, if A is not jointly strictly monotone, then there exists a_1, a_2, b_1, b_2 in the interval $[0,1]$ with $a_1 < a_2$ and $b_1 < b_2$ such that $A(a_1, b_1) \geq A(a_2, b_2)$. By the monotony, we obtain $A(a_1, b_1) = A(a_2, b_2)$. In particular, observe that $a_1 < \frac{a_1+a_2}{2} < a_2$ and $b_1 < \frac{b_1+b_2}{2} < b_2$, hence $A(a_1, b_1) = A(\frac{a_1+a_2}{2}, \frac{b_1+b_2}{2}) = A(a_2, b_2)$.

We consider the following fuzzy subsets of G:

$$\mu(x) = \begin{cases} a_2 & \text{if } x = e \\ \frac{a_1+a_2}{2} & \text{if } x \neq e \end{cases} \qquad \eta(x) = \begin{cases} b_2 & \text{if } x = e \\ \frac{b_1+b_2}{2} & \text{if } x \neq e \end{cases}$$

We know by Proposition 1 that μ and η are fuzzy subgroup of G. Since $\mu(x) < \mu(y)$ if and only if $\eta(x) < \eta(y)$ and $supp\ \mu = G = supp\ \eta$, we have $\mu \sim_M \eta$. Finally, we prove that $A(\mu, \eta) \notin [\mu]_{\sim_M}$. Fix $x \in G$ with $x \neq e$ and suppose $A(\mu, \eta) \in [\mu]_{\sim_M}$. Since $\mu(e) > \mu(x)$, we have $A(\mu, \eta)(e) > A(\mu, \eta)(x)$, but

$$A(\mu(e), \eta(e)) = A(a_2, b_2) = A\left(\frac{a_1 + a_2}{2}, \frac{b_1 + b_2}{2}\right) = A(\mu(x), \eta(x))$$

this means, $A(\mu, \eta)(e) = A(\mu, \eta)(x)$. Therefore $A(\mu, \eta) \notin [\mu]_{\sim_M}$.   □

Consider a jointly strictly monotone aggregation function A and group G. For each $\mu, \eta \in \mathcal{F}(G)$ with $\mu \sim_I \eta$ ($\mu \sim_{FI} \eta$), $A(\mu, \eta)$ is not always a fuzzy subgroup. However, if we suppose that $A(\mu, \eta)$ is a fuzzy subgroup, we wonder if $A(\mu, \eta)$ belongs to the class $[\mu]_{\sim_I}$ ($[\mu]_{\sim_{FI}}$). The following examples answer the question in negative.

Example 2. Let G be a group and consider $\mu = \eta$ defined by $\mu(x) = \eta(x) = c$ for some $c \in (0, 1)$. Take a jointly strictly monotone aggregation function A satisfying that $A(c, c) \neq c$, for example $A = T_P$, that is the product t-norm. We have $A(\mu, \eta)$ is the fuzzy subgroup

$$A(\mu, \eta)(x) = A(\mu(x), \eta(x)) = A(c, c) = c^2$$

for all $x \in G$. Trivially, $\mu \sim_I \eta$. But the range of $A(\mu, \eta)$ is different to the range of μ, hence $A(\mu, \eta) \not\sim_I \mu$.

Example 3. Let G be the Klein group. Let us consider the following fuzzy subgroups

$$\mu(x) = \begin{cases} 1 & \text{if } x \in \{(0,0), (1,0)\} \\ 0.6 & \text{if } x \in \{(0,1), (1,1)\} \end{cases} \qquad \eta(x) = \begin{cases} 1 & \text{if } x \in \{(0,0), (1,1)\} \\ 0.4 & \text{if } x \in \{(0,1), (1,0)\} \end{cases}$$

We have that $\mu \sim_{FI} \eta$. Consider the following function $A : [0,1]^2 \longrightarrow [0,1]$ defined by

$$A(x, y) = \begin{cases} \frac{\min\{x,y\}}{4} & \text{if } (x, y) \in [0, 1)^2 \\ \frac{y}{2} & \text{if } x = 1 \text{ and } y \neq 1 \\ \frac{x}{3} & \text{if } y = 1 \text{ and } x \neq 1 \\ 1 & \text{if } (x, y) = (1, 1) \end{cases}$$

We have that $A(0, 0) = 0$ and $A(1, 1) = 1$. Moreover, A satisfies that if $x_1 \leq x_2$ and $y_1 \leq y_2$, then $A(x_1, y_1) \leq A(x_2, y_2)$. Therefore, A is an aggregation function. In addition, A is jointly strictly monotone by construction. $A(\mu, \eta)$ is defined by

$$A(\mu, \eta)(x) = \begin{cases} 1 & \text{if } x = (0, 0) \\ 0.2 & \text{if } x = (1, 0) \text{ or } x = (1, 1) \\ 0.1 & \text{if } x = (0, 1) \end{cases}$$

Since $|\text{Im } \mu| = 2 \neq 3 = |\text{Im } A(\mu, \eta)|$, we conclude that $A(\mu, \eta) \not\sim_{FI} \mu$.

5 Concluding Remarks

The authors showed some results about the preservation of properties of fuzzy subgroups under the equivalence relation introduced by Dixit. In this paper, we consider three other relevant equivalent relations and we continue the work started in [2]. We have studied the relationship among the four equivalence relations on fuzzy subgroups. We have obtained a complete implication diagram about them which is valid when the given group is cyclic.

Acknowledgement. The authors acknowledge the financial support of the Spanish Ministerio de Economía y Competitividad (Grant TIN2014-59543-P and Grant MTM 2016-79422-P). Carlos Bejines also thanks the support of the Asociación de Amigos of the University of Navarra.

References

1. Anthony, J.M., Sherwood, H.: Fuzzy groups redefined. J. Math. Anal. Appl. **69**, 124–130 (1979)
2. Bejines, C., Chasco, M.J., Elorza, J., Montes, S.: On the preservation of an equivalence relation between fuzzy subgroups. In: Kacprzyk, J., Szmidt, E., Zadrożny, S., Atanassov, K.T., Krawczak, M. (eds.) IWIFSGN/EUSFLAT 2017. AISC, vol. 641, pp. 159–167. Springer, Cham (2018). https://doi.org/10.1007/978-3-319-66830-7_15
3. Beliakov, G., Pradera, A., Calvo, T.: Aggregation Functions: A Guide for Practitioners. Studies in Fuzziness and Soft Computing, vol. 221. Springer, Heidelberg (2007). https://doi.org/10.1007/978-3-540-73721-6
4. Calvo, T., Mayor, G., Mesiar, R.: Aggregation Operation. New Trends and Applications. Physica-Verlag, Heidelberg (2002)
5. Das, P.S.: Fuzzy groups and level subgroups. J. Math. Anal. Appl. **84**, 264–269 (1981)
6. Dixit, V.N., Kumar, R., Ajmal, N.: Level subgroups and union of fuzzy subgroups. Fuzzy Sets Syst. **37**, 359–371 (1990)
7. Fodor, J., Kacprzyk, J.: Aspect of Soft Computing, Intelligent Robotics and Control. Studies in Computational Intelligence, vol. 241. Springer, Heidelberg (2009). https://doi.org/10.1007/978-3-642-03633-0
8. Grabisch, M., Marichal, J., Mesiar, R., Pap, E.: Aggregation functions: means. Inf. Sci. **181**, 1–22 (2011)
9. Iranmanesh, A., Naragui, H.: The connection between some equivalence relations on fuzzy subgroups. Iran. J. Fuzzy Syst. **8**(5), 69–90 (2011)
10. Jain, A.: Fuzzy subgroup and certain equivalence relations. Iran. J. Fuzzy Syst. **3**, 75–91 (2006)
11. Li, S.Y., Chen, D.G., Gu, W.X., Wang, H.: Fuzzy homomorphisms. Fuzzy Sets Syst. **79**(2), 235–238 (1996)
12. Mordeson, J.N., Bhutani, K.R., Rosenfeld, A.: Fuzzy Groups Theory. Springer, Heidelberg (2005). https://doi.org/10.1007/b12359
13. Murali, V., Makamba, B.: On an equivalence of fuzzy subgroups I. Fuzzy Sets Syst. **123**, 259–264 (2001)
14. Murali, V., Makamba, B.: On an equivalence of fuzzy subgroups II. Fuzzy Sets Syst. **136**, 93–104 (2003)

15. Murali, V., Makamba, B.: On an equivalence of fuzzy subgroups III. Int. J. Math. Sci. **36**, 2303–2313 (2003)
16. Ray, S.: Isomorphic fuzzy groups. Fuzzy Sets Syst. **50**, 201–207 (1992)
17. Rosenfeld, A.: Fuzzy groups. J. Math. Anal. Appl. **35**, 512–517 (1971)
18. Negoita, C.V., Ralescu, D.A.: Applications of Fuzzy Sets to Systems Analysis, pp. 54–59. Wiley, New York (1975)
19. Zhang, Y.: Some properties on fuzzy subgroups. Fuzzy Sets Syst. **119**, 427–438 (2001)

A PageRank-Based Method to Extract Fuzzy Expressions as Features in Supervised Classification Problems

Pablo Carmona[1]([✉]), Juan Luis Castro[2], Jesús Lozano[1], and José Ignacio Suárez[1]

[1] Industrial Engineering School, University of Extremadura, Badajoz, Spain
{pablo,jesuslozano,jmarcelo}@unex.es
[2] Department of Computer Science and Artificial Intelligence, University of Granada, Granada, Spain
castro@decsai.ugr.es

Abstract. This work presents a new ranking method inspired on PageRank to reduce the dimensionality of the feature space in supervised classification problems. More precisely, as it relies on a weighted directed graph, it is ultimately inspired on TextRank, a PageRank based method that adds weights to the edges to express the strength of the connections between nodes. The method is based on dividing each original feature used to describe the training set into a set of fuzzy predicates and then ranking all of them by their ability to differentiate among classes in the light of this training set. The fuzzy predicates with the best scores can be then used as new features, replacing the original ones. The novelty of the proposal relies on being an approach halfway between feature selection and feature extraction approaches, being able to improve the discrimination ability of the original features but preserving the interpretability of the new features in the sense that they are fuzzy expressions. Preliminary results supports the suitability of the proposal.

Keywords: Fuzzy logic · Supervised classification · Ranking methods
Feature selection · Feature extraction · PageRank · TextRank

1 Introduction

In many contexts where a huge amount of information needs to be analyzed, the overwhelming processing requirements can be mitigated by evaluating the usefulness of each pieze of information for the goal we are pursuing and selecting the best ones. In supervised classification problems, where the data is expressed through a set of labeled examples (each example represented by a set of values for a fixed set of features and its class label), those techniques fits with the concept of dimensionality or feature reduction.

Two main approaches are used to reach this goal: feature extraction and feature selection [1,5,6]. While the former is based on projecting the original

© Springer Nature Switzerland AG 2018
F. Herrera et al. (Eds.): CAEPIA 2018, LNAI 11160, pp. 154–163, 2018.
https://doi.org/10.1007/978-3-030-00374-6_15

feature space into a new feature space, the latter directly selects a subset of the best original features using some ranking method to evaluate their relevance.

The feature reduction gained by both approaches not only suppose an improvement of the computational efficiency, but often it also improves the classification performance increasing its generalization ability. However, each approach has its own drawbacks. On the one hand, the new feature space obtained through feature extraction is usually less interpretable than the original feature space since the physical meaning of the original features is lost. On the other hand, feature selection is confined to the original features, whose ability to discriminate among classes is expected, in general, to be lower than a set of new features specifically designed to improve this ability.

In this work, we propose a new ranking method for feature reduction inspired by PageRank [8], a well-known ranking algorithm initially used by Google to rank web pages and widely extended to other contexts [4,7]. This algorithm relies on a directed graph with scores associated with its nodes that is iteratively traversed by a random walker making the scores evolve. Once a convergence condition is reached, the final scores are used to rank the web pages.

One of the multiple extensions of this algorithm is TextRank [7], a ranking model applied to text processing, whose main difference with respect to PageRank consist of including weights into the edges of the graph. This weighting allows to express the strength of the connection between text units.

Our PageRank-based proposal also uses a weighted directed graph and, therefore, it is ultimately based on TextRank. The nodes represent fuzzy expressions and each directed edge has a weight that assess the degree of information gained by adding the expression in the head node to the expression in the tail node. This weight is based on the concept of uncertainty measured through the entropy of fuzzy expressions in the light of the dataset.

2   Notation

In the context of supervised classification problems, we consider a set of d classes $\mathcal{C} = \{c_1, \ldots, c_d\}$, n features $\mathcal{F} = \{f_1, \ldots, f_n\}$ and m examples $\mathcal{E} = \{e_1, \ldots, e_m\}$, each example with the form $e_k = ([x_1^k, \ldots, x_n^k], c^k)$, where x_i^k is its value for the feature f_i and c^k is its class label. Each feature f_i is associated with a fuzzy variable X_i whose fuzzy domain is denoted as $\widetilde{\mathcal{X}_i} = \{LX_i^1, \ldots, LX_i^{p_i}\}$, being p_i the number of fuzzy values associated with the variable and being LX_i^j the linguistic label of its jth fuzzy value.[1]

3   PageRank and TextRank Models

The TextRank model is a graph-based ranking algorithm for text processing inspired by PageRank [8], a well-known ranking algorithm used by Google to rank web pages.

---

[1] In case of categorical features, each value of the feature will be represented by a fuzzy singleton.

PageRank shift the paradigm of web page ranking from analyzing the local information contained in a page to analyzing the global information contained on the whole web. This shift is based on computing the importance of that page by taking into account the links a page receive (backward links) and the importance of the pages linking to it. To achieve that, a directed graph is built where nodes represent web pages and directed edges represent the links from one page to another. Next, an iterative process allows to transfer the importance of the web pages connected by links until a convergence condition is reached. Finally, each node will have a score proportional to its importance.

A more formal definition is the following: Let $G = (\mathcal{N}, \mathcal{A})$ be a directed graph with a set of nodes $\mathcal{N}$ representing web pages and a set of directed edges (arrows) $\mathcal{A}$ representing links between web pages. Let $In(N_i)$ and $Out(N_i)$ the set of nodes that point to and are pointed from node N_i, respectively. Then, the score of a node N_i is defined as:

$$S(N_i) = (1 - d) + d \cdot \sum_{N_j \in In(N_i)} \frac{1}{|Out(N_j)|} \cdot S(N_j), \tag{1}$$

where $d \in [0, 1]$ is a damping factor that models the probability of randomly jumping from one node to another and $|\cdot|$ is the cardinality of a set.

The algorithm starts with arbitrary scores for the nodes and iteratively computes the new scores from (1) until a convergence condition is reached. Usually, this condition is expressed in terms of the difference between scores of two consecutive iterations, being satisfied when this difference is smaller than a threshold τ.

TextRank is an extrapolation of PageRank where the nodes represent text units and the edges represent some relation between them. However, it also extends the PageRank model by considering weighted graphs, where each edge from N_i to N_j is associated with a weight w_{ij} that represents the strength of the connection. In order to integrate these weights into the model, the score function (1) is replaced with the following one:

$$S(N_i) = (1 - d) + d \cdot \sum_{N_j \in In(N_i)} \frac{w_{ji}}{\sum_{V_k \in Out(N_j)} w_{jk}} \cdot S(N_j), \tag{2}$$

Since our proposal is also based in a weighted graph, it is ultimately an extension of the TextRank model.

4   Our Feature Ranking Method

The proposed ranking method falls into a category halfway between the feature selection and the feature extraction approaches. This is due to the fact that the underlying idea is not to rank the original features but to decompose each of them on a subset of new features in the form of fuzzy expressions and to rank these new features in order to select the best ones. Concretely, each original feature f_i is decomposed into p_i fuzzy predicates $P_i^j : X_i$ is LX_i^j and, once the ranking method is applied, the fuzzy predicates with highest scores are selected as new features.

This allows, on the one hand, to modify the original features in order to improve the ability to distinguish among classes and, on the other hand, to have an interpretable description of these new features, in the sense that they are fuzzy expressions.

4.1   The Graph

The graph contains a node for each possible fuzzy predicate $P_i^j : X_i$ is LX_i^j, obtaining a graph with p_i nodes for each feature f_i. Therefore, the graph will contain a total of $\sum_{i=1}^n p_i$ nodes.

Besides, initially the graph is fully connected, although after the calculation of the weights described in the next section and according to (2), all edges with a weight equal to 0 are as if they did not exist.

4.2   The Weights of the Edges

A weight is assigned to each edge that measures the usefulness of adding the fuzzy expression contained in the target node to the fuzzy expression contained in the source node in the light of the dataset. Concretely, a weight w_{AB} is assigned to the edge from node A to node B that measures the usefulness of adding the fuzzy predicate $P_B = P_{i_B}^{j_B}$ to the fuzzy predicate $P_A = P_{i_A}^{j_A}$, defined as

$$w_{AB} = U(P_A) - U(P_A, P_B), \tag{3}$$

where U is an uncertainty measure. In other words, w_{AB} measures the decrease of uncertainty when adding P_B to P_A.

The univariate version of the uncertainty measure, $U(P)$, is based on the entropy associated with the presence and the absence of the fuzzy predicate P:

$$U(P) = \rho(P) \cdot H(P) + \rho(\neg P) \cdot H(\neg P), \tag{4}$$

where $\rho(P)$ and $\rho(\neg P)$ represent the ratio of examples satisfying P and $\neg P$, and $H(P)$ and $H(\neg P)$ are the entropy associated with P and $\neg P$, respectively.

Concretely, given a fuzzy predicate $P_i^j : X_i$ is LX_i^j, $\rho(P_i^j)$ is defined as:

$$\rho(P_i^j) = \frac{\displaystyle\sum_{e_k \in \mathcal{E}} \mu_{LX_i^j}(x_i^k)}{m}, \tag{5}$$

whereas the distribution of probability among classes associated with $H(P_i^j)$ is calculated as:

$$p(c) = \frac{\displaystyle\sum_{e_k \in \mathcal{E}_c} \mu_{LX_i^j}(x_i^k)}{\displaystyle\sum_{e_k \in \mathcal{E}} \mu_{LX_i^j}(x_i^k)}, \tag{6}$$

being $\mathcal{E}_c$ the subset of examples within class c. In the case of $\rho(\neg P_i^j)$ and $H(\neg P_i^j)$ in (4), the expression $\mu_{LX_i^j}(x_i^k)$ is replaced by $1 - \mu_{LX_i^j}(x_i^k)$ in (5) and (6).

Regarding the bivariate version of the uncertainty measure, $U(P_A, P_B)$, it is analogously based on the entropy associated with the presence and the absence of the fuzzy predicates P_A and/or P_B. Namely:

$$U(P_A, P_B) = \rho(P_{A \wedge B}) \cdot H(P_{A \wedge B}) + \rho(P_{A \wedge \neg B}) \cdot H(P_{A \wedge \neg B}) + \cdots$$
$$\cdots + \rho(P_{\neg A \wedge B}) \cdot H(P_{\neg A \wedge B}) + \rho(P_{\neg A \wedge \neg B}) \cdot H(P_{\neg A \wedge \neg B}), \quad (7)$$

where the functions ρ and H keep the same meaning.

Concretely, given the fuzzy predicates $P_{i_A}^{j_A} : X_{i_A}$ is $LX_{i_A}^{j_A}$ and $P_{i_B}^{j_B} : X_{i_B}$ is $LX_{i_B}^{j_B}$, $\rho(P_{A \wedge B})$ is defined as:

$$\rho(P_{A \wedge B}) = \frac{\displaystyle\sum_{e_k \in \mathcal{E}} \mu_{LX_{i_A}^{j_A}}(x_{i_A}^k) \cdot \mu_{LX_{i_B}^{j_B}}(x_{i_B}^k)}{m}, \quad (8)$$

whereas the distribution of probability among classes associated with $H(P_{A \wedge B})$ is calculated as:

$$p(c) = \frac{\displaystyle\sum_{e_k \in \mathcal{E}_c} \mu_{LX_{i_A}^{j_A}}(x_{i_A}^k) \cdot \mu_{LX_{i_B}^{j_B}}(x_{i_B}^k)}{\displaystyle\sum_{e_k \in \mathcal{E}} \mu_{LX_{i_A}^{j_A}}(x_{i_A}^k) \cdot \mu_{LX_{i_B}^{j_B}}(x_{i_B}^k)}. \quad (9)$$

In the case of the fuzzy predicates $P_{A \wedge \neg B}$, $P_{\neg A \wedge B}$, and $P_{\neg A \wedge \neg B}$ in (7), the expressions $\mu_{LX_{i_A}^{j_A}}(x_{i_A}^k)$ and $\mu_{LX_{i_B}^{j_B}}(x_{i_B}^k)$ are accordingly replaced by $1 - \mu_{LX_{i_A}^{j_A}}(x_{i_A}^k)$ and $1 - \mu_{LX_{i_B}^{j_B}}(x_{i_B}^k)$ in (8) and (9).

Additionally, in order to grasp the individual potential of each fuzzy expression, an autoreference is also added to each node with a weight that measures the usefulness of adding the predicate P_i^j it represents to the *void* predicate $P_i^\emptyset : X_i$ is *True*. Concretely, for each node A, this weight is

$$w_{\emptyset A} = U(P_\emptyset) - U(P_\emptyset, P_A). \quad (10)$$

Since $\mu_{True}(x) = 1$ for all x, on one hand,

$$U(P_\emptyset) = \rho(P_\emptyset) \cdot H(P_\emptyset) + \rho(P_{\neg\emptyset}) \cdot H(P_{\neg\emptyset}) = 1 \cdot H(P_\emptyset) + 0 \cdot H(P_{\neg\emptyset}) = H(P_\emptyset), \quad (11)$$

and, on the other hand,

$$U(P_\emptyset, P_A) = \rho(P_{\emptyset \wedge A}) \cdot H(P_{\emptyset \wedge A}) + \rho(P_{\emptyset \wedge \neg A}) \cdot H(P_{\emptyset \wedge \neg A}) + \cdots$$
$$\cdots + \rho(P_{\neg\emptyset \wedge A}) \cdot H(P_{\neg\emptyset \wedge A}) + \rho(P_{\neg\emptyset \wedge \neg A}) \cdot H(P_{\neg\emptyset \wedge \neg A})$$
$$= \rho(P_A) \cdot H(P_A) + \rho(P_{\neg A}) \cdot H(P_{\neg A}) + \cdots$$
$$\cdots + 0 \cdot H(P_{\neg\emptyset \wedge A}) + 0 \cdot H(P_{\neg\emptyset \wedge \neg A}) = U(P_A), \quad (12)$$

resulting that this weight is finally reduced to

$$w_{\emptyset A} = H(P_\emptyset) - U(P_A). \quad (13)$$

After the ranking method is applied, the score of each node gives a measure of the information provided by the fuzzy expression it represents in the light of the dataset. Then, the best k fuzzy expressions (i.e., the nodes with the highest k scores) are selected as new features and each example is transformed into a new one as follows: for each selected fuzzy predicate $P_i^j : X_i$ is LX_i^j, the value x_i^k of the kth example is changed by its membership value to P_i^j, that is, by $\mu_{LX_i^j}(x_i^k)$.

5 Experimental Results

5.1 Datasets

In order to validate the suitability of the proposal, the ranking method was applied to three well known classification datasets from UCI repository [2]: Iris, Wine and Wisconsin Diagnostic Breast Cancer (WDBC). Table 1 shows the main information for these datasets.

Table 1. Main information of the used datasets.

Dataset	Features	Classes	Data per class
Iris	4	3	$50 + 50 + 50$
Wine	13	3	$59 + 71 + 48$
WDBC	30	2	$212 + 357$

5.2 Experimental Setup

We used the same fuzzy partition with 5 triangular membership functions for every feature. Regarding the ranking method, the parameter d in (2) is usually set to 0.85 [8] and we also used this value. The threshold τ related with the convergence condition was set to 0.00001.

For each run of the ranking method, the dataset was randomly split in training and test sets with a distribution of 80% and 20%, respectively, using a stratified sampling to balance the number of examples from each class. Then, the method was applied to the training set and the classification performance was measured analyzing the accuracy obtained when applying to the test set the implementation from Scikit-learn [9] of a linear support vector classifier. For each dataset, the method was run 10 times and the averaged accuracy was obtained.

5.3 Analysis of the Results Obtained with the Proposed Method

In order to illustrate the results obtained with the proposed ranking method, Table 2 shows the score of each fuzzy predicate after applying to the Iris dataset the ranking method in one of the runs, where the labels VS, S, M, L and

VL stand for *Very small*, *Small*, *Medium*, *Large* and *Very large*, respectively. The last column shows, for each original feature, the sum of the scores of the fuzzy predicates corresponding to this feature, and can be considered as a basic estimation of the global importance of the feature. We can observe from this column that the ranking method gives the highest importance to the feature *Petal width*, followed by *Petal length*, *Sepal length* and *Sepal width*. Despite the scores varied among runs, the same order of importance for these original features was obtained in all the runs.

Table 2. Scores of the fuzzy predicates after the ranking method converged (Iris dataset).

Features	Labels					Sum
	VS	S	M	L	VL	
Sepal length	0.724	1.096	0.714	0.732	0.519	3.785
Sepal width	0.251	0.651	0.208	0.671	0.222	2.003
Petal length	2.108	0.693	1.807	1.506	0.710	6.824
Petal width	2.555	0.507	2.041	1.232	1.052	7.388

Moreover, the three fuzzy predicates with highest scores *Petal width is Very small* (2.555), *Petal length is Very small* (2.108), and *Petal width is Medium* (2.041) would indicate that not only the original features *Petal width* and *Petal length* are the more important ones, but also that we can focus only on their fuzzy regions *Very small* and *Medium*.

Besides, in order to analyze the suitability of this feature ranking, the averaged classification accuracy was obtained for the k-best and k-worst new features (fuzzy predicates), for $k \in [1, \sum_{i=1}^{n} p_i]$, once the datasets have been properly transformed into the new feature space. The results shown in Fig. 1 support the appropriateness of the ranking, where it can be appreciated that the difference between the k-best and the k-worst features increases as k decreases in all the three datasets. Besides, as an illustrative example, the three Iris predicates mentioned above were the ones with the highest scores in all the runs, achieving an average accuracy of 0.96.

5.4   Comparison with Other Feature Selection Methods

In this section, our proposal is compared with the results obtained with three feature selection methods: Fisher Score [3], ReliefF [11], and Minimum Redundancy Maximum Relevance (MRMR) [10]. For these three methods the implementation from Scikit-feature feature selection repository [6] was used. Our method is named in this section as Fuzzy Feature Ranking (FFrank) method.

It must be stressed that the goal of the FFrank method is to rank fuzzy predicates and not the original features. Therefore, in order to allow the comparison with other feature selection methods, the method was adapted using the basic

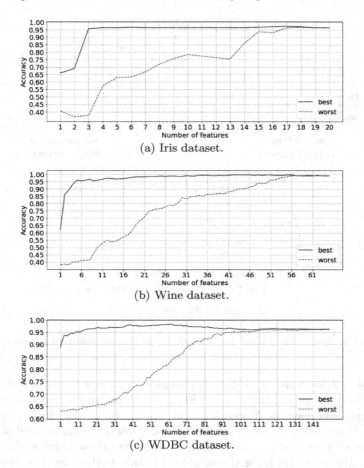

(a) Iris dataset.

(b) Wine dataset.

(c) WDBC dataset.

Fig. 1. Evolution of accuracy along the k-best vs. k-worst new features (fuzzy predicates).

approach of transforming the predicate scores into feature scores by adding up the scores of the predicates pertaining to the same original feature (as in the last column of Table 2).

Figure 2 shows the evolution of the averaged accuracy of each feature selection method along the k-best features, for $k \in [1, n]$. The results of the adapted FFrank method seem to be quite robust (despite it is a feature selection adaptation of the original method) and support the appropriateness of the approach, providing averaged accuracies better or worse than the other methods depending on the number of selected features and the dataset used.

Again, it is important to emphasize that, even in the cases where other feature selection methods improve the adapted FFrank method, the original FFrank method still could provide a benefit. As an illustrative example, selecting two

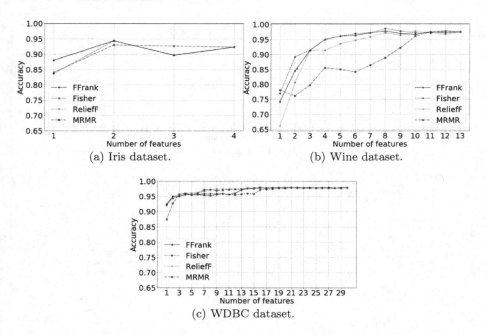

(a) Iris dataset.   (b) Wine dataset.

(c) WDBC dataset.

Fig. 2. Evolution of accuracy with different feature selection methods.

features ($k = 2$) with the Wine dataset provides an averaged accuracy of 0.89 and 0.84 for Fisher Score and the adapted FFrank method, respectively, since Fisher Score selects the features *Flavanoids* and *Proline* in all the runs whereas the adapted FFrank method selects *Od280/od315_of_diluted_wines* instead of *Proline* in 5 of the 10 runs (for $k = 3$, both methods selects these three features in all the runs). However, applying the original FFrank method to this dataset the averaged accuracy for only 5 predicates is 0.96, where 7 of the 10 runs extracted the same following fuzzy predicates: *Alcohol is Small*, *Flavanoids is Very small*, *Flavanoids is Medium*, *Color_intensity is Very small*, and *Proline is Large*. This result indicates that is more advantageous to consider only some fuzzy regions of the features *Flavanoids* and *Proline* along with some other regions of the features *Alcohol* and *Color_intensity* than considering only the original features *Flavanoids* and *Proline* selected by Fisher Score.

6 Conclusions

This work proposes a novel ranking method whose goal is to reduce the dimensionality of the feature space in supervised classification problems. It is presented an approach consisting in splitting the original features into a set of fuzzy expressions in the form of fuzzy predicates and then ranking them in order to select the best ones. The approach can be regarded as falling in a category halfway between feature selection and feature extraction, with the advantage of maintaining the physical meaning of the original features while allowing to translate them to a new feature space focused on improving the classification performance.

In order to validate the suitability of the proposal, the method was applied to three well known datasets and the classification performance was analyzed in terms of classification accuracy when using a linear support vector classifier. Besides, a version of the proposed method adapted to select the original features was compared with three features selection methods. The results endorse the appropriateness of both the global scores of the original features obtained by summing the scores of theirs associated fuzzy predicates and the ranking of the fuzzy expressions.

Acknowledgments. This work has been co-funded by the grants to research groups, IB16048 Proyect (Government of Extremadura), and the NanoSen-AQM Project (SOE2/P1/E0569).

References

1. Cai, J., Luo, J., Wang, S., Yang, S.: Feature selection in machine learning: a new perspective. Neurocomputing **300**, 70–79 (2018)
2. Dheeru, D., Karra Taniskidou, E.: UCI machine learning repository (2017). http://archive.ics.uci.edu/ml
3. Duda, R.O., Hart, P.E., Stork, D.G.: Pattern Classification, 2nd edn. Wiley, New York (2000)
4. Gupta, P., Goel, A., Lin, J., Sharma, A., Wang, D., Zadeh, R.: WTF: the who to follow service at Twitter. In: Proceedings of the 22nd International Conference on World Wide Web - WWW 2013. ACM Press (2013)
5. Khalid, S., Khalil, T., Nasreen, S.: A survey of feature selection and feature extraction techniques in machine learning. In: 2014 Science and Information Conference. IEEE, August 2014
6. Li, J., Cheng, K., Wang, S., Morstatter, F., Trevino, R.P., Tang, J., Liu, H.: Feature selection. ACM Comput. Surv. **50**(6), 1–45 (2017)
7. Mihalcea, R., Tarau, P.: TextRank: bringing order into texts. In: Proceedings of the Conference on Empirical Methods in Natural Language Processing - EMNLP 2004, July 2004
8. Page, L., Brin, S., Motwani, R., Winograd, T.: The pagerank citation ranking: bringing order to the web. Technical report 1999–66, Stanford InfoLab, November 1999. http://ilpubs.stanford.edu:8090/422/. Previous number = SIDL-WP-1999-0120
9. Pedregosa, F., et al.: Scikit-learn: machine learning in Python. J. Mach. Learn. Res. **12**, 2825–2830 (2011)
10. Peng, H., Long, F., Ding, C.: Feature selection based on mutual information criteria of max-dependency, max-relevance, and min-redundancy. IEEE Trans. Pattern Anal. Mach. Intell. **27**(8), 1226–1238 (2005)
11. Robnik-Šikonja, M., Kononenko, I.: Theoretical and empirical analysis of ReliefF and RReliefF. Mach. Learn. **53**(1), 23–69 (2003)

A Universal Decision Making Model for Restructuring Networks Based on Markov Random Fields

Julia García Cabello[1](✉)[iD] and Enrique Herrera-Viedma[2][iD]

[1] Department of Applied Mathematics, University of Granada,
Granada, Spain
cabello@ugr.es
[2] Department of Computer Science and Artificial Intelligence,
University of Granada, Granada, Spain
viedma@decsai.ugr.es

Abstract. The process of re-structuring physical networks is often based on local demographics. However, there are major variations across countries when defining demographics according to "local" parameters, which hinders the export of methodologies based on local specifications. This paper presents a *universal* decision making model for re-structuring networks aimed at working on a global basis since local parameters has been replaced by "internationally accepted" notions thereby allowing cross-border correlations. This a first step towards the globalization of demographic parameters which would also be fruitful in other disciplines where demographics play a role.

Importantly, the model variables can be replaced/expanded as needed thereby providing a decision making tool that can be applied to a wide range of contexts.

Keywords: Universal decision model
Spatial stochastic processes (Markov and Gibbs random fields)

1 Introduction

The process of re-structuring physical networks is often based on local demographics (see for instance, (Abbasi 2003) or (Ioannou et al. 2007), where authors presented a decision support system for reconfiguring networks, based specifically on information about local demographics). However, there are major variations across countries when defining demographics according to "local" specifications. As a matter of fact, local demography has to be managed carefully due to the significant variations when defining it through local parameters. For instance, the distinction between- urban and rural areas is growing *fuzzy*: while the main criteria to define them commonly include population size/density and availability of some support services such as secondary schools and hospitals, the combination of criteria applied can vary greatly. Actually, even the population

F. Herrera et al. (Eds.): CAEPIA 2018, LNAI 11160, pp. 164–173, 2018.
https://doi.org/10.1007/978-3-030-00374-6_16

thresholds used across countries can be different: "Scientists from different disciplines diverge when defining these zones (rural/urban) or their limits; they even often mention the zones without any definition. This practice excludes comparison between studies" [sic], (see André et al. 2014) In light of this situation, this means that cross-border comparisons are difficult for all methodologies based on local parameters.

With respect to work published in the literature, authors could not find any studies that deal with the problem of globalising demographic parameters into a unified approach. However, there are several studies that address the restructuring of physical networks, which is approached from different perspectives in the current literature. One perspective treats site selection according to certain pre-established criteria. The problem of selecting the best place for a new site can be also viewed as part of the more general problem of restructuring the network. In all cases, a full range of mathematical techniques is used. In (Miliotis et al. 2002), mathematical programming was used to present a method for reorganising the network by combining geographical information systems (GIS) representing local geographical/social attributes- with demand-covering models. In (Ruiz-Hernandez et al. 2015), authors presented a restructuring model by using integer 0-1 programming while other approaches are based on Fuzzy Cognitive Maps (FCMs), see (Glykas et al. 2005) where authors generate a hierarchical and dynamic network of interconnected financial knowledge concepts by using FCMs.

In this paper the authors design a *universal* decision-making model for restructuring networks under a pre-fixed criterion. The term *universal* means here that the proposed decision-making model is aimed at working on a *global* basis since local demographic parameters have been replaced by "internationally accepted" notions, thereby avoiding the aforementioned fuzziness of local specifications. As a result, cross-border correlations should be now allowed. Also, a global approach will be beneficial from many standpoints: for instance, in order to decrease costs when replacing several local approaches by a universal one. A further benefit of decoupling the theoretical foundations from any local geographical premise is that it widens the decision-model scope of application. Actually, the proposed model applies to all contexts which may be viewed either as real or imaginary networks as long as they comply with some slight requirements.

Specifically, our objective is to create a universal tool which assists in deciding where to (re)locate a new node of the network according to some pre-fixed criterion. This will rely on a joint probability distribution. The use of probability functions in decision-making models is not novel since Bayesian-based frameworks traditionally have been based on them. However, the main disadvantage of Bayesian networks is that they are data-intensive, as long as they require sufficient input in order to derive the probabilistic relationships. Unlike Bayesian methods, our approach relies on a probability distribution expressly stated just in terms of *a few significant nodes*: the cliques. Besides the theoretical decision-making methodology, a method that redesigns the bank branch network according to branch size is formulated, intended for illustrative purposes. This example is carried out for the banking sector, where mergers and acquisitions make necessary to frequently restructure the bank branch network, and it is a perfect example of how local demographics influence the selected criterion for re-structuring (for further details, see Sect. 4). Importantly, this new methodology applies to many other contexts. As a matter of fact, replacing classical standards with "internationally accepted" notions would also be fruitful in any contexts where demographics play a role, thus enabling other disciplines to fully enjoy the benefits of the globalisation of demographic parameters.

The remainder of the paper is organised as follows. Section 2 sets out the background on graphs, graphical models, Markov and Gibbs random fields. Section 3 focuses on developing the global decision-model. In Sect. 4, the theoretical methodology is fully applied to the banking context. Finally, Sect. 5 concludes the paper.

2   Background: Graphs, Graphical Models, Gibbs and Markov Random Fields

Recall that a graph in discrete mathematics is a set of vertices (or nodes) and a collection of edges, each connecting a pair of vertices, represented by $G = (V, E)$, V for vertices and E for edges. An *undirected graph* is a graph where all the edges are bidirectional. In contrast, a graph where the edges point in a single direction is called a directed graph. Two vertices u and v are *adjacent*, $u \sim v$, if $(u, v) \in E$. A *maximally connected subgraph* is a connected subgraph of a graph to which no vertex can be added and it still be connected. A *clique* C in a graph $G = (V, E)$ is a subset C of the set of vertices V, $C \subset V$, in such a way that C consists of a single node or for every par of vertices $u, v \in C$ must be that $u \sim v$. That is, cliques in a graph are maximally connected subgraphs. Graphical models are a further step towards. They can be considered as *spatial stochastic processes* understood as those collections of random variables $\{X_v \mid v \in V\}$ which take a value X_v for each location v over the region of interest V. Such spatial stochastic processes are also called *random fields on* V. The two most common types of graphical models are Bayesian and Markov networks (also called Markov random Fields). The main difference between them is the underlying graph, Bayesian networks are based on a directed graph whereas Markov random fields use an undirected graph.

Consider a finite collection of random variables $X = \{X_v\}$ taking values in a finite set V, $X = \{X_v \mid v \in V\}$ ($V = \{1, 2, \ldots, n\} \times \{1, 2, \ldots, n\}$). We denote $P[X]$ to the joint distribution of this finite collection of variables: $P[X] = P[\{X_v = x_v \mid v \in V\}] = \{P[X_v = x_v] \mid v \in V\}$. We say that $P[X]$ is a Gibbs distribution relative to the graph $G = (V, E)$ if it can be expressed as a product of some functions on the cliques in G (called clique potentials):

$$P[X] = \frac{1}{Z} \exp\left[-\sum_{C \in \mathcal{C}} f(C)\right] \tag{1}$$

where $C \in \mathcal{C} \subset V$ denotes cliques and Z is a normalisation constant.

For a collection of random variables $X = \{X_v \mid v \in V\}$ we say that X is a Markov random field relative to $G = (V, E)$ so long as the full conditional distribution of X depends only on the neighbours, according to some definition of "neighbourhood". To this end, both concepts, *direction and proximity*, need to be specified. Let $V = \{1, 2, \ldots, n\} \times \{1, 2, \ldots, n\}$ be the set of nodes of a finite gride. Each $v \in V$ may be also called *site*. The *neighbourhood of a site* (i, j), written $N(i, j)$, is commonly defined as $N(i, j) = \{(i-1, j), (i+1, j), (i, j-1), (i, j+1)\}$ where one may take $(0, j) = (n, j), (n+1, j) = (1, j), (i, 0) = (i, n), (i, n+1) = (i, 1)$. For instance, the neighbourhood of $(1, 1)$, $N(1, 1)$, is shown in Fig. 1. Thus, the Markov property can be now be defined as follows:

$$P[X_v = x_v | X_{V-\{v\}} = x_{V-\{v\}}] = P[X_v = x_v | X_{N(v)} = x_{N(v)}].$$

Also, the concept of neighbourhood of a site allows to consider the notion of *clique*: a *clique c* is a set of sites where any pair of elements $c_i, c_j \in c$ hold that $c_i \in N(c_j)$ and $c_j \in N(c_i)$. Finally, the relationship between the Gibbs and Markov Random Fields is given by the Hammersley-Clifford theorem:

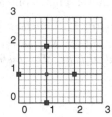

Fig. 1. Neighbourhood of $(1, 1)$

Theorem 1 (Hammersley-Clifford theorem). *According to the positivity condition, X is a Gibbs random field relative to an undirected graph G is and only if X is a Markov random field relative to G.*

3 The Universal Decision Making Model

This section is aimed at developing a decision making model which would work on a universal basis for re-structuring those networks which comply with some requirements (MRF's). For this, the primary objective is to replace local demographic specifications by global ones.

3.1 Creation of a Universal Network $(X_v, N_v)_{v \in V}$

Towards the Universality. In designing a universal tool which assists to re-locate a new node according to some pre-fixed criterion, we start from any situation that may be viewed as a spatial network (V, E), either a real or an imaginary one. Real networks consist of nodes which are physically connected amongst them like networks of motorways or railway lines. On the contrary, by imaginary networks we mean those scenarios whose nodes consist of objects like individuals or groups of persons which are connected by non-tangible links like friendship (social networks) or business relationships (markets). Thus, we proceed as follows:

(i) *Select a random variable X according to the criterion for re-structuring.* For instance, if we are to re-structure the bank branch network according to branch size (*size* is the selected criterion), we must choose a random variable which best represents the branch size (see Sect. 4 for further details).

(ii) *Construct a universal network $(X_v, N_v)_{v \in V}$.* For this, we identify each grid node $v \in V$ with the random variable X_v (which should have been selected according to the pre-fixed criterion, see step a) thereby providing a collection of random variables $X = \{X_v \mid v \in V\}$. Further, X_v may be extensively detailed by means of a collection of features $x_v^k, k = 1, \ldots, n$: i.e., $v \sim X_v \simeq (x_v^1, x_v^2, \ldots, x_v^n)^t$ for each $v \in V$, where both vector coordinates themselves (i.e., the features $x_v^k, k = 1, \ldots, n$) and number of them can be freely selected. Next, key notions must be decoupled from any geographical premise. This is one as follows:

Definition 1. *Consider nodes $X_{v_i} = (x_{v_i}^1, \ldots, x_{v_i}^n)^t$, $X_{v_j} = (x_{v_j}^1, \ldots, x_{v_j}^n)^t$, $i \neq j$, and let $\overline{X_{v_i}}$ be the mean value of the random variable over some fixed interval of time. The distance between such two nodes X_{v_i}, X_{v_j}, $i \neq j$, written d_{ij}, is*

$$d_{ij} = d(v_i, v_j)(= d(X_{v_i}, X_{v_j})) = +\sqrt{\sum_{k=1}^n (\overline{x^k}_{v_i} - \overline{x^k}_{v_j})^2}. \tag{2}$$

Definition 2. *The neighbourhood of a node $v_i \in V$, N_{v_i}, is defined as*

$$N_{v_i} = \{v_j \in V \text{ such that } d(v_i, v_j) \leq k, \quad k \in \mathbb{R}, k \neq 0\},$$

where the degree of similarity (i.e., the benchmark k) should be specified for each particular case.

Note that by defining neighbourhood of a node, the edges of the network have been specified as well: for $v_j \in N(v_i)$ an edge (v_i, v_j) will join them if $d(v_i, v_j) \leq k \neq 0$. Hence, these definitions lead to a network $(X_v, N_v)_{v \in V}$, detached from any local premises. We refer to $(X_v, N_v)_{v \in V}$ as universal network.

We may now state what cliques are in the network $(X_v, N_v)_{v \in V}$.

Proposition 1 (What cliques are). *A clique C consists of those nodes c_1^i, c_2^i with a high degree of similarity with respect to the random variable X_v, where the term "high" means that $d(c_1^i, c_2^i) \leq k_i$.*

Proof. Since nodes in a clique are in their corresponding neighbourghoods, thus $d(c_1^i, c_2^i) \leq \min\{k_1, k_2\} \leq k_1, k_2$. □

Remark 1. The choice of the random variable X makes cliques appear as different groups of nodes, thus simulating different categorising practices. That is, each choice of X leads to a re-configuration of the network. Also note that the degree of similarity will vary for each C_i, i.e., nodes $c_1^i, c_2^i \in C_i$ may be similar with degree k_i while $c_1^j, c_2^j \in C_j$ will be similar with different degrees of similarity $k_i \neq k_j$ whenever $i \neq j$ (e.g., the Internet, which is a set of routers linked by physical cables with *different* lengths).

(iii) *Show that the universal network* $(X_v, N_v)_{v \in V}$ *is a MRF.* This property does not hold in general but it strongly depends on the selection of random variable X we make. In next section, where a case study shall be developed, we will see that it is not a too restrictive assumption.

Applying the MRF Theory. Whenever the universal network $(X_v, N_v)_{v \in V}$ is a MRF, following theorem holds:

Theorem 2. *The universal network joint probability distribution may be expressed in terms of cliques:* $P[X] = \frac{1}{Z} \exp\left[-\sum_{C \in \mathcal{C}} f(C)\right]$, *for some functions* f *on cliques* C, *provided that* $(X_v, N_v)_{v \in V}$ *is a MRF.*

Proof. It is straightforward according to Theorem 1.

Functions f may vary depending on each particular context: this fact would allow sensitivity tests in order to find which best suit each particular situation. Furthermore, the practical value of this result is that it provides a simple way of specifying the joint probability which is explicitly given on terms of a few significant nodes (cliques).

3.2 How to Identify the best Site for a New Node

Suppose that the universal network (X_v, N_v) is divided into subnets $S^i(X_v, N_v)$ where this partition offers different scenarios for locating a new node v^*:

$$(X_v, N_v) = \cup_{i=1} S^i(X_v, N_v).$$

Consider now that such subnets are its cliques, $(X_v, N_v) = \cup_{i=1} C^i(X_v, N_v)$. Note that the following property holds true:

Theorem 3. *Cliques on a MRF, considered as subnetworks, are MRF themselves.*

Proof. Since the Markov property is based on the concept of neighbourhood, and cliques are subsets of some node' neighbourhood, thus the result follows. $\square$

The decision making model (for identifying the best site for a new node in the network under a pre-fixed criterion) consists of following steps:

Step 1. The universal network (X_v, N_v) provided it is a MRF should be divided into its own cliques $C^i(X_v, N_v)$ in such a way that this partition offers different scenarios for locating a new node v^*:

$$(X_v, N_v) = \cup_{i=1} C^i(X_v, N_v).$$

Step 2. Considered such subnetworks together with the new node v^* as one of its nodes:

$$C^i_{v^*}(X_v, N_v) = C^i(X_v, N_v) \cup \{v^*\}.$$

Step 3. The joint distributions of such subnetworks $\{P[C_{v*}^i(X_v, N_v)]\}_i$ are computed according to Theorem 2. Once these numerical scores are compared, they allow to make decisions about the most suitable locations.

Step 4. From all the outputs, select the most convenient one according to pre-established criteria. Such criteria may take many forms including minimising costs, minimising the distances between the existing facilities, etc.

4   Case Study: Restructuring the Bank Branch Network

This section is devoted to applying previous results to the banking context. Thus, a decision making model to restructure the branch network depending on branch size (as the selected criterion), shall be formulated now. To this end, *branch cash holdings* (i.e., the total amount of cash allowed in a branch) was selected as the general random variable X since it is a main determinant of branch

Fig. 2. The branch network

size. Moreover, this is a perfect example of the need for a universal decision making tool since local demographics strongly influence branch size. Actually, there exists a deep relationship between branch size and local demographics: branch size depends on branch cash transactions -number and amounts- while branch cash transactions depend on branch customers' needs for cash, which are strongly related to local demographics (a heavy retail stores area will require much more cash than a heavy industrial area where firms do not deal with much cash) (Fig. 2).

The application of the core model is based on previously proving that the bank branch network is a Markov random field for the branch cash holdings as selected random variable. To this end, let CH stand for the cash holdings while BN denotes the Bank Branch Network. There are two stochastic processes associated with the random variable CH. Firstly, $\{CH^n \mid n \in \mathbb{N}\}$ represents the *temporal* stochastic process of cash holdings' movements through time, where n is the time unit, (García Cabello 2017). And secondly, $\{CH_b \mid b \in BN\}$ denotes the *spatial* stochastic process (or random field on BN) where b represents a branch in the network BN. In order to express BN as a universal network following Sect. 3.1, we shall identify each branch b with the random variable which represents its cash holdings, CH_b.

Hence, the set of random variables $CH = \{CH_b \mid b \in BN\}$ is an *undirected* graph with nodes the corresponding branch cash holdings CH_b for each branch b, while the (undirected) edges indicate financial similarities between branches, to be specified as a particular instance of Definition 2. Specifically:

Definition 3. *Each branch $b \in BN$ is represented by the vector (n_b, v_b) with*

n_b *the number of branch transactions at b,*

v_b *the volume of branch transactions allowed at b.*

Remark 2. Recall that the number of vector coordinates as well as the coordinates themselves can be freely selected.

Definition 4. *The distance between two branches $b_i = (n_{b_i}, v_{b_i})$, $b_j = (n_{b_j}, v_{b_j})$, $i \neq j$, is the Euclidean distance between the vectors once the random variables are evaluated in the corresponding mean values over some interval of time:*

$$b_{ij} = d(b_i, b_j) = +\sqrt{(\overline{n_{b_i}} - \overline{n_{b_j}})^2 + (\overline{v_{b_i}} - \overline{v_{n_j}})^2}.$$

Proposition 2. *The neighbourhood of a branch $b_i = (n_{b_i}, v_{b_i})$, $N(b_i)$, consists of branches with the same(or quite the same) size.*

Proof. The neighbourhood of a branch $b_i = (n_{b_i}, v_{b_i})$, $N(b_i)$, consists of similar branches b_j such that

$$N(b_i) = \{b_j \in BN \text{ such that } b_{ij} \leq k\},$$

As branch *size* mainly depends on branch cash transactions -number and amounts involved- then the result follows. □

Remark 3. The choice of the random variable X (CH is this example) makes cliques appear as different groups of branches, simulating different branch managers' practices. While merging branches according to similar sizes is a common practice, there are others such as configuring the branch network through multi-location operations in order to diversify business and hedge against risks. Thus, it is noticeable that the proposed decision making model allows to different reconfigurations of the branch network as part of its setting options.

Now, the main result is:

Theorem 4. *With the above definition, branch network' cash holdings $\{CH_b \mid b \in BN\}$ are a Markov Random Field.*

Proof. Since the neighbourhood of a branch $b_i = (n_{b_i}, v_{b_i})$, $N(b_i)$, consists of all branches b_j with same (very similar) size, then the Markov property must be shown to hold true:

$$P[CH_b = c_b | CH_{BN-\{b\}} = c_{BN-\{b\}}] = P[CH_b = c_b | CH_{N(B)} = c_{N(b)}].$$

Theorem 5. *The joint distribution of the branch network' cash holdings $CH = \{CH_b \mid b \in BN\}$ can be expressed as a product of clique potentials in BN, say V_C. That is, if we denote $CH_C = \{CH(c) \mid c \in C\}$, where C is a clique in BN, we require functions V_C such that the joint distribution of CH, $P[CH]$, takes the form*

$$P[CH] = \frac{1}{Z} e^{\frac{-1}{T} \sum_{c \in C} V_c(CH_C)}, \tag{3}$$

Proof. The joint distribution of CH, $P[CH]$, has the alternate form $P[CH] = \frac{1}{Z} \prod_{c \in C} e^{\frac{-1}{T} V_c(CH_C)} = \frac{1}{Z} e^{\frac{-1}{T} \sum_{c \in C} V_c(CH_C)}$.

5 Conclusions

This paper has designed and presented a theoretical model for reconfiguring networks (either real or imaginary ones) as long as they are structured as MRF's. Its main advantage is that it is intended for *working on a universal basis* when replacing local demography by internationally accepted standards. Thus, it is a potential solution to the fuzziness that exists when defining demographics according to "local" specifications, which hinders the export of methodologies based on local parameters. The versatility of the procedure allows to expand/replace all model variables. As a result, a decision-making tool for restructuring networks is provided with a wide range of application areas. As an illustrative example, a case study for the banking sector has been developed by formulating a decision-making method to (re-)locate a bank branch. However, this new methodology can also apply to supermarkets, petrol stations, or other businesses with networks.

We are also studying some possible extensions of our decision making model by using fuzzy tools for managing linguistic information, see (Li et al. 2017) and new proposals of multicriteria decision making models based on incomplete information, see (Ureña et al. 2015).

The theoretical structure designed in this paper can be translated into computational terms by means of algorithms because, besides being a language for formulating models, graphical models inherit the excellent computational properties of graphs. This could be a solution when numerically-valued examples are attempted, since a huge quantity of output data has to be managed. Thus, to implement the procedure into an algorithm should provide an easy-to-handle system which are useful for conducting the selection procedure. Such computational version of the proposed method is a future research project within a foreseeable period of time.

Acknowledgments. This study was funded by FEDER funds under grant TIN2016-75850-R. Moreover, financial support from Proyectos de Excelencia 2012 Junta de Andalucía "Mecanismos de resolución de crisis: cambios en el sistema financiero y efectos en la economía real" (P12-SEJ-2463) is gratefully acknowledged.

References

Abbasi, G.Y.: A decision support system for bank location selection. Int. J. Comput. Appl. Technol. **16**, 202–210 (2003). https://doi.org/10.1504/IJCAT.2003.000326

André, M., Mahy, G., Lejeune, P., Bogaert, J.: Toward a synthesis of the concept and a definition of the zones in the urban-rural gradient. Biotechnol. Agron. Soc. **18**(1), 61–74 (2014)

García Cabello, J.: The future of branch cash holdings management is here: new Markov chains. Eur. J. Oper. Res. **259**(2), 789–799 (2017). https://doi.org/10.1016/j.ejor. 2016.11.012

Ioannou, G., Mavri, M.: Performance-net: a decision support system for reconfiguring a bank's branch network. Omega-Int. J. Manag. S **35**, 190–201 (2007). https://doi. org/10.1016/j.omega.2005.05.007

Glykas, M., Xirogiannis, A.G.: A soft knowledge modeling approach for geographically dispersed financial organizations. Soft Comput. **9**, 579–593 (2005). https://doi.org/ 10.1007/s00500-004-0401-8

Li, C.-C., Dong, Y., Herrera, F., Herrera-Viedma, E., Martínez, L.: Personalized individual semantics in computing with words for supporting linguistic group decision making. an application on consensus reaching. Inf. Fusion **33**(1), 29–40 (2017). https://doi.org/10.1016/j.inffus.2016.04.005

Miliotis, P., Dimopoulou, M., Giannikos, I.: A hierarchical location model for locating bank branches in a competitive environment. Int. Trans. Oper. Res. **9**(5), 549–565 (2002). https://doi.org/10.1111/1475-3995.00373

Ruiz-Hernandez, D., Delgado-Gomez, D., Lopez-Pascual, J.: Restructuring bank networks after mergers and acquisitions: a capacitated delocation model for closing and resizing branches. Comput. Oper. Res. **62**, 316–324 (2015). https://doi.org/10.1016/ j.cor.2014.04.011

Ureña, M.R., Chiclana, F., Morente-Molinera, J.A., Herrera-Viedma, E.: Managing incomplete preference relations in decision making: a review and future trends. Inf. Sci. **302**(1), 14–32 (2015). https://doi.org/10.1016/j.ins.2014.12.061

Fuzzy Information and Contexts for Designing Automatic Decision-Making Systems

Maria Teresa Lamata, David A. Pelta, and José Luis Verdegay[(✉)]

Department of Computer Science and Artificial Intelligence,
Universidad de Granada, 18014 Granada, Spain
{mtl,dpelta,verdegay}@decsai.ugr.es

Abstract. The replacement of people by Automatic Decision-making Systems (ADS) has become a threat today. However, it seems that this replacement is unstoppable. Thus, the need for future and current ADS to perform their tasks as perfectly as possible is, more than a necessity an obligation. Hence, the design of these ADS must be carried out in accordance with the theoretical models on which they are to be built. From this point of view, this paper considers the classic definition of General Decision Making Problem and introduces two new key elements for building ADS: the nature of the information available and the context in which the problem is being solved. The new definition allows to cover different models and decision and optimization problems, some of which are presented for illustrative purposes.

Keywords: Fuzzy information · Decision making problems · Contexts

1 Introduction

In our day to day we are constantly faced with decision making, which can manifest itself in many ways, from checking our mobile phones to deciding that we are going to wear and even thinking that we are going to eat today or tomorrow for example.

Thus, while we make an average of 35,000 decisions per day, a recent report from Huawei [7] shows that we are aware of less than 1% of the decisions we make every day, in fact, we ignore 99.74% of the decisions we make.

It is obvious that the substantial advances that have occurred in the last decade in Machine Learning have produced systems that compete with the behavior of people in situations that pose challenges, are ambiguously raised or require high doses of ability. Sometimes they even surpass us.

These systems, which in what follows here will be called Automatic Decision Systems (ADS), when managed with intelligent automation techniques, can increase, and in some cases replace, through completely autonomous systems, the capacity to act, to make decisions, of the human beings. The crucial fact is that this "substitution" of functions could produce, sooner rather than later, massive losses of jobs, as well as the disqualification of the people who performed them, so that:

(a) the ethical issues related to the behavior of the ADS must include in their technological design, to facilitate more than the "risk factors" or restrictions, their performance as driving forces of innovation [5], and

© Springer Nature Switzerland AG 2018
F. Herrera et al. (Eds.): CAEPIA 2018, LNAI 11160, pp. 174–183, 2018.
https://doi.org/10.1007/978-3-030-00374-6_17

(b) in the event of the substitution of functions, that does not produce misfunctions, that is, that the corresponding ADS act exactly as the human decision maker in turn, reproducing and improving its behavior and trying to avoid the inescapable and unpredictable failures that human beings can have when making decisions, especially when they have to be taken in unknown contexts [16].

Aware of these two demands, on the ethical character and the correct functioning of the ADS, "Informatics Europe", which represents to the European Union the research and academic community in Europe, has recently published an important document [4] that collects the policies that must be taken into account in order to achieve a balanced and effective development of the ADS in our Society. In a very summarized way, these recommendations are the following:

1. Establish means, measures and standards to assure that ADS are fair.
2. Ensure that Ethics remain at the forefront of, and integral to, ADS development and deployment.
3. Promote value-sensitive ADS design
4. Define clear legal responsibilities for ADSs use and impacts.
5. Ensure that the economic consequences of ADS adoption are fully considered.
6. Mandate that all privacy and data acquisition practices of ADS deployers be clearly disclosed to all users of such systems.
7. Increase public funding for non-commercial ADS-related research significantly.
8. Foster ADS-related technical education at the University level.
9. Complement technical education with comparable social education.
10. Expand the public's awareness and understanding of ADS and its impacts

Given the importance that ADSs have and will have in our daily lives, these recommendations mark the way in which we will have to develop future advances in this matter and highlight the need to incorporate ADSs into the possibility of having an intelligent behavior, the ability to consider ethical aspects and the recognition of qualitative properties in the solutions they propose. This requires therefore that these three scenarios (intelligent behavior, ethical aspects and qualitative properties) be included among the elements that must be considered in any decision process that is going to be automated, that is, that these three elements must be present in the models of the decision problems to be solved [11]. As it is well known, usually a classic problem of decision supposes a single decision-maker, that the available information is of probabilistic nature and that the rewards, the results of the different actions of the decision maker, can be ordered according to a criterion that he himself establishes and that allows to make the best decision among the possible [18], that is, a classic General One-person Decision Making problem is defined by a quartet $(X, A, C, \leq)$ in which,

(1) X is a space of alternatives, actions or possible decisions which contains at least two elements (if only one single possible action exists then there is no problem of choice).
(2) A is the set of states of nature that describes the environment in which the problem is posed, i.e., that the decision-maker does not control and therefore they condition the results of their actions.

(3) C is the set of results, rewards or consequences, each of which is associated to a pair constituted by an alternative and a state of nature, and
(4) ≤ is a relationship that establishes the preferences of the decision-maker on the possible results that can be obtained [12, 14].

As it is clear, in this definition there is no reference to the two most important elements for the construction of an ADS: information available and context (concept different from that of environment) in which the decision process is developed.

Therefore, taking into account the Smithson's Taxonomy [1] and [15] to characterize the different types of information, and following the line that has already been advanced in [8–10] in a descriptive way, what we propose in this paper is a new definition of the General One-person Decision Making problem that takes into account those two essential elements, I and K, where:

- I is a characterization of the available information that indicates, in terms of knowledge, what we know about the elements taking part into the problem.
- K is a framework of behaviour that establishes the rules (ethical, sustainable, purely competitive, etc.) to guide the decision-maker in the selection of alternatives.

Thus, for the analysis, design and exploitation of ADSs we will have to consider the sextet (X, A, C, ≤ , I, K), with clear meaning from above.

From this point of view the aim of this paper is to describe the new Decision Making Problems that can appear according to the nature of the information available I and the context K assumed. Because of the limit on the number of pages we will not take into account other possible generalizations, as for instance the case of considering multiple decision makers, with which our scenario would be that of Group Decision Making (GDM) or multiple criteria, with what we would be in the case of Multiple Criteria Decision Making (MCDM). Consequently, in next section the case of having an incomplete information will be analyzed, focusing on the difference between a probabilistic information and a fuzzy information. Then, in Sect. 3, different contexts shall be presented and its main characteristics described. Finally in Sect. 4 some examples are presented.

2   Information Available

In any decision making problem one of the following two circumstances occurs: all the elements taking part into the problem are perfectly known or not. In the first case one says that the information is complete and in the second case that the information is incomplete. When the information is complete, the decision problem usually becomes an optimization one, since what we are looking for is finding the best option among a perfectly known, defined and limited set of possibilities. However, when the information is not complete, the nature of this information conditions the solutions, so it is essential to classify it so that the solution approaches that are going to be used are the most appropriate.

We will consider in what follows the Taxonomy of Ignorance by Smithson [15], of which we will only comment on the aspects that interest us in this work. This taxonomy first distinguishes between passive and active ignorance. Passive ignorance involves areas that we are ignorant of, whereas active ignorance refers to areas we ignore. He uses the term 'error' for the unknowns encompassed by passive ignorance and 'irrelevance' for active ignorance. The key issue is that the taxonomy demonstrates that there are multiple kinds of unknowns, many, if not all, of which will be inherent in any complex problem (Fig. 1).

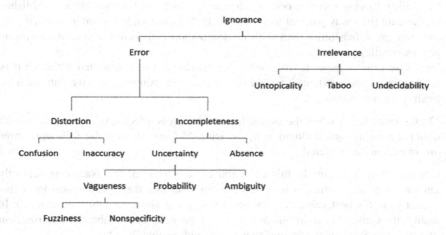

Fig. 1. Taxonomy of ignorance

In terms of 'incompleteness', the taxonomy first distinguishes between what one calls "incompleteness in degree" or 'uncertainty', and "incompleteness in kind" or 'absence'. Let us focus on "Uncertainty". As in Decision Making the term "uncertainty" is used to refer to an environment, to avoid confusion in this paper, we will replace it with "incertitude". Incertitude refer to partial information and it is subdivided into three categories: vagueness, probability and ambiguity. In short, vagueness relates to a range of possible values on a continuum (it is between 6 and 8); probability refers to random (she will win the prize with a probability of 0.01); and ambiguity refers to a finite number of different possibilities (we can travel to France, to China, to Cuba).

The essential difference between Vagueness and Ambiguity is graduality, in the sense that one information can be more accurate than another. Thus, for purposes of decision-making problems, one can suppose that Vagueness and Ambiguity represent the same type of information and consequently they can have the same Environments. Thus in what follows we will talk only about Vagueness. Hence if the information is vague, it may be because there is Fuzziness in the data [20] or because these are not specified at all, that is, there is Nonspecificity. In the first case, the management of the information would be modeled by Fuzzy Sets. In the second, the Nonspecificity assumes that handling such data requires more precision, a better definition, to be able to operate effectively (in the 20th century, in Europe, etc.). But if they are more specified, they could be handled with the same tools as the vague type data. Thus, the different Environments that we can find in the case that the available information is of a vague nature are:

- Certainty Environment: It arises when there is certainty about the values that the data will take, an essential nuance with respect to certitude. For example, we can know that it will rain tomorrow (certainty), which does not match to be certain about something: 10.4 l/m2 were collected.
- Possibility Environment: It occurs when a distribution of possibility is established on the data (in a way parallel to a probability distribution, but more associated with the concept of feasibility than with randomization and therefore without verifying any axiomatic).
- Imprecise Environment: It arises when the data are not well specified, although they are known to some extent. It would be the case, for example, to have data such as: fresh, pleasant, warm, etc.

On the other hand, when the available information is affected by randomness, with no doubt the most studied situation in Decision Making Theory, the following three Environments are considered:

- Certitude Environment: In this case, the current State of the Nature is perfectly known. In general terms, it is in this environment that the Optimization Problems appear (find the best value in a set, according to certain pre-established criteria). In reality, the Certitude Environment is only a degenerate case of the risk environment (there is one State of the Nature that arises with probability 1).
- Risk Environment: It appears when the decision maker only knows the probabilities of occurrence of the different states. The essential tool, and more generally used, to solve the problems that are associated with this environment is the Criterion of the Expected Utility.
- Uncertainty Environment: Only the list of possible States of the Nature is known, without further information about its occurrence. In this environment one can apply decision rules so well known as Laplace, Maxmin, etc. among many.

Once explained the different environments that we can consider in the case of having incomplete information of vague type or random type (other cases would overflow the objectives of this paper), we will present in the following section the most important contexts that we can find in a decision process.

3 The Contexts

A context is a set of rules, often established in the form of logical predicates, which establish the qualitative characteristics that the decisions we have to choose to solve our problem must have. The most known contexts are the classics of Ethic, Concurrence and Theoretical [14]:

- Ethic Framework, which can appear in decision processes that take place in very specific and professional contexts, such as the legal, military, medical or, in general, any other where the final decision is subject to the compliance of certain "code". In this framework, it is not only a matter of making decisions adjusted to certain moral behaviors, but rather that decisions are tailored to interests that conform to ethical codes.

- Concurrence Framework, which is ideal for decision processes in which several decision makers compete to achieve a result that is the best possible for each one at the expense of the damage that their decision may cause to others. These are situations mainly associated with games in which what a player wins is what his opponent loses.
- Theoretical frame, which is the one that considers when the only interest associated with the decision process is its modeling and study of properties.

But, in addition to these classic contexts there are others that have emerged associated with the development of the Digital Society that deserve to be mentioned. Specifically, new contexts that can condition our decisions may be the following:

- Adversarial Decision Making Context, which occurs when the decisions we make are known to our adversaries. Decision-making in the presence of adversaries poses difficulties inherent to a situation that sometimes requires resorting to sub-optimal decisions in order to confuse the adversaries. This type of situation clearly arises in the military field, but also in areas such as surveillance of perimeters, development of computer games, design of intelligent systems for personnel training, cyber crime, etc. In general terms, an adversary is an entity whose benefits (in some sense) are inversely proportional to ours. This adversary is capable of altering our benefits by taking certain actions and, in addition, can observe our actions/decisions having the opportunity to learn our pattern of behavior.
- Emergency Framework, which occurs when there are exceptional circumstances (for example catastrophes, accidents,...) and in which the best possible decision must be made among the available ones, which are usually not all of the possible ones. In these cases, a good solution strategy, inspired by the design of precomputing algorithms, can be the protocolization of the problem.
- Induced Framework, that is to say, if some psychological (nudge) factors are considered [16] we could assume an Induced Framework, as the deliberated design of a framework to make decisions can help the decision maker to take better decisions. The previous personalization (the so called induction) of the Framework in which we have to make decisions is usually a very convenient mechanism for effective decision making.
- Sustainability Framework, associated with what is meant by sustainable decisions in a specific ecosystem. As it is well known [2] for a decision to be sustainable has to meet the expectations of the moment in which it is taken, that is, to be optimal in some sense that establishes the decision maker, and at the same time not to compromise the choices made on the problem in question can be taken in the future. It makes perfect sense therefore to consider the Sustainability Frameworks to make decisions that, adhering to the needs of the problem in question, allow us to be able to solve the same problem once it is presented again, without being conditioned by the previous decisions.
- Dynamic Framework, in which at the moment in which the best possible decision is taken, the conditions that have led to that decision change and may cause that first

optimal decision no longer is. This framework is typical of Transportation, Investment and Management Problems [19], but it is becoming more frequent in social networks.

– The Corporate Social Responsibility (CSR) is an economical concept to conduct companies based on the management of the impacts that their activity generates on their customers, employees, shareholders, local communities, the environment and on society in general [13]. In this way a CSR Framework is understood as an action way through which the decision maker can voluntarily integrate social and environmental dimensions into the consequences of his actions [9].

– Context of Stress, nowadays, balancing work and family is getting more demanding, finding time for oneself is difficult, and the demands of work are overwhelming; life is becoming more stressful. Stress has become one of the factors that decision makers must contend with in most life-or-death situations. In business, stress can be detrimental to the success of managers when making key decisions. Making strategic decisions is the most critical component of an executive's job. Though executive decisions usually have very important consequences, executives must make decisions of high quality regardless of the situations and conditions they face, knowing that executives making decisions under stress in limited time and resources or uncertainty may be forced to narrow their alternatives [6]. Decision making under stress can have disastrous consequences.

4   Some Examples

The new proposed definition of a General Decision Problem gives raise to a wide and diverse set of possibilities that will have to be exploited and developed in the future. A detailed explanation is out of the scope of this paper, but here we can illustrate and explore how the definition works in a specific case.

For the sake of illustration let us consider two key examples of the potential of this definition focusing on the nature of the information (examples on how different solutions may appear depending on the context used may be found in [10]).

First, let suppose the case of one-decision making problem $(X, A, C, \leq, I, K)$ that is specified by the following elements:

– The set of alternatives X is defined by

$$X = \{x \in R^n / g_i(x) \leq 0, \text{ for } i = 1, \ldots, m\}$$

– The set A of States of the Nature is defined by only one state
– The consequences are defined by means of a function, the so called objective function,

$$f: R^n \to R$$

– According to $f(\cdot)$ the consequences are ranked as,

$$\forall x_1, x_2 \in X, x_1 \geq x_2 \Leftrightarrow f(x_1) \geq f(x_2)$$

- The information available is assumed fuzzy, and
- The context K is the Theoretical one. Particularly, and for illustration purposes, we will suppose that K is a cone.

Under these hypotheses it becomes evident that the solution of the decision making problem will come from solving the following optimization problem,

$$
\begin{aligned}
\text{Maximize:} \quad & f(x) \\
\text{Subject to:} \quad & g_i(x) \leq 0, i = 1, \ldots, m \\
& x \in K
\end{aligned}
\tag{1}
$$

If K is defined as the positive orthant, (1) becomes a conventional Mathematical Programming problem,

$$
\begin{aligned}
\text{Maximize:} \quad & f(x) \\
\text{Subject to:} \quad & g_i(x) \leq 0, i = 1, \ldots, m; x \geq 0
\end{aligned}
\tag{2}
$$

But, as we are assuming that the information has a fuzzy nature, depending on the parameters affected for this nature, we can get different problems,

(a) Fuzziness is in the objective function, (1) easily becomes

$$
\begin{aligned}
\text{Maximise:} \quad & f(x)^f \\
\text{Subject to:} \quad & g_i(x) \leq 0, i = 1, \ldots, m; \ x \geq 0
\end{aligned}
\tag{3}
$$

where usually $f(x)^f$ is characterized for having its coefficients defined as fuzzy numbers.

(b) Fuzziness is in the set of alternatives, that is one has a set of fuzzy constraints defining this set, then (1) becomes

$$
\begin{aligned}
\text{Maximise:} \quad & f(x) \\
\text{Subject to:} \quad & g_i(x) \leq^f 0, \ i = 1, \ldots, m; x \geq 0
\end{aligned}
\tag{4}
$$

(c) Fuzziness is in the parameters defining the set of alternatives, which obviously also supposes to have fuzzy constraints, then

$$
\begin{aligned}
\text{Maximise:} \quad & f(x) \\
\text{Subject to:} \quad & g_i(x)^f \leq^f 0, \ i = 1, \ldots, m; \ x \geq 0
\end{aligned}
\tag{5}
$$

(d) Fuzziness is in all the coefficients taking part in the problem, we get the more general fuzzy optimization problem,

$$
\begin{aligned}
\text{Maximise:} \quad & f(x)^f \\
\text{Subject to:} \quad & g_i(x)^f \leq^f 0, \ i = 1, \ldots, m; \ x \geq 0
\end{aligned}
\tag{6}
$$

As it may be patent, the best studied case of all these six models is the so-called Fuzzy Linear Programming problems, which appear when the functions defining (3)-(6) are linear ones. In this case there are very well known algorithms solving the corresponding problems [3]. However it is worth noting that this linear case is not he most usual in practice. Thus for instance if the Set of Consequences X is defined by discrete variables, Combinatorial Fuzzy Optimization problems may appear and then one needs to resort to fuzzy sets based heuristic algorithms [17].

Second, let now consider the case in which the elements of the General Decision Making problem are defined as above but with the novelty that instead of only one decision maker there are multiple ones. Then Fuzzy Multicriteria Decision Making problems (FMCDM) are to be considered. In this case the different contexts that we have defined before can provide new models of FMCDM. In particular, an immediate case related to the optimization problems that we have described above occurs when on $(X, A, C, \leq, I, K)$ we have the same definitions than above, except for the fact that we now have p decision makers.

Thus if the p decision makers want to find a solution that maximizes their profits, what we have is a Multiobjective Optimization problem, in which each objective corresponds to one of the decision makers, that is, we have an objective given by a p-dimensional vector,

$$[f_1(x), f_2(x), \ldots, f_p(x)] \tag{7}$$

From the fuzzy sets and systems point of view, in case of having objective functions defined by means of fuzzy coefficients, that is, a MCDM problem with fuzzy objectives, this is one of the most important problems to be analyzed. In fact, if this is the case, (7) becomes

$$[f_1(x), f_2(x), \ldots, f_p(x)]^f \tag{8}$$

where the symbol $[\ldots]^f$ stands for fuzzy coefficients in the different objective functions. As it is well known, fuzzy coefficients are usually modelled by means of fuzzy numbers but, if this is the case and each objective corresponds to one different decision maker, then it makes sense that each decision maker wants to compare the fuzzy quantities that shall appear in solving the problem by different comparison criteria of fuzzy numbers. Studying solution method for the different models that could appear in such a case is beyond the objectives and limits of this paper so we leave it as an important goal to meet in a next paper.

5   Conclusions

In this paper we wanted to show the importance of introducing the concept of context and information in the definition of a decision process. In consequence we have shown how a great catalog of different new models of decision problems appears. In order to illustrate how this new definition of decision problem works as a general framework, we have shown a few examples that, although in the literature they seem to be part of different theories or fields of study, are now clearly seen as particular models of the general model presented.

Acknowledgement. This paper has been partially supported by the projects TIN2014-55024-P and TIN2017-86647-P (both including FEDER funds) from the Spanish Ministry of Economy and Competitiveness.

References

1. Bammer, G., Smithson, M.: The nature of uncertainty. In: Bammer, G., Smithson, M. (eds.) Uncertainty and Risk: Multidisciplinary Perspectives, pp. 289–303. Earthscan (2009)
2. Brundtland Report: Our Common Future. United Nations (1987)
3. Ebrahimnejad, A., Verdegay, J.L.: Fuzzy Sets-Based Methods and Techniques for Modern Analytics. Studies in Fuzziness and Soft Computing, vol. 364. Springer, Cham (2018). https://doi.org/10.1007/978-3-319-73903-8
4. European Recommendations on Machine-Learned Automated Decision Making: When Computers Decide. Informatics Europe & EUACM (2018)
5. Cebrián, J.L.: The estate of the planet. EL PAÍS, 22 April 2018. (in Spanish)
6. Hejase, H.J., Hamdar, B., Hashem, F., Bou, S.R.: Decision making under stress: an exploratory study in lebanon. MENA Sci. **3**(12), 1–16 (2017)
7. http://www.huawei.com/es/press-events/news
8. Lamata, M.T., Pelta, D.A., Verdegay, J.L.: Optimisation Problems as Decision Problems: The case of fuzzy Optimisation Problems. Inf. Sci. **460**, 377–388 (2018). https://doi.org/10.1016/j.ins.2017.07.035
9. Lamata, M.T., Verdegay, J.L.: On new frameworks for decision making and optimization. In: Gil, E., Gil, E., Gil, J., Gil, M.Á. (eds.) The Mathematics of the Uncertain. SSDC, vol. 142, pp. 629–641. Springer, Cham (2018). https://doi.org/10.1007/978-3-319-73848-2_58
10. Lamata, M.T, Pelta, D.A., Verdegay, J.L.: The role of the context in decision and optimization problems. In: Marsala, C., Lesot, M.J. (eds.) A Fuzzy Dictionary of Fuzzy Modelling. Studies in Fuzziness and Soft Computing. Springer, Heidelberg (in press)
11. Lin, P., Abney, K., Jenkins, R.: Robot Ethics 2.0 (From Autonomous Cars to Artificial Intelligence). Oxford University Press, Oxford (2017)
12. Lindley, D.V.: Making Decisions. Wiley, New York (1971)
13. Observatory of CSR. http://observatoriorsc.org/la-rsc-que-es/
14. Rios, S.: Decision analysis. ICE Ediciones, Madrid (1976). (in Spanish)
15. Smithson, M.: Ignorance and Uncertainty: Emerging Paradigms. Springer, New York (1989). https://doi.org/10.1007/978-1-4612-3628-3
16. Thaler, R., Sunstein, C.: Nudge, Improving decisions about health, wealth and happiness. Yale University Press, New Haven (2008)
17. Verdegay, J.L.: Fuzzy Sets based Heuristics for Optimization. Studies in Fuzziness and Soft Computing, vol. 126. Springer, New York (2010)
18. Von Neumann, J., Morgenstern, O.: Theory of Games and Economic Behavior. Wiley, New York (1944)
19. Yankelevich, D. https://www.linkedin.com/pulse/prediciendo-un-futuro-predicho-daniel-yankelevich?trk=hp-feed-article-title-like
20. Zadeh, L.A.: Fuzzy sets. Inf. Control **8**, 338–353 (1965)

Evolutionary Algorithms

Distance-Based Exponential Probability Models for Constrained Combinatorial Problems

Josu Ceberio[1(✉)], Alexander Mendiburu[1], and Jose A. Lozano[1,2]

[1] University of the Basque Country UPV/EHU,
Paseo Manuel Lardizabal 1, 20018 Donostia, Spain
josu.ceberio@ehu.eus
[2] Basque Center for Applied Mathematics (BCAM),
Alameda Mazarredo 14, 48009 Bilbao, Spain

Abstract. Estimation of Distribution Algorithms (EDAs) have already demonstrated their utility when solving a broad range of combinatorial problems. However, there is still room for methodological improvement when approaching problems with constraints. The great majority of works in the literature implement repairing or penalty schemes, or use ad-hoc sampling methods in order to guarantee the feasibility of solutions. In any of the previous cases, the behavior of the EDA is somehow denaturalized, since the sampled set does not follow the probability distribution estimated at that step. In this work, we present a general method to approach constrained combinatorial optimization problems by means of EDAs. This consists of developing distance-based exponential probability models defined exclusively on the set of feasible solutions. In order to illustrate this procedure, we take the 2-partition balanced Graph Partitioning Problem as a case of study, and design efficient learning and sampling methods to use distance-based exponential probability models in EDAs.

Keywords: Constraint · Estimation of Distribution Algorithm
Distance-based exponential model · Graph Partitioning Problem

1 Introduction

A Combinatorial Optimization Problem (COP), $\mathbf{P} = (\Omega, f)$, consists of a domain of solutions Ω also known as *search space* (finite or infinite countable), and an *objective function* f that is defined as

This work has been partially supported by the Research Groups 2013–2018 (IT-609-13) programs (Basque Government), and TIN2016-78365R and TIN2017-82626R (Spanish Ministry of Economy, Industry and Competitiveness). Jose A. Lozano is also supported by BERC 2014–2017 and Elkartek programs (Basque government) and Severo Ochoa Program SEV-2013-0323 (Spanish Ministry of Economy and Competitiveness).

© Springer Nature Switzerland AG 2018
F. Herrera et al. (Eds.): CAEPIA 2018, LNAI 11160, pp. 187–197, 2018.
https://doi.org/10.1007/978-3-030-00374-6_18

$$f : \Omega \to \mathbf{R}$$
$$x \mapsto f(x)$$

Given an instance of the problem, the aim is to find a solution $x \in \Omega$ such that f is maximized (or minimized).

The work on computers and intractability by Garey and Johnson [6] showed that many COPs are NP-*hard*, denoting the difficulty of finding the optimum solution of a problem of this type (motivated principally by the size of Ω). In this sense, literature on evolutionary computation has proposed a broad range of heuristic and metaheuristic algorithms that are able to provide acceptable solutions in reasonable computation times. Among those algorithms, Estimation of Distribution Algorithms (EDAs) [11,16] have proved to be a powerful evolutionary algorithm for solving either artificial or real-world COPs [3,13].

EDAs are a type of population-based evolutionary algorithm designed for solving combinatorial and numerical optimization problems. Based on machine learning techniques, at each iteration, EDAs learn a probabilistic model from a subset of the most promising solutions, trying to explicitly express the dependencies between the different variables of the problem. Then, by sampling the probabilistic model learnt in the previous step, a new population of solutions is created. The algorithm iterates until a certain stopping criterion is met, such as a maximum number of iterations, homogeneity of the population, or a lack of improvement in the solutions.

Among different types of COPs, there is one where there is room for methodological improvement in the design of EDAs. We refer to problems with constraints (also known as restrictions). In general, a n dimensional constrained COP, or from now on, *constrained problem* [10] consists of (following the same notation as for COPs)

$$\begin{aligned}
&\text{minimizing} \ \ f(\mathbf{x}), \quad \mathbf{x} = (x_1, \ldots, x_n) \in \Omega \\
&\text{subject to,} \ \ g_i(\mathbf{x}) \leq 0, \ i = 1, \ldots, r \\
&\qquad\qquad\ \ h_j(\mathbf{x}) = 0, \ j = r+1, \ldots, m
\end{aligned}$$

where g and h are linear functions that are used to describe, respectively, inequality and equality constraints. Some relevant examples of constrained problems are the knapsack problem [12], capacitated arc routing problems [5] or graph partitioning problem [15].

Constrained problems introduce a challenging characteristic in the definition of the solutions in the search space: feasibility. A solution $\mathbf{x}$ is *feasible* if it holds all the constraints (functions g and h), otherwise it is *unfeasible*. It must be noted that the codification used to describe the solutions may induce solutions that are unfeasible. For instance, if the codification used to formalize solutions of size n is the binary coding, then the set of all solutions that fit in the codification are all the binary vectors of size n (2^n in total). Nonetheless, many of those solutions might be unfeasible, and thus, procedures that discriminate between feasible and unfeasible solutions are needed within the algorithms.

This point represents a serious drawback for EDAs, since these algorithms learn a probability distribution defined on the whole set of solutions induced by the codification (either feasible or unfeasible solutions), and then the new solutions are sampled from that distribution. As a consequence, both type of solutions are usually generated. In this sense, in the case of EDAs, in order to hold the feasibility of the solutions, three trends have been followed: (1) repair unfeasible solutions, (2) use penalty functions to punish unfeasible solutions[1], or (3) implement sampling procedures that update the probability distribution while constructing the solutions [18]. When following any of these trends, the behavior of EDAs is somehow denaturalized as the obtained sample of solutions does not follow the probability distribution defined at that iteration.

Works in the literature with regard to the design of EDAs for solving permutation problems have shown that introducing probability models defined exclusively on the space of feasible solutions (the set of all permutations of a given size) can be a step forward in terms of performance [2]. Inspired by that work, recently, Ceberio et al. [4] proposed a lattice-based probability model defined exclusively on the set of feasible solutions of the 2-partition balanced Graph Partitioning Problem (GPP). Despite its novelty, the proposed approach is ad-hoc for that problem, and can hardly be extended to other problems.

Based on the work of Irurozki [8] on probability models for permutation spaces, in this manuscript we address constrained problems in the framework of EDAs following a more general research line: design and implement new distance-based exponential probability models that define a probability distribution on the set of feasible solutions. In order to design such types of models, there are three key aspects related to the model to consider, and it is desirable to give computationally accurate, and efficient, implementations: (1) calculate the probability of any x in Ω (feasible set), (2) given a set $\{\mathbf{x}^1, \ldots, \mathbf{x}^n\}$ of solutions, estimate the parameters of the probability model, and (3) sample solutions given the parameters of the probability model.

With illustrative purposes, we consider the GPP as a case of study (in Sect. 2). Then, in Sect. 3, one by one we address the three key-aspects enumerated above. In Sect. 4, we carried out a set of experiments to evaluate the performance of the proposed EDA. Lastly, conclusions and future works are listed in Sect. 5.

2   Graph Partitioning Problem

The Graph Partitioning Problem (GPP) is the problem of finding a k-partition of the vertices of a graph while minimizing the number (or sum of weights) of the edges between sets [15]. Among its many variants, in this paper we considered the balanced 2-partition case, i.e., a solution for the problem groups the vertices into two sets of equal size. So, any solution is encoded as a binary vector $\mathbf{x} \in \{0,1\}^n$, where x_i indicates the set to which vertex i is assigned, n stands for the size of the problem, and is subject to the restriction of having the same number of zeros

---

[1] Trends (1) and (2) have been proposed for a broad range of algorithms that include EDAs.

and ones, i.e., $\sum_{i=1}^{n} x_i = n/2$. Taking this restriction into account, the search space of solutions Ω is composed of $\binom{n}{n/2}$ binary vectors.

Given a weighted undirected graph $G = (V, E)$ with $n = |V|$ vertices, and the weights w_{ij} between each pair of vertices in G, the objective function is calculated as the total weight of the edges between the sets (sum of the weights of the edges in the cut), and it is denoted as

$$f(\mathbf{x}) = \sum_{i=1}^{n} \sum_{j=1}^{n} x_i(1 - x_j)w_{ij}$$

Figure 1 introduces a GPP instance of $n = 8$, where two solutions are illustrated. As we can see, $\mathbf{x}^1$ is better than $\mathbf{x}^2$ since the sum of the weights of the edges of the cut (in bold) is lower: $f(\mathbf{x}^1) = 1 + 3$ and $f(\mathbf{x}^2) = 2 + 2 + 1$.

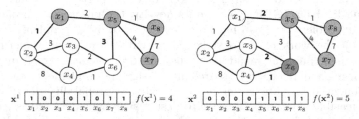

Fig. 1. Example of $n = 8$ instance of GPP with two different solutions $\mathbf{x}^1$ and $\mathbf{x}^2$.

The GPP is a very well-known constrained problem and has been studied extensively. When the problem is balanced (as in our case), the problem is NP-hard [1]. This problem is present in many real world applications [14], in fields such as tracking areas in cellular networks [19] or image segmentation [7].

3   Distance-Based Exponential Model

Let us consider a distance-based exponential probability model P defined on a finite (or infinite countable) domain of solutions Ω. Under this model, the probability value of every solution in the domain, $x \in \Omega$, is calculated as

$$P(x) = \frac{e^{-\theta d(x, \bar{x})}}{\psi(\theta)} \tag{1}$$

where $\psi(\theta)$ is the normalization constant, θ denotes the spread of the distribution, and $d(x, \bar{x})$ is the distance of x to a reference solution $\bar{x}$.

This model assigns every solution x a probability that decays exponentially with respect to its distance to $\bar{x}$. When θ equals 0, the model assigns equal probability to every solution in Ω. The larger the value of θ, the more concentrated the distribution becomes around the reference solution.

As previously mentioned, in order to introduce a model of this type in EDAs to deal with constrained problems, there are three key aspects that are mandatory to be solved efficiently. First, we need to calculate the probability of any given solution in Ω, i.e., compute Eq. 1, in closed form. Next, given a set of solutions, it is necessary to calculate the parameters of the model: θ and $\bar{x}$. In this work, we decided to calculate their maximum likelihood estimators (MLE). Finally, given the parameters of the model, we need to be able to obtain a sample of solutions that follows the inferred probability distribution.

It is worth remarking that, as EDAs are iterative algorithms, and learning/sampling procedures are repeated innumerable times in that framework, developing low complexity methods for each procedure is critical for the general feasibility of the algorithm. For the sake of illustrating the approach presented in this manuscript, we considered the balanced 2-partition GPP as a case of study, and we individually address each of the three key aspects described above.

3.1   A Case of Study: The Graph Partitioning Problem

The first decision to make is the election of the distance-metric d, which directly depends on the codification of solutions. As described in Sec. 2, the solutions of the 2-balanced GPP are codified as binary vectors of size n with an equal number of zeros as ones. Among binary vectors, the most natural distance-metric is the Hamming distance, i.e., given two solutions x and y, the Hamming distance is calculated as $d_H(x,y) = \sum_{i=1}^{n} \mathbf{1}_{[x(i) \neq y(i)]}$. It is worth noting that the codification employed to describe the solutions is highly redundant. For instance, the solutions $x^1 = (0,0,0,1,1,1)$ and $x^2 = (1,1,1,0,0,0)$ are equal. One is a relabeling of the other, but in terms of the Hamming distance, they are at maximum distance from each other. Therefore, we define a new distance for this case that is based on the Hamming distance.

Definition 1. *Let x be a solution for a 2-balanced GPP problem, and $\neg x$ is the negated or complementary solution of x, then the distance-metric $d(x, \bar{x})$ is calculated as*

$$d(x, \bar{x}) = \min\{d_H(x, \bar{x}), d_H(\neg x, \bar{x})\}$$

Under this metric, the maximum distance K to which a solution can be defined is $\lfloor n/2 \rfloor$. The minimum distance is 0 (when x or $\neg x$, are equal to $\bar{x}$). Solutions can only be at paired distances in order to hold the constraint of the balanced 2-partition case. The number of solutions at distance k (being k even) is $\binom{n/2}{k/2}^2$. Note that each possible solution can be coded with two different binary vectors (one the negation of the other).

Computing the Probability Value. Given a solution x and the parameters θ and $\bar{x}$, computing $P(x)$ is given by an efficient computation of the normalization constant $\psi(\theta)$. This function, which is roughly computed as

$$\psi(\theta) = \sum_{y \in \Omega} e^{-\theta d(y, \bar{x})},$$

considering the previous observations, can be reformulated as follows:

$$\psi(\theta) = \sum_{l=0}^{K/2} \binom{n/2}{l}^2 e^{-\theta 2l}$$

Learning. Given a set $\mathbf{x} = \{x^1, \ldots, x^N\}$ of solutions, the learning step consists of estimating the parameters of the proposed model: $\bar{x}$ and θ. In this work, we decided to estimate the MLE of the parameters. To that end, we define the likelihood expression as

$$L(\theta, \bar{x}|\mathbf{x}) = \prod_{i=1}^{N} \frac{e^{-\theta d(x_i, \bar{x})}}{\psi(\theta)},$$

The aim is to find the parameters that maximize that function. To that end, we apply the logarithm, as calculus is more simple, but the MLE parameters are not affected. So we have the function

$$\log L(\theta, \bar{x}|\mathbf{x}) = -\theta N \bar{d} - N \log \left[\sum_{l=0}^{K/2} \binom{n/2}{l}^2 e^{-\theta 2l} \right] \tag{2}$$

where $\bar{d} = \frac{1}{N} \sum_{i=1}^{N} d(x_i, \bar{x})$.

The term related to the estimation of $\bar{x}$ is independent to that of θ, so by minimizing $\sum_{i=1}^{N} d(x_i, \bar{x})$, the likelihood is maximized. Thus, first we define the computation of $\bar{x}$ as

$$\bar{x} = \arg \min_{x \in \Omega} \sum_{i=1}^{N} d(x_i, x)$$

Such an estimation is an optimization task itself. Therefore, we propose estimating $\bar{x}$ as in [3]. First, we calculate the average values at each position individually in the N solutions, and next, the $n/2$ positions with largest values in the resulting vector are assigned 1, and the rest 0.

Once the $\bar{x}$ is estimated, the MLE of the spread parameter θ is calculated. In order to find the maximum value of θ that maximizes Eq. 2, we compute its derivative and equate it to 0:

$$-N\bar{d} + N \frac{\sum_{l=0}^{K/2} \binom{n/2}{l}^2 2l e^{-\theta 2l}}{\sum_{l=0}^{K/2} \binom{n/2}{l}^2 e^{-\theta 2l}} = 0,$$

and finally,

$$\sum_{l=0}^{K/2} \binom{n/2}{l}^2 (2l - \bar{d}) e^{-\theta 2l} = 0. \tag{3}$$

θ cannot be calculated exactly from Eq. 3 and, thus, we propose using numerical methods, such as Newton-Raphson [3] to estimate it.

Sampling. Once the parameters of the model have been estimated, the next step in EDAs is to sample a set of solutions that follow the probability distribution defined by the parameters calculated in the previous step. In this case, we propose using a distance-based sampling algorithm based on the following two statements [8]: (1) every solution has the same number of solutions at each distance, and (2) every solution at same distance from $\bar{x}$ has same probability.

First, a distance $k = 2l$ at which generate a solution is sampled. The probability under the proposed model to generate a solution at distance k is

$$P(k) = \sum_{x|d(x,\bar{x})=k} P(x) = \binom{n/2}{k/2}^2 \frac{e^{-\theta k}}{\psi(\theta)}$$

Once k has been decided, we generate uniformly at random one solution among those at distance k from the reference solution $\bar{x}$. To that end, we choose u.a.r. $k/2$ zeros and $k/2$ ones, and the chosen items are bit-flipped.

4 Experiments

The experimental study has been organized in two parts. First, the feasible ranges to estimate the θ parameter are calculated. Then, a performance evaluation of the EDA on a benchmark of instances of GPP is carried out.

4.1 Feasible Ranges for θ

In order to incorporate the model to the framework of EDAs, and avoid scenarios in which the proposed model is not suitable for optimization, the spread parameter θ needs to be controlled in a range of feasible values. To that end, we calculate for different $n = \{124, 250, 500, 1000\}$, in the range $[0,10]$ for θ with step size 0.1, the probability assigned to the reference solution. Results are introduced in Fig. 2.

Without performing previous experiments, for each problem size, we set the lower and upper bounds that assign to $\bar{x}$ the probabilities 10^{-9} and 0.1 respectively. The values are presented in Table 1.

4.2 Performance Analysis

In order to obtain some hints about the performance of the proposed exponential probability model (abbreviated as **Exp**) for solving the GPP, we carried some experiments on the set of 22 GPP instances [9]. In addition, with comparison purposes, we included three EDAs in the analysis: the univariate marginal distribution algorithm (UMDA) [16], the tree EDA (Tree) [17] and the Lattice EDA [4] (Lattice). The UMDA assumes that the n-dimensional joint probability distribution factorizes as a product of n univariate and independent probability distributions[2]. The second EDA, Tree, estimates, at each iteration, a dependency-tree

---

[2] Despite this strong assumption, due to its good general performance and low time consumption it is appropriate to include UMDA as baseline in the experimentation.

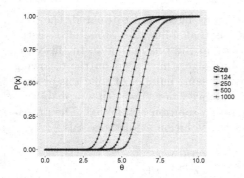

Table 1. Lower and upper bounds for the estimation of θ parameters.

n	θ_{lower}	θ_{upper}
124	1.6	3.5
250	2.4	4.2
500	3.1	4.9
1000	3.8	5.6

Fig. 2. Probability assigned to $\bar{x}$ for different n and θ.

which encodes conditional dependencies between the problem variables. Finally, in [4], the authors of the paper made the first attempt of developing EDAs for constrained COPs, and proposed a square lattice probability model to optimize the GPP.

As this is a preliminary work, we considered using the same parameter settings as in [4]:

- Population size is set to $10\,n$.
- $5\,n$ solutions are selected to learn the probabilistic model.
- At each iteration, $10\,n$ solutions are sampled.
- A maximum number of $100\,n^2$ evaluations are considered as the stopping criterion.

Equally, the initial population is generated uniformly at random.

Particular Settings of the Exponential EDA. A preliminary analysis of the quality of the reference solution estimated at the learning step pointed out that the proposed method, although it provides MLE of the reference solution $\bar{x}$, it is not competitive. In this sense, we decided to set the best solution of the population as $\bar{x}$.

In addition, as sampling at distance 0 provides the reference solution, we avoid sampling at this distance. Finally, due to convergence and diversity considerations, the solutions that do not exist in the population are only accepted when sampling.

Results. Results are presented in Table 2 as Average Relative Percentage Deviation (ARPD) and Average Execution Times (AET) of 10 executions of each algorithm-instance pair.

Looking at the results, we can conclude that with regard to the ARPD, Tree is the best performing algorithm followed by Lattice and Exp. Finally, UMDA

Table 2. ARPD and AET results for Johnson's benchmark instances. The results in bold denote the approach with the lowest ARPD and AET. (*) denotes the instances in which 10^6 evaluations were used instead of $100n^2$.

Instance	ARPD					AET			
	Best	Exp	Lattice	UMDA	Tree	Exp	Lattice	UMDA	Tree
$G_{124,0.02}$	*13*	0,32	0,32	0,61	**0,19**	**33**	145	57	464
$G_{124,0.16}$	*449*	0,04	0,02	0,05	**0,01**	**32**	144	60	496
$G_{250,0.01}$	*31*	0,40	0,33	0,49	**0,20**	**511**	2835	879	19543
$G_{250,0.02}$	*118*	0,10	0,07	0,14	**0,06**	**506**	2698	882	22610
$G_{250,0.04}$	*360*	0,04	0,04	0,10	**0,03**	**515**	2271	802	21963
$G_{250,0.08}$	*830*	0,05	**0,01**	0,05	**0,01**	**507**	2187	818	20652
$G_{500,0.005}$	*61*	0,38	0,30	0,40	**0,08**	**7768**	64873	12760	771427
$G_{500,0.01}$	*234*	0,14	0,09	0,21	**0,07**	**7948**	54905	14025	758556
$G_{500,0.02}$	*642*	0,06	**0,03**	0,11	**0,03**	**7918**	58294	13992	833603
$G_{500,0.04}$	*1754*	0,05	**0,02**	0,06	**0,02**	**7612**	44498	14366	768596
$G_{1000,0.0025}{}^{*}$	*131*	1,16	2,96	3,20	**0,74**	**1420**	2832	234724	250136
$G_{1000,0.005}{}^{*}$	*496*	**0,52**	1,22	1,28	0,88	**1395**	2775	229806	276535
$G_{1000,0.01}{}^{*}$	*142*	**0,28**	0,56	0,66	0,62	**1358**	2859	239480	256820
$G_{1000,0.02}{}^{*}$	*3450*	**0,18**	0,35	0,40	0,39	**1349**	2775	237858	252978
$U_{500,0.05}$	*23*	1,34	1,17	1,89	**0,57**	7754	58430	14541	721097
$U_{500,0.1}$	*61*	1,50	1,05	1,12	**0,57**	**7698**	45947	13065	745215
$U_{500,0.2}$	*185*	0,89	0,56	0,87	**0,44**	**7767**	43988	12518	745311
$U_{500,0.4}$	*412*	0,81	0,41	0,38	**0,28**	**7728**	48326	12867	764009
$U_{1000,0.05}{}^{*}$	*77*	4,82	**1,62**	12,83	2,39	**1364**	2938	2757	268978
$U_{1000,0.1}{}^{*}$	*170*	5,09	**1,67**	11,67	3,73	**1349**	3451	2734	271070
$U_{1000,0.2}{}^{*}$	*352*	5,23	**1,67**	10,58	4,94	**1322**	3379	2648	263100
$U_{1000,0.4}{}^{*}$	*862*	4,09	**1,53**	3,24	2,29	**1322**	3348	2494	264821

is the worst performing algorithm. Note that Exp obtained the best result in three of the largest instances while it is by far the fastest algorithm for all sizes considered.

5   Conclusions and Future Work

In this work we proposed using distance-based exponential probability models defined exclusively on the set of feasible solutions. In this sense, we took as a case of study the 2-balanced Graph Partitioning Problem, and designed step by step a probability model to be used in the framework of EDAs. In addition, some experiments to fine-tune the probability model have also been carried out.

Finally, in order to evaluate the performance of the proposed algorithm, some benchmark experiments were carried out. Results revealed that the proposed methodology is a promising research line for future work.

In this work, we used an heuristic method to calculate the reference solution $\bar{x}$ as it is a cheap algorithm to approach the estimation problem. However, we have no intuition about its precision, in this sense, for the future, we should (1) perform experiments to estimate its accuracy, and (2) investigate other estimation methods for MLE of $\bar{x}$.

Regarding the experimental study, in future research, we aim to provide a more general experimental study in which the parameters of the compared algorithms are fined-tuned, and are evaluated on a larger benchmark of instances. We also consider to hybridize Exp in order to perform a fair comparison of the algorithms.

References

1. Bui, T.N., Jones, C.: Finding good approximate vertex and edge partitions is NP-hard. Inf. Process. Lett. **42**(3), 153–159 (1992)
2. Ceberio, J.: Solving permutation problems with estimation of distribution algorithms and extensions thereof. Ph.D. thesis, University of the Basque Country, December 2014
3. Ceberio, J., Irurozki, E., Mendiburu, A., Lozano, J.A.: A distance-based ranking model estimation of distribution algorithm for the flowshop scheduling problem. IEEE Trans. Evol. Comput. **18**(2), 286–300 (2014)
4. Ceberio, J., Mendiburu, A., Lozano, J.A.: A square lattice probability model for optimising the graph partitioning problem. In: 2017 IEEE Congress on Evolutionary Computation (CEC), pp. 1629–1636. IEEE (2017)
5. Dror, M.: Arc Routing: Theory Solutions and Applications. Springer, Heidelberg (2012). https://doi.org/10.1007/978-1-4615-4495-1
6. Garey, M.R., Johnson, D.S.: Computers and Intractability: A Guide to the Theory of NP-Completeness. W. H. Freeman & Co., New York (1979)
7. Grady, L., Schwartz, E.L.: Isoperimetric graph partitioning for image segmentation. IEEE Trans. Pattern Anal. Mach. Intell. **28**(3), 469–475 (2006)
8. Irurozki, E.: Sampling and learning distance-based probability models for permutation spaces. Ph.D. thesis, University of the Basque Country, November 2014
9. Johnson, D.S., Aragon, C.R., McGeoch, L.A., Schevon, C.: Optimization by simulated annealing: an experimental evaluation. part i, graph partitioning. Oper. Res. **37**(6), 865–892 (1989)
10. Kusakci, A.O., Can, M.: Constrained optimization with evolutionary algorithms: a comprehensive review. Southeast Eur. J. Soft Comput. **1**(2) (2012)
11. Larrañaga, P., Lozano, J.A.: Estimation of Distribution Algorithms: A New Tool for Evolutionary Computation. Kluwer Academic Publishers (2002)
12. Martello, S., Toth, P.: Knapsack Problems: Algorithms and Computer Implementations. Wiley Inc., Hoboken (1990)
13. Mendiburu, A., Miguel-Alonso, J., Lozano, J.A., Ostra, M., Ubide, C.: Parallel EDAs to create multivariate calibration models for quantitative chemical applications. J. Parallel Distrib. Comput. **66**(8), 1002–1013 (2006)

14. Menegola, B.: A Study of the k-way graph partitioning problem. Ph.D. thesis, Institute of Informatics, Federal University of Rio Grande Do Sul (2012)
15. Mezuman, E., Weiss, Y.: Globally optimizing graph partitioning problems using message passing. In: Proceedings of the 15th (AISTATS), pp. 770–778 (2012)
16. Mühlenbein, H., Paaß, G.: From recombination of genes to the estimation of distributions I. Binary parameters. In: Voigt, H.-M., Ebeling, W., Rechenberg, I., Schwefel, H.-P. (eds.) PPSN 1996. LNCS, vol. 1141, pp. 178–187. Springer, Heidelberg (1996). https://doi.org/10.1007/3-540-61723-X_982
17. Pelikan, M., Tsutsui, S., Kalapala, R.: Dependency trees, permutations, and quadratic assignment problem. Technical report, Medal Report No. 2007003 (2007)
18. Santana, R., Ochoa, A.: Dealing with constraints with estimation of distribution algorithms: the univariate case. In: Second Symposium on Artificial Intelligence. Adaptive Systems. CIMAF 99, La Habana, pp. 378–384 (1999)
19. Toril, M., Luna-Ramírez, S., Wille, V.: Automatic replanning of tracking areas in cellular networks. IEEE Trans. Vehic. Technol. 62(5), 2005–2013 (2013)

Studying Solutions of the p-Median Problem for the Location of Public Bike Stations

Christian Cintrano[1](✉), Francisco Chicano[1], Thomas Stützle[2], and Enrique Alba[1]

[1] E.T.S. Ingeniería Informática, University of Málaga Andalucía Tech,
Bulevar Louis Pasteur 35, 29071 Málaga, Spain
{cintrano,chicano,eat}@lcc.uma.es
[2] Universite Libre de Bruxelles, CoDE, IRIDIA, Av. F. Roosevelt 50,
1050 Brussels, Belgium
stuetzle@ulb.ac.be

Abstract. The use of bicycles as a means of transport is becoming more and more popular today, especially in urban areas, to avoid the disadvantages of individual car traffic. In fact, city managers react to this trend and actively promote the use of bicycles by providing a network of bicycles for public use and stations where they can be stored. Establishing such a network involves the task of finding best locations for stations, which is, however, not a trivial task. In this work, we examine models to determine the best location of bike stations so that citizens will travel the shortest distance possible to one of them. Based on real data from the city of Malaga, we formulate our problem as a p-median problem and solve it with a variable neighborhood search algorithm that was automatically configured with irace. We compare the locations proposed by the algorithm with the real ones used currently by the city council. We also study where new locations should be placed if the network grows.

Keywords: Bike station location · p-median problem
Variable neighborhood search

1 Introduction

Driving a vehicle through the city is an increasingly difficult and annoying task. A large number of traffic jams at different times of the day, the rising cost of fuel, and the rising level of pollution in cities are some of the biggest problems for citizens. For all these reasons, more and more citizens are looking for alternative and sustainable ways to move around the city. Bicycles are a good way to get around: they are clean and environmentally friendly, beneficial to the health, and help to avoid getting trapped in traffic jams. As a result, an increasing number of people use bicycles as their main means of transport in the city. Municipalities have become aware of this trend and try to promote their use by providing the

© Springer Nature Switzerland AG 2018
F. Herrera et al. (Eds.): CAEPIA 2018, LNAI 11160, pp. 198–208, 2018.
https://doi.org/10.1007/978-3-030-00374-6_19

necessary infrastructure, such as areas to place them and bike lanes. In addition, numerous initiatives have been taken by local authorities and private companies to promote the so-called bike-sharing. To manage the distribution of bicycles, stations are usually set up where bicycles can be picked up/dropped off. But finding the best location for those facilities is not a trivial job.

In this work, we tackle the public bike stations location problem (PBSLP). To do so, we formulate the problem as a classical problem of localization, namely the p-median problem. The p-median problem tries to identify, given a set of locations and customers and distances between locations and customers, a subset of locations of size p such that the total distance of travel of customers to the closest location is minimized. In other words, the goal is to identify locations so that the average distance of customers to bike stations is minimized. While we are not the first to consider this problem to a possible model for the PBSLP [17], we study here different variants of the problem including different distance metrics. This classic formulation is accompanied by real data from the city of Malaga, Spain, which allows us to test our proposal and variants of the p-median problem as close as possible to the reality of the city. Finally, an advantage of this modeling is that the p-median problem is well-studied and many efficient and effective algorithms are available for it. We further enhance the performance of some popular p-median algorithms by an automatic algorithm design process.

The rest of this article is organized as follows: Sect. 2 presents the formulation of the problem. Section 3 describes the selected optimization algorithm. Section 4 analyzes the main results of our work using as a real scenario the city of Malaga. We discuss related work in Sect. 5 and conclude in Sect. 6.

2 The p-median Problem

The p-median problem is one of the most-studied NP-hard discrete location problems [6,14]. The problem can be formulated as follows. Given a set of customers N and a set of possible facility locations F, the p-median problem asks to allocate p facilities to the set of available locations F while minimizing the weighted sum of the distances between the customers and their closest facility. Formally, the optimization problem is defined as:

$$\min \sum_{i=1}^{|N|} w_i \min_{j \in L} d_{ij}, \qquad (1)$$

where $L \subseteq F$, $|L| = p$, w_i is the weight of customer i, and d_{ij} is the distance between customer i and facility j. If we have $w_i = 1$, $i = 1, \ldots, |N|$, that is, all weights are one, we have the unweighted version of the problem.

In our study the customers are the citizens and the facilities are the bicycle stations that can be placed on different street segment of the city. We selected this formulation of the problem for two reasons: (i) the model is easy to understand and implement and (ii) the p-median problem is a classical location problem that has been well-studied in the scientific literature [6]. From the PBSLP perspective, the p-median may serve as a good proxy to identify interesting locations and relevant possible distributions of the bicycle stations across an urban area.

While there are other formulations of the problem [2,10,12], these may in part require various types of information. Differently, the p-median problem requires rather little information (only the distance matrix and possibly the weights). Although this may seem like a limitation, it is relatively simple to add additional information, either by pre-processing the weights or distances, (e.g., by considering the slopes of the streets) or by adding terms in the formulation itself (e.g., by adding capacity information related to the number of bikes in each bicycle parking site). Hence, in the present work we define a baseline for future works that will take these advanced characteristics into account.

3   Algorithm

The algorithm, that we use in this article, is based on the variable neighborhood search (VNS) algorithm for the p-median problem presented by Mladenovic and Hansen [15]. It was selected as the base for our development because the VNS algorithms and variants of it have been used in localization problems, and, in particular, for the p-median problem reaching very good performance [1,16,18].

While there are a number of VNS variants for the p-median problem, we have developed a component-wise implementation of such algorithms. In particular, we use algorithmic components that have been used for the original, basic VNS algorithm [15] (OVNS), and two more recent variants proposed in [7] (BVNS and DVNS). In order to find an improved version of the VNS, we use an automatic algorithm configuration stage through the use of the iterated racing procedure implemented by the irace package [13]. We used a budget (number of configurations) of 5000 and, as training instances, we chose instances of the TSPLIB[1] library. The locations of the cities correspond to the locations of the customers and the possible facilities. As in the practical problem we tackle the number of p with small values, we used here values of $p \in \{10, 20, 30\}$ for each instance of the set of training instances. For the configuration process we considered the following components together with associated numerical parameters:

Neighborhood Model. We used two models to select the (k) neighbors: nearest points (NEAR) and the division of the space into four quadrants from the point and selecting (if possible) an equal number of neighbors in each quadrant (QUAD).

Local Search. We apply a local search procedure to the initial solution (localsearch1) and in each iteration of the algorithm (localsearch2). The possible local search algorithms are: Fast Interchange [20], IALT [8] (Laux parameter), and IMP [7].

Shake. We use two different solution modifications: either we exchange n facilities with randomly chosen other ones in the neighborhood (NEIGHBOR) or we select then randomly in the whole set of facilities (RAND).

---

[1] TSPLIB instances: http://comopt.ifi.uni-heidelberg.de/software/TSPLIB95/tsp/.

Initial Solution. We have four procedures to generate the initial solution: random (RAND), START [7], the best of 100 random solutions (100RAND), and the best of performing IMP to 100 different random solutions (100IMP).

Table 1 presents the parameters information used by irace. The last column shows the best values found by irace. These values form the parameters of the VNS algorithm used in the experiment. While we leave a detailed evaluation of the configured VNS algorithm to previously proposed VNS algorithms to an extended version of the paper, in what follows we focus on the usage of the configured algorithm to study the location of public bike stations in Malaga.

Table 1. Parameters for the automatic configuration process. Given are the parameter, its type, its domain of possible values, the condition under which the parameter is relevant (conditional parameters) and the value the parameter takes in the automatically obtained configuration.

Parameter	Type[a]	Range	Condition	Configured
Algorithm	c	$(OVNS, DVNS, BVNS)$		DVNS
Localsearch1	c	(FI,IALT,IMP)		IMP
Laux1	i	$(1,50)$	localsearch1 = "$IALT$"	—
Localsearch2	c	(FI,IALT,IMP)	algorithm $\in$ ("$DVNS$", "$BVNS$")	FI
Laux2	i	$(1,50)$	localsearch2 = "$IALT$"	—
Generation	c	$(RAND, START, 100RAND, 100IMP)$		100IMP
Theta	r	$(0,1)$	generation = "$START$"	—
Shake	c	$(RAND, NEIGHBOR)$		$NEIGHBOR$
Neighborhood	c	$(NEAR, QUAD)$		$QUAD$
k	i	$(1, 40)$		40
Kmayus	i	$(10, 100)$	algorithm $\in$ ("$DVNS$", "$BVNS$")	92
m	r	$(0, 1)$	algorithm = "$DVNS$"	0.12
Lambda	r	$(1, 5)$	algorithm = "$DVNS$"	4.30

[a] c: categorical, i: integer, r: real;

4 Experimental Study

As mentioned, our study focuses on the usage of the p-median problem to suggest possible solutions that may be implemented in the real world to tackle the PBSAP. For this reason, the use of actual data is really important in this work. We have decided to use the city of Malaga as a study object as it has a wide variety of open data (see municipality open data website: http://datosabiertos. malaga.eu/) and has already a functioning shared bikes system with 23 public accessible stations, which will allow us to compare our proposed solutions by the p-median problem to the actual situation.

The p-median problem needs two sets of points: customers and facilities. The customers are the citizens. As positions where these are located, we have chosen the centers of the different neighborhoods (363) of the city (we have excluded six neighborhoods at the outskirts of Malaga) and the weights are the real number of inhabitants in each of the neighborhoods. All possible street segments of the city have been chosen as possible locations for the facilities, that is, the bike stations. This makes a total of 33,550 possible locations. Figure 1 shows the layout of the customers concerned and the bike stations in the city, where the red points correspond to the centers of the neighborhoods and the clue points to the current location of the 23 public bike stations. We should note there that the ratio between the number of clients and facilities we have in this instance is not common in the existing studies in the p-median problem; hence, further studies with instances with such ratios may be an interesting direction for future work.

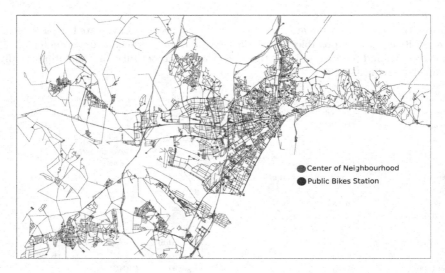

Fig. 1. Public bike stations in the city of Malaga (blue points) and the center of each neighborhood (red points). (Color figure online)

We next present the main results of our case study and compare them with the real scenario. We used real data and the tuned VNS to optimize each instance. The algorithm was running 60 CPU seconds and reports the best solution found in each of the trials. The implementation was done in C++, compiled with flag -O3. The computation platform used is a desktop computer with one Intel i5-4460 processor running at 3.2 GHz, and 8 GB memory. We have carried out 100 runs of the algorithm to statistically compare the results obtained.

4.1 Comparison with Real Scenario

In our study we have taken into account different levels of realism to analyze its impact on the proposed solutions. We have used two types of distance between customers and stations: straight-line Euclidean distance and the shortest path through the city streets (calculated using the Dijkstra's algorithm), we refer to this last one as real distance. We have also taken into account the population density in each neighborhood by weighing the problem using the number of citizens and compare it to how solutions would look if these weights were not taken into account. This gives us four scenarios to study, corresponding to weighted or unweighted customers locations with Euclidean or real distances. We used the same number of stations that are currently in use in Malaga, that is, $p = 23$. Figure 2 shows the empirical cumulative distribution of the percentage of improvement in the objective function value of each scenario when compared to the result of calculating the objective function for the actual location of the 23 bike stations in Malaga (evaluated according to the corresponding scenario).

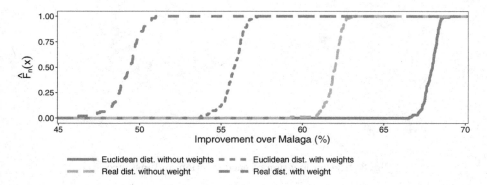

Fig. 2. Empirical cumulative distribution of the improvement of our solutions in each scenario, compared to current real location of bike stations in Malaga. For each scenario the current solution of Malaga is evaluated with the corresponding objective function.

The versions without weights show larger improvements, with a median of 68% (Euclidean distance) and 62% (real distance), than the versions with weights, where the improvement was less: 56% (Euclidean distance) and 49% (real distance). This results can be explained because by not taking the population size of each neighborhood into account, the stations become spread across the city, reaching more areas closer to the borders, reducing the distance to each neighborhood. However, the population size should be taken into account, as has apparently been done in the current solution implemented in Malaga, where the stations are concentrated in the central part of the city, which is also the most densely populated area. Even so, the p-median versions with weights obtain substantial improvements of about 50% improvement over the base scenario.

After analyzing the quality of the results, we can see in Fig. 3 the geographical distribution in each instance of the problem. As we expected, the weighted versions put more stations in the central area of the city. However, all of them offer good coverage of the main neighborhoods, so that each citizen has a reasonably nearby station to use the service.

Finally, bike-sharing systems should try to minimize the distance users need to travel to their nearest station. Therefore, for each solution found, we

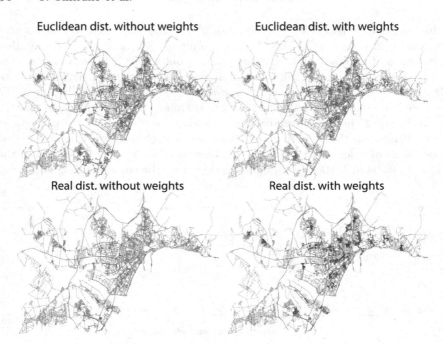

Euclidean dist. without weights   Euclidean dist. with weights

Real dist. without weights   Real dist. with weights

Fig. 3. Station location of each solution and instance.

Table 2. Distance traveled per inhabitant to the nearest station in the different scenarios, evaluated as real weighted distance. The minimum values are marked in bold.

Scenario	Distance (m)			
	Min	Max	Median	Mean
D. Euclidean without weights	877.3	984.2	933.3	933.1
D. Euclidean with weights	761.4	865.5	802.2	802.4
D. Real without weights	853.7	951.7	902.9	902.3
D. Real with weights	**728.2**	**801.2**	**751.3**	**754.1**
Malaga	1485.9			

calculated the average distance that the citizens must travel to their nearest station. Table 2 shows this distance information for each scenario. In general, our solutions reduce the city council's solution by 500 m (on average per person even in the worst case). As expected, the real distance with weights obtained the best results when taking this distance into account during the optimization process. It is interesting to note that using weights in the Euclidean distance case reduces the average distance walked more than when changing the distance computation from Euclidean to real distances. This tells us that using demographic

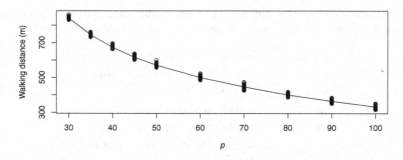

Fig. 4. Traveled distance for each new value of p.

information is more important than the type of distance used, if one wants to improve the quality of service to the user.

4.2 Increasing Number of Stations

As it is is often costly to make changes to urban infrastructure, it is not always possible to move the city infrastructure. So, instead of that, we will study now the optimal locations if new stations are to be installed in the city. Using the most realistic scenario, weights and real distances, we have considered different values of p, while keeping the 23 stations in the city of Malaga as fixed in each solution. As a result, in each of the following experiments, a number of $p - 23$ bike stations is added to the current solution already existing. We have performed 100 iterations of the algorithm for $p \in \{30, 35, \ldots, 50, 60, \ldots, 100\}$. Table 3 summarizes the results obtained. With only seven additional stations, one already may obtain a substantial improvement of 43% when compared to the actual situation in Malaga. More additional stations lead obviously to additional improvements, with a steep decrease in average distances to travel for small increments and to diminishing returns when adding a large number of stations. The traveled distance is also plotted in Fig. 4. The slope of this curve is quite steep up to $p = 50$ stations and smoother thereafter. As a summary, this analysis indicates that increasing the stations in the city considering optimized locations could greatly improve the quality of service to citizens.

Comparing this data with our more realistic scenario, real distance with weights, we see that the average distance traveled (751.3) is even better than increasing Malaga's stations to 30 (being even better in some cases to $p = 35$). The power of VNS to solve this problem underscores the usefulness of algorithms to solve real problems and to consider optimization aspects upfront.

5 Related Work

In bicycle sharing systems there are multiple problems to address such as predicting the filling of stations [19], the location-allocation of bicycles [3,5], routes

Table 3. Walking distance for different numbers of additional stations; the number of additional stations is $p - 23$, where p is the number given in the first column.

Scenario	Distance (m)			
	Min	Max	Median	Mean
Malaga (23)	1485.9			
30	830.87	860.92	839.93	840.82
35	732.89	761.39	744.62	745.47
40	663.80	696.49	672.87	674.19
45	602.65	637.48	615.93	616.75
50	557.29	599.86	570.82	571.52
60	485.66	523.55	499.93	500.35
70	426.08	473.54	447.19	446.74
80	387.16	417.81	400.71	400.51
90	351.08	384.28	364.40	364.45
100	316.26	351.99	332.33	332.77

for users or the transfer of bicycles [9], etc. There are also complete solutions that take into account multiple aspects of bike sharing systems [11]. However, they are very complex solutions that require large amounts of information that are not always available.

If we focus on the optimal location of the stations we find solutions such as those proposed in [2]. In this paper, the authors use real data and machine learning techniques to find the best places to position the stations. A similar approach to the above is used in [12], where they use New York City, which has a large network of bike stations. In [17] a comparison between two models for the station location problem is presented. The authors compare the p-median and the MCLP. However, they do not give details about the algorithm that is used. In [4] the authors try to find the best locations so as to ensure the availability to collect/deposit the bicycles, taking into account the demand (at peak times) and the possible routes between stations made by users. These works involve a demand on the use of the bike-sharing system. However, they do not take into account potential users who could use the system if it were closer to them.

Our work uses demographic data to bring the system closer to all users (which has not been taken into account in other work). In addition, existing solutions use a custom formulation, making it difficult to compare them with other state-of-the-art solutions. We formulate the problem of locating bike stations as a p-median which allows us to use a solid base for our studies, as well as to enrich ourselves with all the advances made in the p-median problem.

6   Conclusions

In this work we have modeled the public bike stations location problem as a classic problem of location, the p-median problem. To solve it we used a VNS algorithm that was automatically configured with irace. We have studied different scenarios: Euclidean and real distances (calculated by Dijkstra's algorithm), and whether or not to consider population densities. As might be expected, the most realistic scenario (real distances and weights) reported the best results, and all configurations performed much better than the Malaga public bike-sharing system. Finally, we have studied where ideally locations should be added if the current public bike system is extended by new stations. Already with few additional stations, we have obtained substantial reductions in the average distance of customers to bike stations; overall the improvements have been in the range of 43–77% with respect to the current situation in Malaga.

As future work we want to test our model in bigger cities like Madrid or New York, try other heuristic algorithms and integer linear programming solvers, as CPLEX; and variants of the p-median problem such as minimizing the distance to customers while maximizing the distance between facilities or the capacitated version. It is also interesting include more realistic data such as the bike trips, cycle lanes, traffic patterns, point of interest in the city (schools, hospitals, museums, etc.), or stops of other public transport, to promote multimodal transport.

Acknowledgements. This research was partially funded by the University of Málaga, Andalucía Tech, the Spanish MINECO and FEDER projects: TIN2014-57341-R, TIN2016-81766-REDT, and TIN2017-88213-R. C. Cintrano is supported by a FPI grant (BES-2015-074805) from Spanish MINECO.

References

1. Avella, P., Boccia, M., Salerno, S., Vasilyev, I.: An aggregation heuristic for large scale p-median problem. Comput. Oper. Res. **39**(7), 1625–1632 (2012)
2. Chen, L., et al.: Bike sharing station placement leveraging heterogeneous urban open data. In: Proceedings of the 2015 ACM International Joint Conference on Pervasive and Ubiquitous Computing - UbiComp 2015, pp. 571–575. ACM Press, NY (2015)
3. Chen, Q., Liu, M., Liu, X.: Bike fleet allocation models for repositioning in bike-sharing systems. IEEE Intell. Transp. Syst. Mag. **10**(1), 19–29 (2018)
4. Chen, Q., Sun, T.: A model for the layout of bike stations in public bike-sharing systems. J. Adv. Transp. **49**(8), 884–900 (2015)
5. Chira, C., Sedano, J., Villar, J.R., Cámara, M., Corchado, E.: Urban bicycles renting systems: modelling and optimization using nature-inspired search methods. Neurocomputing **135**, 98–106 (2014)
6. Dantrakul, S., Likasiri, C., Pongvuthithum, R.: Applied p-median and p-center algorithms for facility location problems. Expert Syst. Appl. **41**(8), 3596–3604 (2014)
7. Drezner, Z., Brimberg, J., Mladenović, N., Salhi, S.: New heuristic algorithms for solving the planar p-median problem. Comput. Oper. Res. **62**, 296–304 (2015)

8. Drezner, Z., Brimberg, J., Mladenović, N., Salhi, S.: New local searches for solving the multi-source Weber problem. Ann. Oper. Res. **246**(1–2), 181–203 (2016)
9. Hu, S.R., Liu, C.T.: An optimal location model for a bicycle sharing program with truck dispatching consideration. In: 17th International IEEE Conference on Intelligent Transportation Systems (ITSC), pp. 1775–1780. IEEE, October 2014
10. Kloimüllner, C., Raidl, G.R.: Hierarchical clustering and multilevel refinement for the bike-sharing station planning problem. In: Battiti, R., Kvasov, D.E., Sergeyev, Y.D. (eds.) LION 2017. LNCS, vol. 10556, pp. 150–165. Springer, Cham (2017). https://doi.org/10.1007/978-3-319-69404-7_11
11. Lin, J.R., Yang, T.H., Chang, Y.C.: A hub location inventory model for bicycle sharing system design: formulation and solution. Comput. Ind. Eng. **65**(1), 77–86 (2013)
12. Liu, J., et al.: Station site optimization in bike sharing systems. In: 2015 IEEE International Conference on Data Mining, pp. 883–888. IEEE, November 2015
13. López-Ibáñez, M., Dubois-Lacoste, J., Cáceres, L.P., Birattari, M., Stützle, T.: The irace package: iterated racing for automatic algorithm configuration. Oper. Res. Perspect. **3**, 43–58 (2016)
14. Megiddot, N., Supowits, K.J.: On the complexity of some common geometric location problems. SIAM J. Comput. **13**(1), 182–196 (1984)
15. Mladenović, N., Hansen, P.: Variable neighborhood search. Comput. Oper. Res. **24**(11), 1097–1100 (1997)
16. Mladenović, N., Brimberg, J., Hansen, P., Moreno-Pérez, J.A.: The p-median problem: a survey of metaheuristic approaches. Eur. J. Oper. Res. **179**(3), 927–939 (2007)
17. Park, C., Sohn, S.Y.: An optimization approach for the placement of bicycle-sharing stations to reduce short car trips: an application to the city of Seoul. Transp. Res. Part A: Policy Pract. **105**, 154–166 (2017)
18. Reese, J.: Methods for Solving the p-Median Problem: An Annotated Bibliography (2006)
19. Singhvi, D., et al.: Predicting Bike Usage for New York City's Bike Sharing System (2015)
20. Whitaker, R.A.: A Fast algorithm for the greedy interchange for large-scale clustering and median location problems. INFOR: Inf. Syst. Oper. Res. **21**(2), 95–108 (1983)

An Empirical Validation of a New Memetic CRO Algorithm for the Approximation of Time Series

Antonio Manuel Durán-Rosal[1(✉)], Pedro Antonio Gutiérrez[1],
Sancho Salcedo-Sanz[2], and César Hervás-Martínez[1]

[1] Department of Computer Science and Numerical Analysis, University of Córdoba,
Rabanales Campus, Albert Einstein Building, 14071 Córdoba, Spain
{aduran,pagutierrez,chervas}@uco.es
[2] Department of Signal Processing and Communications, Universidad de Alcalá,
Madrid, Spain
sancho.salcedo@uah.es

Abstract. The exponential increase of available temporal data encourages the development of new automatic techniques to reduce the number of points of time series. In this paper, we propose a novel modification of the coral reefs optimization algorithm (CRO) to reduce the size of the time series with the minimum error of approximation. During the evolution, the solutions are locally optimised and reintroduced in the optimization process. The hybridization is performed using two well-known state-of-the-art algorithms, namely Bottom-Up and Top-Down. The resulting algorithm, called memetic CRO (MCRO), is compared against standard CRO, its statistically driven version (SCRO) and their hybrid versions (HCRO and HSCRO, respectively). The methodology is tested in 15 time series collected from different sources, including financial problems, oceanography data, and cardiology signals, among others, showing that the best results are obtained by MCRO.

Keywords: Time series size reduction · Segmentation
Coral reefs optimization · Memetic algorithms

1 Introduction

Time series analysis has been very important for several decades from the point of view of statistical and traditional methodologies [21]. Currently, time series

Supported by the projects TIN2017-85887-C2-1-P, TIN2017-85887-C2-2-P, TIN2014-54583-C2-1-R, TIN2014-54583-C2-2-R and TIN2015-70308-REDT of the Spanish Ministry of Economy and Competitiveness (MINECO), and FEDER funds (FEDER EU). Antonio M. Durán-Rosal's research has been subsidised by the FPU Predoctoral Program of the Spanish Ministry of Education, Culture and Sport (MECD), grant reference FPU14/03039.

© Springer Nature Switzerland AG 2018
F. Herrera et al. (Eds.): CAEPIA 2018, LNAI 11160, pp. 209–218, 2018.
https://doi.org/10.1007/978-3-030-00374-6_20

data mining and machine learning have become an important research field, motivated by the increasing computational capabilities. This kind of problems can be found in many fields of science and engineering applications, including hydrology, financial problems, climate, etc. Time series data mining includes a wide range of tasks, such as the reconstruction of missing values [6], forecasting [2] or segmentation [15], among others.

Due to the great amount of data which can be collected from different resources in different periods of time, it is necessary to develop algorithms with the aim of reducing the number of points of the corresponding time series. This problem can be tackled by time series segmentation or approximation algorithms. Essentially, time series segmentation consists in dividing the time series into a set of specific points, trying to satisfy different objectives. It is often a prerequisite for other time series tasks (classification, clustering, motif detection, forecasting, etc.). There are two well-known objectives in a segmentation-type time series problem: The first one is to discover useful patterns of observed similarities along time [4] or to detect important points in the series [14]. The second objective considered in time series segmentation problems is related to simplifying the raw time series with the minimum information loss. This second group of algorithms are focussed on reducing time series size, alleviating the difficulty of processing, analysing or mining complete time series databases. They usually select a subset of points by optimizing a given approximation error, where this approximation is made by using interpolations of the subset of points. In this context, Keogh et al. [11] proposed several algorithms using piecewise linear approximations (PLA), including Top-Down, Bottom-Up, Sliding Window and SWAB (a combination of Sliding Windows and Bottom-Up) methodologies. Other works have been recently presented with the same objective [22]. Also, PLA representation can be changed by other alternatives, such as piecewise aggregate approximation (PAA) or the adaptive piecewise constant approximation (APCA) [1]. Finally, there are works that combine both points of view, ie.e, the objective is to find a set of points with minimum error and also resulting in useful patterns [8].

This paper is focused on the second group of time series segmentation methods, specifically on PLA representation algorithms. We propose a novel modification of a powerful metaheuristic, which is called coral reefs optimization (CRO) [17]. CRO is a bioinspired evolutionary algorithm with simulates the processes in real coral reefs. It has been applied to different optimization problems with a great performance, such as energy problems, telecommunications, and vehicle routing [16]. In general, all evolutionary algorithms are able to find high-quality areas using a population of individuals. However, the main disadvantage is that they are poor when looking for the precise optimum in these areas. To prevent this problem, authors in [7] proposed to further optimize the best solution found by the CRO with a local search procedure, resulting in an algorithm called HCRO. In this paper, we propose to use a different technique, which is called memetic hybridization, where the combination is not only performed to the best solution obtained by the CRO in the last generation, but also in different parts of the evolution, reintroducing optimized solutions in the population. This algorithm is called memetic CRO (MCRO). We test the MCRO algorithm in 15 time series collected from different sources, showing that the proposed methodology outperforms the remaining state-of-the-art CRO algorithms.

The rest of the paper is organized as follows: Sect. 2 defines the problem of the time series approximation algorithm, while Sect. 3 describes the CRO algorithm. The time series used, the performed experiments and the statistical discussion of the results are presented in Sect. 4. Finally, the paper is concluded in Sect. 5.

2   Problem Definition

As stated before, we try to reduce the number of points of a given time series $Y = \{y_i\}_{i=1}^{N}$ by obtaining a set of K segments defined by $K - 1$ cut points $(t_1 < t_2 < \cdots < t_{K-1})$. Considering a linear interpolation between every two consecutive cut points, the error of the approximation needs to be minimized with the aim to represent the time series with the minimum information loss. In this way, the cut points t are extracted from all time indexes, obtaining the set of segments $\mathcal{S} = \{s_1, s_2, \ldots, s_K\}$, where $s_1 = \{y_1, \ldots, y_{t_1}\}$, $s_2 = \{y_{t_1}, \ldots, y_{t_2}\}, \ldots, s_K = \{y_{t_{K-1}}, \ldots, y_N\}$. It is important to mention that the cut points are always included in two segments: the last point of the previous segment and the first point of the next one. Given that the search space is too large, we apply CRO algorithms. Figure 1 shows an example of a time series segmentation with 8 cut points, together with the corresponding 9 segments.

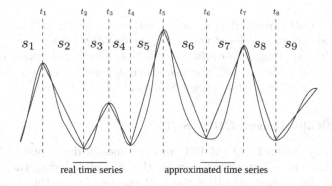

Fig. 1. Example of time series segmentation

3   Coral Reef Optimization Algorithms

In this section, previous versions of the CRO algorithm, the proposed memetic CRO and the different adaptations for time series approximation problem are described.

3.1   Basic CRO

CRO [18] is a bio-inspired algorithm for optimization based on the real processes that happen in a coral reef. The reef is represented by a matrix $\mathcal{C}$, where each position $\mathbf{C}_{ij}$ represents a coral (i.e. a solution of the problem). Firstly, the coral reef is initialised with some empty positions (F_{free}), and each coral is encoding depending on the problem. After that, the following evolution is made:

Asexual Reproduction: corals asexually reproduce by budding or fragmentation. A random coral from a set of the best corals (F_a) is selected and modified to settle in the new coral reef.

External Sexual Reproduction: a percentage F_b of the corals are selected, and with these corals, the algorithm selects pairs of parents. To generate each new larva, two broadcast spawners are selected and a crossover operator or any other exploration strategy is applied.

Internal Sexual Reproduction: hermaphrodite corals mainly reproduce by brooding. This reproduction can be any kind of mutation mechanism and takes place on a fraction of corals of $1 - F_b$.

Settlement: when the reproduction is performed the new larvae try to settle in the reef. For each new larva, a random position is generated (i, j). If the position is empty, the larva will be introduced in the coral reef, if it is occupied, it will be introduced only if it is better than the existent coral.

Depredation: corals can die during the evolution, that is, a little percentage of the worst corals (F_d) are deleted with a low probability (P_d).

   Finally, after applying the evolution the best solution obtained by CRO is optimised by using a local search process, which depends on the problem tackled (see Sect. 3.4), resulting in that we called hybrid CRO (HCRO).

3.2   Statistically-Driven CRO (SCRO)

The Statistically-driven CRO (SCRO) was specifically developed in [7] for the problem of time series segmentation with the aim of reducing the number of points in the time series, which is the problem tackled in this work. One of the problems of evolutionary algorithms is the adjustment of parameter values. SCRO is a self-adaptive version of CRO, where the user does not need to specify any parameter. Let f_i be the fitness value of coral $\mathbf{c}_i$ (we use linear notation instead of matrices for simplicity), and let $\bar{f}_j$, S_{f_j} be the mean and the standard deviation of the corals of generation j, respectively. The following modifications are performed to guide the evolution:

Initialization: the coral is randomly generated. Then, those corals whose fitness verifies $f_{i0} \notin (\bar{f}_0 - S_{f_0}, 1]$ are deleted.

Asexual Reproduction: F_a is substituted by considering only those corals whose fitness satisfies $f_{ij} \in (\bar{f}_j + S_{f_j}, 1]$.

External Sexual Reproduction: the use of the percentage F_b of corals is avoided by considering the corals whose fitness verifies $f_{ij} \in (\bar{f}_j - S_{f_j}, 1]$.

Internal Sexual Reproduction: the rest of corals, i.e. those whose fitness value satisfies $f_{ij} \in [0, \bar{f}_j - S_{f_j}]$, are used instead of $1 - F_b$.

Settlement: the settlement procedure is performed in the same way than in standard CRO.

Depredation: instead of applying a probability depredation P_d to a percentage of corals F_d, SCRO deletes those corals with fitness $f_{ij} \in [0, \bar{f}_j - 2S_{f_j}]$.

Finally, as in CRO, a hybrid version of SCRO is proposed (HSCRO) by applying a local search to the best solution.

3.3   Proposed Memetic CRO (MCRO)

As we stated before, a way to prevent evolutionary algorithms from not finding the precise optimum is the application of different local search processes during the evolution. In this work, we propose to use a memetic strategy to improve the optimization performed by standard CRO, called memetic CRO (MCRO). The memetic strategies consist in applying local search at the beginning or during the evolution and introducing improved individuals in the population. There are different combination strategies, for instance, Lamarckian learning, Baldwinian learning, or partial Lamarckianism [12]. In our case, we consider the following strategy: firstly the coral reef is initialised using a F_{free} percentage of unoccupied positions. Then the algorithm randomly selects a 50% of corals, which are applied a local search, and the optimised corals replace the original ones. Once the evolution process has started, the best coral (the fittest one) of the population is locally optimised and reintroduced in the reef by the settlement process. This is done four times during the evolution (in generations $G/4$, $G/2$, $3G/4$ and G, where G is the maximum number of generations).

3.4   CRO Algorithms for Time Series Approximation

In this section, we describe how to adapt the CRO algorithms to the specific problem of the time series approximation.

Coral Encoding: each coral $\mathbf{c}_i$ is an array of binary values with length N. That is, each position l of the coral stores a 1 if it represents a cut point, a 0 otherwise. In this way, $\sum_{l=1}^{N} c_{i,l} = K - 1$.

Initialization: the coral is initialised randomly with a predefined number of cut points and with a percentage F_{free} of free positions in the coral.

External Sexual Reproduction: the algorithm randomly selects pairs of parents without replacement from a F_b percentage of the reef. For each pair of parents (c_1, c_2), the algorithm chooses a time index (t_x). Then, those positions $t_i \geq t_x$ where $c_{1,t_i} = 1 \wedge c_{2,t_i} = 0$ are interchanged, and also those where $c_{1,t_i} = 0 \wedge c_{2,t_i} = 1$. However, in order to maintain the same number of cut points in the offspring, the maximum number of interchanges allowed is the minimum between the number of positions where the first condition is fulfilled and the number of positions where the second one is fulfilled.

Internal Sexual Reproduction: $1 - F_b$ corals are mutated by an operator which consists in performing a cyclic rotation of the coral to the left or to the right.

Local Search: the solutions are optimised with a local search, based on the combination of two well-known traditional algorithms, Bottom-Up and Top-Down. Bottom-Up consists in merging the two adjacent segments with the lowest cost in each iteration, while Top-Down starts from one single segment and iteratively partitions the time series using the point that produces the highest decrease of error. We use a local search method which removes a percentage of cut points using the Bottom-up strategy and then adds the same number of cut points with the Top-Down process.

Fitness Function: the fitness function is based on the minimization of approximation error. The error for the l-th point of the i-th coral is:

$$e_l(\mathbf{c}_i) = (y_l - \hat{y}_l(\mathbf{c}_i)), \tag{1}$$

where y_l is the real value of the time series, and $\hat{y}_l$ is the approximation resulting from the interpolation coded in individual $\hat{y}_l(\mathbf{c}_i)$. Then, the root mean squared error is calculated:

$$\mathrm{RMSE}(\mathbf{c}_i) = \sqrt{\frac{1}{N} \sum_{l=1}^{N} e_l^2(\mathbf{c}_i)}. \tag{2}$$

To have a maximization problem, the fitness function is calculated as $f_i = 1/(1 + \mathrm{RMSE}(\mathbf{c}_i))$, $f_i \in (0, 1]$. It is important to mention that, in this paper, we consider different computational tricks from a related problem (optimal polygonal approximation [20]), such as precomputing different terms needed for obtaining the error of each candidate segment. These tricks greatly improve the computational time of the algorithm with respect [7].

4   Experiments

This section describes in detail the time series used, experiments performed and the statistical discussion of the results obtained.

4.1   Time Series

In our experiments, we have used 15 time series with different length and collected from a variety of sources. The time series are obtained from different areas of application: cardiology data [13] (Arrhythmia), wave height time series [10] (B41043, B41044, B46001 and B46075), financial data (BBVA, DEUTSCHE, IBEX, SAN PAOLO and SO_GENERAL, available at https://es.finance.yahoo. com/), and synthetic time series previously used for classification tasks [3] (Donoho-Johnstone, Hand Outlines, Mallat, Phoneme and StarLightCurves). The length of the time series varies from 2048 points of Donoho-Johnstone to 9000 points of Arrhythmia time series.

4.2   Experimental Setting

We evaluate the performance of the proposed MCRO algorithm against CRO, SCRO and their hybrid versions HCRO and HSCRO, respectively. CRO, HCRO and the proposed MCRO are configured with the same parameters than in [19], while the SCRO and HSCRO do not need specific parameters to be adjusted. All algorithms are run 30 times with different seeds due to their stochasticity. The population size (size of the coral reef) is set to 100, while the stop condition is a maximum number of evaluations. This number of evaluations is established based on the length of each time series, N, by considering the value $3.5N$. The results are evaluated in error terms (RMSE), and a posterior statistical test is used to check the existence of differences.

4.3   Results and Discussion

The results are shown in Table 1, where each column represents the mean and the standard deviation of an algorithm in the 30 runs for each dataset. The datasets are sorted alphabetically. The mean rank for each algorithm is also included considering a 1 for the best algorithm in each dataset and a 5 for the worst one. As can be seen, the hybrid versions, HCRO and HSCRO, are better than their evolutionary versions (improving the mean rank from 4.333 and 4.667 to 2.433 and 2.567 respectively), which shows that the local search (combination of the Bottom-Up and Top-Down strategies) is suitable for time series approximation. Analysing all the algorithms, the proposed MCRO outperforms the rest of methods in all databases, considerably reducing the error. Furthermore, the lowest standard deviations are also obtained by MCRO, which shows the robustness of the method and that the results do not depend on the initialization. The second best performing methods are HCRO and HSCRO. However, HSCRO does not involve the adjustment of specific parameters. These results agree with those in [7], where it was statistically validated that there are no differences between these algorithms.

To determine the existence of differences between the algorithms, two statistical tests have been used. First, a Friedman test [9] has been run using the RMSE rankings. With a level of significance $\alpha = 0.05$, the test rejects the null-hypothesis

Table 1. Error approximation results (RMSE) and mean rankings ($\bar{r}$) for all the algorithms

Algorithm	CRO (Mean ± SD)	HCRO (Mean ± SD)	SCRO (Mean ± SD)	HSCRO (Mean ± SD)	MCRO (Mean ± SD)
Arrhythmia	0.084 ± 0.003	*0.051 ± 0.002*	0.083 ± 0.004	*0.051 ± 0.002*	**0.036 ± 0.001**
B41043	0.450 ± 0.009	0.387 ± 0.006	0.451 ± 0.007	*0.386 ± 0.006*	**0.347 ± 0.004**
B41044	0.453 ± 0.008	*0.378 ± 0.008*	0.452 ± 0.006	0.380 ± 0.007	**0.341 ± 0.004**
B46001	1.088 ± 0.012	*0.971 ± 0.009*	1.091 ± 0.010	0.973 ± 0.007	**0.906 ± 0.007**
B46075	1.132 ± 0.011	*1.016 ± 0.009*	1.138 ± 0.011	1.019 ± 0.008	**0.949 ± 0.009**
BBVA	0.382 ± 0.012	0.313 ± 0.006	0.382 ± 0.013	*0.312 ± 0.006*	**0.278 ± 0.004**
DEUTSCHE	2.275 ± 0.083	*1.840 ± 0.041*	2.292 ± 0.075	1.842 ± 0.052	**1.630 ± 0.018**
Donoho-Johnstone	2.779 ± 0.058	*2.508 ± 0.061*	2.812 ± 0.079	2.529 ± 0.056	**2.322 ± 0.024**
Hand Outlines	0.018 ± 0.002	*0.006 ± 0.000*	0.017 ± 0.002	*0.006 ± 0.000*	**0.004 ± 0.000**
IBEX	262.688 ± 8.176	203.276 ± 6.292	263.200 ± 8.447	*201.785 ± 4.106*	**176.267 ± 1.575**
Mallat	0.270 ± 0.012	0.159 ± 0.007	0.270 ± 0.016	*0.158 ± 0.007*	**0.101 ± 0.002**
Phoneme	0.974 ± 0.012	*0.882 ± 0.007*	0.981 ± 0.014	0.883 ± 0.008	**0.833 ± 0.006**
SAN PAOLO	0.130 ± 0.005	*0.106 ± 0.002*	0.131 ± 0.004	0.107 ± 0.003	**0.094 ± 0.001**
SO Genéralé	2.580 ± 0.082	2.100 ± 0.044	2.531 ± 0.072	*2.084 ± 0.043*	**1.849 ± 0.024**
StarLightCurves	0.051 ± 0.005	*0.023 ± 0.001*	0.054 ± 0.005	*0.023 ± 0.001*	**0.017 ± 0.000**
Mean Rankings ($\bar{r}$)	4.333	2.433	4.667	2.567	1.000

Bold face and *italics* for the best and the second methods, respectively.

that states that the differences are not significant, with a confidence interval of $C_0 = (0, F_{0.05} = 2.54)$ and a F-distribution statistical value $F^* = 22.19$. Consequently, the choice of the algorithm is a statistically significant factor. Based on this rejection, we use the Holm post-hoc test to compare the five algorithms to each other. Holm test is a multiple comparison procedure that considers a control algorithm (CA), in this case, MCRO, and compares it with the remaining methods (for more information see [5]). The Holm test results for $\alpha = 0.05$ can be seen in Table 2, using the corresponding p and α^*_{Holm} values. From the results of this test, it can be concluded that the MCRO algorithm obtains a significantly higher ranking of RMSE when compared to the remaining two algorithms, which justifies the proposal.

Table 2. Results of the Holm test using MCRO as control algorithm (CA) when comparing its average RMSE to those of CRO, HCRO, SCRO, and HSCRO: corrected α values, compared methods and p-values, all of them ordered by the number of comparison (i). CA results statistically better than the compared algorithm are marked with (*).

i	CA:MCRO $\alpha^*_{0.05}$	RMSE Algorithm	p_i
1	0.013	SCRO	0.000 (*)
2	0.017	CRO	0.000 (*)
3	0.025	HSCRO	0.007 (*)
4	0.050	HCRO	0.013 (*)

5   Conclusions

In this paper, we have proposed a new memetic coral reef optimization algorithm (MCRO) for size reduction of time series with minimum approximation error. The algorithm finds a set of indexes for performing linear interpolations between every two consecutive points. The memetic strategy is based on performing a local search at the beginning of the evolution and repeating the local search in specific generations. Two-well known algorithms (Bottom-Up and Top-Down) are used for the local search. The proposed methodology has been tested on 15 time series collected from different sources, and it has been compared against other state-of-the-art CRO algorithms, such as CRO, HCRO, SCRO, HSCRO.

The results show that the algorithm with the lowest information loss is the proposed MCRO, reducing the error drastically with respect to the other algorithms. Also, the standard deviation of MCRO is the lowest one in almost all datasets. Finally, a statistical test corroborates the existence of statistical differences between MCRO and the rest of algorithms.

Future research includes the adaptation of the different CRO algorithms to other tasks, such as numerical or real functions minimization.

References

1. Chakrabarti, K., Keogh, E., Mehrotra, S., Pazzani, M.: Locally adaptive dimensionality reduction for indexing large time series databases. ACM Trans. Datab. Syst. (TODS) **27**(2), 188–228 (2002)
2. Chen, M.Y., Chen, B.T.: A hybrid fuzzy time series model based on granular computing for stock price forecasting. Inf. Sci. **294**, 227–241 (2015)
3. Chen, Y., et al.: The UCR time series classification archive, July 2015. www.cs.ucr.edu/~eamonn/time_series_data/
4. Chung, F.L., Fu, T.C., Ng, V., Luk, R.W.: An evolutionary approach to pattern-based time series segmentation. IEEE Trans. Evol. Comput. **8**(5), 471–489 (2004)
5. Demšar, J.: Statistical comparisons of classifiers over multiple data sets. J. Mach. Learn. Res. **7**(Jan), 1–30 (2006)
6. Durán-Rosal, A., Hervás-Martínez, C., Tallón-Ballesteros, A., Martínez-Estudillo, A., Salcedo-Sanz, S.: Massive missing data reconstruction in ocean buoys with evolutionary product unit neural networks. Ocean Eng. **117**, 292–301 (2016)
7. Durán-Rosal, A.M., Gutiérrez, P.A., Salcedo-Sanz, S., Hervás-Martínez, C.: A statistically-driven coral reef optimization algorithm for optimal size reduction of time series. Appl. Soft Comput. **63**, 139–153 (2018)
8. Durán-Rosal, A.M., Gutiérrez, P.A., Martínez-Estudillo, F.J., Hérvas-Martínez, C.: Simultaneous optimisation of clustering quality and approximation error for time series segmentation. Inf. Sci. **442**, 186–201 (2018)
9. Friedman, M.: A comparison of alternative tests of significance for the problem of m rankings. Ann. Math. Stat. **11**(1), 86–92 (1940)
10. National Buoy Data Center: National Oceanic and Atmospheric Administration of the USA (NOAA) (2015). http://www.ndbc.noaa.gov/
11. Keogh, E., Chu, S., Hart, D., Pazzani, M.: Segmenting time series: a survey and novel approach. In: Data mining in time series databases, pp. 1–21 (2004)

12. Martínez-Estudillo, A.C., Hervás-Martínez, C., Martínez-Estudillo, F.J., García-Pedrajas, N.: Hybridization of evolutionary algorithms and local search by means of a clustering method. IEEE Trans. Syst. Man Cybern. Part B (Cybern.) **36**(3), 534–545 (2005)
13. Moody, G., Mark, R.: The impact of the MIT-BIH arrhythmia database. Eng. Med. Biol. Mag. IEEE **20**(3), 45–50 (2001)
14. Nikolaou, A., Gutiérrez, P.A., Durán, A., Dicaire, I., Fernández-Navarro, F., Hervás-Martínez, C.: Detection of early warning signals in paleoclimate data using a genetic time series segmentation algorithm. Clim. Dyn. **44**(7–8), 1919–1933 (2015)
15. Pérez-Ortiz, M., et al.: On the use of evolutionary time series analysis for segmenting paleoclimate data. Neurocomputing. (2017, in Press)
16. Salcedo-Sanz, S.: A review on the coral reefs optimization algorithm: new development lines and current applications. Prog. Artif. Intell. **6**, 1–15 (2017)
17. Salcedo-Sanz, S., Del Ser, J., Landa-Torres, I., Gil-López, S., Portilla-Figueras, A.: The coral reefs optimization algorithm: an efficient meta-heuristic for solving hard optimization problems. In: Proceedings of the 15th International Conference on Applied Stochastic Models and Data Analysis (ASMDA2013), Mataró, pp. 751–758 (2013)
18. Salcedo-Sanz, S., Del Ser, J., Landa-Torres, I., Gil-López, S., Portilla-Figueras, J.: The coral reefs optimization algorithm: a novel metaheuristic for efficiently solving optimization problems. Sci. World J. **2014** (2014)
19. Salcedo-Sanz, S., Sanchez-Garcia, J.E., Portilla-Figueras, J.A., Jimenez-Fernandez, S., Ahmadzadeh, A.M.: A coral-reef optimization algorithm for the optimal service distribution problem in mobile radio access networks. Trans. Emerg. Telecommun. Technol. **25**(11), 1057–1069 (2014)
20. Salotti, M.: An efficient algorithm for the optimal polygonal approximation of digitized curves. Pattern Recognit. Lett. **22**(2), 215–221 (2001)
21. Zellner, A., Palm, F.: Time series analysis and simultaneous equation econometric models. J. Econom. **2**(1), 17–54 (1974)
22. Zhao, G., Wang, X., Niu, Y., Tan, L., Zhang, S.X.: Segmenting brain tissues from Chinese visible human dataset by deep-learned features with stacked autoencoder. BioMed Res. Int. **2016**, 12 (2016)

An Improvement Study of the Decomposition-Based Algorithm Global WASF-GA for Evolutionary Multiobjective Optimization

Sandra González-Gallardo[1], Rubén Saborido[2(✉)], Ana B. Ruiz[3], and Mariano Luque[3]

[1] Programa de Doctorado en Economía y Empresa, Universidad de Málaga,
C/Ejido 6, 29071 Málaga, Spain
sandragg@uma.es
[2] Department of Computer Science & Software Engineering, Concordia University,
1455 De Maisonneuve Blvd West, Montreal, QC H3G 1M8, Canada
rsain@uma.es
[3] Department of Applied Economics (Mathematics), Universidad de Málaga,
C/Ejido 6, 29071 Málaga, Spain
{abruiz,mluque}@uma.es

Abstract. The convergence and the diversity of the decomposition-based evolutionary algorithm Global WASF-GA (GWASF-GA) relies on a set of weight vectors that determine the search directions for new non-dominated solutions in the objective space. Although using weight vectors whose search directions are widely distributed may lead to a well-diversified approximation of the Pareto front (PF), this may not be enough to obtain a good approximation for complicated PFs (discontinuous, non-convex, etc.). Thus, we propose to dynamically adjust the weight vectors once GWASF-GA has been run for a certain number of generations. This adjustment is aimed at re-calculating some of the weight vectors, so that search directions pointing to overcrowded regions of the PF are redirected toward parts with a lack of solutions that may be hard to be approximated. We test different parameters settings of the dynamic adjustment in optimization problems with three, five, and six objectives, concluding that GWASF-GA performs better when adjusting the weight vectors dynamically than without applying the adjustment.

Keywords: Evolutionary multiobjective optimization
Decomposition-based algorithm · GWASF-GA · Weight vector

1 Introduction

In general, *multiobjective optimization problems* (MOPs) can be defined as:

$$\text{minimize} \quad \{f_1(\mathbf{x}), f_2(\mathbf{x}), \ldots, f_k(\mathbf{x})\}$$
$$\text{subject to} \ \mathbf{x} \in S, \tag{1}$$

© Springer Nature Switzerland AG 2018
F. Herrera et al. (Eds.): CAEPIA 2018, LNAI 11160, pp. 219–229, 2018.
https://doi.org/10.1007/978-3-030-00374-6_21

where $\mathbf{x} = (x_1, x_2, \ldots, x_n)^T$ is the vector of *decision variables*, $S \subset \mathbb{R}^n$ is the *feasible set*, and $f_i : S \rightarrow \mathbb{R}$ for $i = 1, \ldots, k$ ($k \geq 2$) are the *objective functions* to be minimized. All *objective vectors* $\mathbf{f}(\mathbf{x})$ form the *objective set* $Z = \mathbf{f}(S) \subset \mathbb{R}^k$.

In these problems, no solution is available that minimizes all the objectives at the same time, but there is a set of Pareto optimal solutions at which no objective can be improved without worsening, at least, one of the others. Given $\mathbf{y}, \bar{\mathbf{y}} \in Z$, we say that $\mathbf{y}$ *dominates* $\bar{\mathbf{y}}$ if $y_i \leq \bar{y}_i$ for all $i = 1, \ldots, m$, with, at least, one strict inequality. Then, a solution $\mathbf{x} \in S$ is *Pareto optimal* if there does not exist another $\bar{\mathbf{x}} \in S$ so that $\mathbf{f}(\bar{\mathbf{x}})$ dominates $\mathbf{f}(\mathbf{x})$. All Pareto optimal solutions form the *Pareto optimal set* (PS) in the decision space, denoted by E, and the *Pareto optimal front* (PF) in the objective space, denoted by $\mathbf{f}(E)$.

Evolutionary multiobjective optimization (EMO) has been proven to be a good methodology to solve MOPs [1]. Among them, decomposition-based EMO algorithms scalarize the original MOP into a set of single objective optimization problems. The Global Weighting Achievement Scalarizing Function Genetic Algorithm (GWASF-GA) [6] is one of these algorithms, in which an achievement scalarizing function (ASF) [7] based on the Tchebychev distance is used as fitness function to classify the individuals into different fronts. For this classification, two reference points are simultaneously considered: the nadir and the utopian points. The *nadir point* $\mathbf{z}^{\text{nad}} = (z_1^{\text{nad}}, \ldots, z_k^{\text{nad}})^T$ is defined by the worst possible objective function values, i.e. $z_i^{\text{nad}} = \max_{\mathbf{x} \in E} f_i(\mathbf{x})$ ($i = 1, \ldots, k$). To calculate the *utopian point* $\mathbf{z}^{\star\star} = (z_1^{\star\star}, \ldots, z_k^{\star\star})^T$, we need the *ideal point* $\mathbf{z}^{\star} = (z_1^{\star}, \ldots, z_k^{\star})^T$, defined as $z_i^{\star} = \min_{\mathbf{x} \in E} f_i(\mathbf{x}) = \min_{\mathbf{x} \in S} f_i(\mathbf{x})$ ($i = 1, \ldots, k$). Thus, the components of the utopian point are obtained as $z_i^{\star\star} = z_i^{\star} - \epsilon_i$ ($i = 1, \ldots, k$), where $\epsilon_i > 0$ is a small real value. At each generation of GWASF-GA, individuals are classified according to their ASF values calculated using both the nadir and the utopian points, but also depending on a set of weight vectors that determine the search directions for new non-dominated solutions towards the PF (due to the mathematical properties of the ASF [7]). Thus, the weight vectors should preferably define uniformly distributed search directions in the objective space.

Promising results have been provided by GWASF-GA [6]. However, in cases where the PF is discontinuous or has a complex shape, GWASF-GA may not be able to properly approximate the PF, e.g. if different weight vectors are generating similar solutions because they are directing the search towards the same region of the PF. To amend this, we propose to improve the performance of GWASF-GA by dynamically adjusting the weight vectors. The idea is to re-direct some of the search directions taking into account the distribution of the current solutions, so that weight vectors generating solutions in overcrowded areas of the PF are replaced by new ones directing the search towards regions where there are not enough solutions. Since this improvement depends on several parameters, we carry out an experimental study to gain knowledge about their impact on the performance of GWASF-GA when solving several benchmark MOPs.

Next, we motivate the performance improvement suggested for GWASF-GA in Sect. 2 and we describe our proposal in Sect. 3. Section 4 presents and discusses the experiments designed to study the benefits of adjusting the weight vectors and, finally, Sect. 5 concludes this work.

2   Motivation

Minimizing the ASF proposed in [7] over the feasible set means, in practice, to project the reference point used onto the PF in the projection direction given by the inverse components of the weight vector. Owing to this, the weight vectors are key parameters in GWASF-GA since they set the search directions for new non-dominated solutions, either from the nadir or from the utopian point. Thus, they affect the diversity and the convergence of GWASF-GA and, according to [6], the weight vectors used must define projection directions as evenly distributed as possible. Besides, since the front classification is based on the ASF values, each individual in the population produced at each generation is associated either with the nadir or with the utopian point, and with a weight vector.

In practice, a set of weight vectors producing uniformly spread projection directions of the nadir or of the utopian point allows a well-diversified population to be obtained just in case the PF is not complex, such as e.g. a hyper-plane (Fig. 1 (a)). In case the PF is discontinuous (Fig. 1 (b)), the same weight vectors may be generating very similar (i.e. very close) non-dominated solutions, because their projection directions may be pointing towards the same area of the PF, or may be even directing the search for new solutions towards gaps in the PF. This fact does not contribute to the convergence of the algorithm, having areas of the PF where there are not enough solutions.

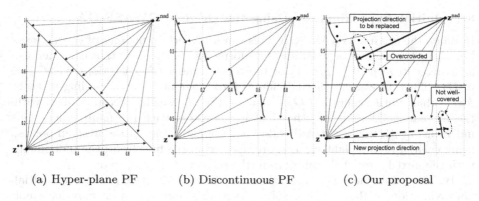

| (a) Hyper-plane PF | (b) Discontinuous PF | (c) Our proposal |

Fig. 1. Projection directions from $\mathbf{z}^{nad}$ and $\mathbf{z}^{**}$ in GWASF-GA.

According to this, to improve the quality (convergence and diversity) of the approximation generated by GWASF-GA, we suggest to re-calculate some of the weight vectors once GWASF-GA has generated a initial approximation of the PF. The new weight vectors generated should be able to fit to the features of the

PF, such as e.g. discontinuities, convexity, etc. To achieve this, one possibility is to re-distribute the weight vectors according to the solutions in the current population. These individuals have survived along the generations and, therefore, are assumed to be the best individuals found so far to approximate the PF.

After performing a number of generations, we suggest to re-orientate the search for new individuals in GWASF-GA as shown in Fig. 1(c). According to the distribution of the solutions in the current population, our proposal is to dynamically replace the weight vectors whose projection directions point towards overcrowded parts (where there are more solutions), by weight vectors helping to approximate regions that are not so well-covered. With this, we try to diversify the solutions generated, helping the algorithm to converge towards regions where the density of solutions is not very high at the moment and that may be hard to be approximated. The convexity of the PF is also considered somehow when adjusting the weight vectors, since more importance is likely to be given to the use of the utopian point if the PF is concave, or of the nadir point if it is convex.

3   Improvement of GWASF-GA Through a Dynamic Adjustment of the Weight Vectors

Next, we describe how to enhance the performance of GWASF-GA. Let G_T denote the total number of generations to be performed. Initially, a representative set W of weight vectors in $[0,1]^k$ is considered. They are generated as suggested in [6] (pp. 321–323) so that the projection directions they define towards the PF are evenly distributed. Let us refer to the number of weight vectors in W as N_μ.

3.1   Initial Approximation Using the Original GWASF-GA

Initially, the original GWASF-GA [6] is run for a certain number of generations. Let G_p be the number of generations performed so. We assume that G_p is defined as a percentage p ($0 < p \leq 1$) of the total number of generations, i.e. $G_p = p \cdot G_T$. Thus, to classify the individuals into different fronts according to their ASF values until generation G_p, a half of the N_μ weight vectors is used with the utopian point and the other half is associated with the nadir point. The individuals with the lowest ASF values for the nadir and for the utopian points, using each of their weight vectors, are selected in the first front; the individuals with the next lowest ASF values form the second front, and so on.

By executing these G_p generations, we let GWASF-GA produce an initial approximation of the PF. The solutions in this approximation may have not converged closely enough to the PF but their distribution gives us an initial insight of the true PF, although regions complicated to converge to may have not still been well-covered at this generation. Thus, the value of G_p should be an intermediate value: if it is very small, the solutions may be still far from the PF and the new weight vectors may not adapt well the convergence to the true shape of the PF; but if G_p is too close to G_T, there may not be enough generations to improve the approximation with the new proposal before the process stops.

3.2 Dynamic Weight Vectors' Adjustment

The adjustment of the weight vectors is aimed at re-orientating some of the projection directions used to converge to the PF. To this aim, as previously explained, useful information about the shape of the PF can be extracted from the distribution of the solutions generated so far. Remind that when, at each generation of GWASF-GA, the individuals which minimize the ASF for each of the weight vectors are selected, in practice, the new approximation gets close to the PF in the search directions given by the inverse components of the weight vectors. Thus, by changing some of the weight vectors as described hereafter, new solutions will be found taking into account new search directions created according to the PF's shape known until this generation.

Let N_a be the number of weight vectors to be re-adapted ($N_a < N_\mu/2$) and $G_a = (1 - p) \cdot G_T$ the number of generations that are left for the dynamical adjustment (i.e. $G_p + G_a = G_T$). In the next G_a generations, we perform n_a adjustments of the weight vectors (with $n_a \in \{1, \dots, G_a\}$). This implies that some of the weight vectors are modified each $E(G_a/n_a)$ generations, starting at generation G_p (where $E(\cdot)$ denotes the integer part of a number). That is, if G_a^r denotes the generation at which the r-th adjustment is done ($r = 1, \dots, n_a$), we have $G_a^r = G_p + (r - 1) \cdot E(G_a/n_a)$. Note that the r-th adjustment of the weight vectors is done as explained hereafter once the generation G_a^r has concluded and before running the generation $G_a^r + 1$. For each $r = 1, \dots, n_a$, let P_a^r denote the population generated at generation G_a^r and W_a^r the new set of weight vectors obtained at the r-th adjustment. Next, we describe the procedure to identify the weight vectors that will be replaced by new ones.

Firstly, at the r-th adjustment, we initialize $W_a^r = W_a^{r-1}$ (with $W_a^1 = W$). To detect which regions of the PF are well-covered by solutions and which are not (according to the current population P_a^r), we need a measurement of the diversity around each individual (in the objective space). For this, we define the *scattering level* of each individual $\mathbf{x} \in P_a^r$, and denote it by $s(\mathbf{x})$, as follows:

$$s(\mathbf{x}) = \prod_{j=1}^k L_2(\mathbf{f}(\mathbf{x}), \mathbf{f}(\mathbf{x}^j)), \qquad (2)$$

where $\mathbf{x}^1, \dots, \mathbf{x}^k \in P_a^r$ are the k solutions with the closest objective vectors to $\mathbf{f}(\mathbf{x})$ regarding the L_2-distance. The objective vectors are assumed to be normalized to avoid scale problems. The higher (respectively, the lower) $s(\mathbf{x})$ is, the less (respectively, the more) crowded is the region where $\mathbf{f}(\mathbf{x})$ lyes. Thus, we can identify areas of the PF that are poorly approximated (with a lack of solutions) by means of the solutions with the highest scattering level, and overcrowded regions as these containing the solutions with the lowest scattering level.

Secondly, the N_a solutions in P_a^r with the lowest scattering level are selected, and we denoted them by $\{\bar{\mathbf{x}}^1, \dots, \bar{\mathbf{x}}^{N_a}\}$. It can be understood that there are close enough individuals of P_a^r around each $\bar{\mathbf{x}}^j$ in the objective space ($j = 1, \dots, N_a$), meaning that the area of the PF where $\mathbf{f}(\mathbf{x})$ lyes has been covered enough at generation G_a^r in comparison to other regions. As previously said, each solution in GWASF-GA is associated with a weight vector, as well as with the nadir or with the utopian point. Then, we can consider that the weight vector corresponding to each $\bar{\mathbf{x}}^j$ is directing the search towards an overcrowded area of the PF, where other weight vectors are also orientating the search towards (those of the solutions around $\bar{\mathbf{x}}^j$). In view of this, the N_a weight vectors corresponding to the solutions $\{\bar{\mathbf{x}}^1, \ldots, \bar{\mathbf{x}}^{N_a}\}$ are the candidates to be replaced by new weight vectors which point towards the least crowded areas (according to P_a^r). Thus, they are removed from W_a^r, in which only $N_\mu - N_a$ weight vectors are left.

Thirdly, to generate the new N_a weight vectors, we identify the N_a solutions in P_a^r with the highest scattering level, referred to as $\{\hat{\mathbf{x}}^1, \ldots, \hat{\mathbf{x}}^{N_a}\}$. The objective vectors of these solutions enable the search to be focused on parts that have not been well-approximated by the population P_a^r, since the new weight vectors $\mu^j = (\mu_1^j, \ldots, \mu_k^j)$ $(j = 1, \ldots, N_a)$ to be introduced into W_a^r are calculated as:

Case (a): If the solution $\hat{\mathbf{x}}^j$ has been obtained in the front classification process as one minimizing the ASF with the utopian point $\mathbf{z}^{\star\star}$, then:

$$\mu_i^j = \tfrac{1}{f_i(\hat{\mathbf{x}}^j) - z_i^{\star\star}} \quad \text{for each } i = 1, \ldots, k. \tag{3}$$

Case (b): In case $\hat{\mathbf{x}}^j$ has been selected in the front classification process as one minimizing the ASF with the nadir point $\mathbf{z}^{\text{nad}}$, then:

$$\mu_i^j = \tfrac{1}{z_i^{\text{nad}} - f_i(\hat{\mathbf{x}}^j)} \quad \text{for each } i = 1, \ldots, k. \tag{4}$$

Once all the new weight vectors are incorporated to W_a^r, new generations of GWASF-GA are run using W_a^r as the set of weight vectors, until the next weight vectors' adjustment needs to be carried out at generation G_a^{r+1}. Note that, to classify the individuals into different fronts, each new μ^j is considered to select individuals with the lowest ASF values either with the utopian point (case (a)) or with the nadir point (case (b)), depending on the way it has been calculated.

The main steps of the proposed improvement for GWASF-GA are in Algorithm 1. Observe that, after the first adjustment, the number of weight vectors used in the ASF for the utopian point is no longer equal to that for the nadir point, so as in the original GWASF-GA. In case the PF is convex, it is likely that a larger amount of weight vectors is automatically assigned to the nadir point, giving more importance to the projection from it. But if it is concave, more new weight vectors may be likely to be associated with the utopian point, meaning that its projection is more suitable.

Algorithm 1. General scheme of the weight vectors' adjustment in GWASF-GA

1: Generate N_μ weight vectors following [6] to form the set W.
2: Set $G_p = p \cdot G_T$, $G_a = (1 - p) \cdot G_T$, and create an initial population P^0 of N individuals.
3: Initialize the adjustment counter $r = 1$.
4: **for** $h = 1, \ldots, G_T$ **do**
5:     Run a new generation h of GWASF-GA to obtain P^h, using the set of weight vectors W.
6:     **if** $h \geq G_p$ and $h = G_p + (r - 1) \cdot E(G_a/n_a)$ **then**
7:         If $h = 1$, initialize $W_a^h = W$. Otherwise, $W_a^r = W_a^{r-1}$.
8:         For every $\mathbf{x} \in P^h$, calculate the scattering level $s(\mathbf{x})$ according to (2).
9:         Let $\{\bar{\mathbf{x}}^1, \ldots, \bar{\mathbf{x}}^{N_a}\}$ be the solutions of P^h with the N_a lowest values of $s(\mathbf{x})$.
10:        For each $j = 1, \ldots, N_a$, delete from W_a^r the weight vector associated to $\bar{\mathbf{x}}^j$.
11:        Let $\{\hat{\mathbf{x}}^1, \ldots, \hat{\mathbf{x}}^{N_a}\}$ be the solutions of P^h with the N_a highest values of $s(\mathbf{x})$.
12:        For each $j = 1, \ldots, N_a$, calculate a new weight vector μ^j according to (3) or (4), depending if $\hat{\mathbf{x}}^j$ has been obtained using $\mathbf{z}^{**}$ or $\mathbf{z}^{\mathrm{nad}}$, respectively. Update $W_a^r = W_a^r \cup \{\mu^j\}$.
13:        Update $r = r + 1$ and $W = W_a^r$.
14:     **end if**
15: **end for**

---

4 Experimental Study

The adjustment of the weight vectors depends on three parameters: the number of adjustments to be performed (n_a), the percentage of generations run before the first adjustment is applied (p), and the number of weight vectors to be adapted (N_a). Next, we analyze different settings for these parameters.

4.1 Experimental Design

A total of 46 test problems are used: 18 with three objectives (DTLZ1-DTLZ4 and DTLZ7 [2], WFG1-9 [3], UF8-10 [9], and LZ09 [4]) and 14 with five and six objectives, respectively (DTLZ1-DTLZ4 and DTLZ7 [2] and WFG1-9 [3]). We set the number of decision variables to $k+4$ for DTLZ1, $k+9$ for DTLZ2-DTLZ4, and $k + 19$ for DTLZ7 (k is the number of objectives). For the WFG problems, the position- and distance-related parameters are $k - 1$ and 10, respectively.

In our analysis, we use a fractional factorial design to investigate the impact of n_a, p, and N_a on the performance of GWASF-GA. We analyze the approximations generated by GWASF-GA when the weight vectors are adjusted using all possible combinations of the following values: $n_a \in \{2, 4, 6\}$, $p \in \{0.6, 0.7, 0.8\}$, and $N_a \in \{5, 20, 25, 30, 50, 60, 75, 100\}$.[1] Thus, we study 72 different configurations to dynamically adjust the weight vectors and we compare their performance against the original GWASF-GA (i.e. without weight vectors' adjustment).

For each algorithm (meaning each version of GWASF-GA with each dynamic adjustment configuration, in addition to the original GWASF-GA), 30 independent runs are executed for each test problem, which implies more than 100,000 runs ($30 \times (72 + 1) \times (18 + 14 + 14)$). In all cases, we use the same evolutionary parameters: 300 individuals (i.e. $N_\mu = 300$ weight vectors), 3,000 generations,

---

[1] These values reported the best results after performing several initial tests. The results are not included due to space limitations, but they are available upon request.

the SBX crossover operator with a distribution index $\eta_c = 20$ and a probability $P_c = 0.9$, and the polynomial mutation operator with a distribution index $\eta_m = 20$ and a probability $P_m = 1/n$, where n is the number of variables.

To run the experiments, we use the implementations of GWASF-GA and of the test problems available in jMetal [5], an object-oriented Java-based framework for multiobjective optimization using metaheuristics. We conduct our experiments in a cluster of 21 computers offering a total of 172 cores and 190 GB of memory. The cluster is managed by HTCondor, a specialized workload management system for compute-intensive jobs.[2] We use HTCondor because it provides a job queuing mechanism, scheduling policy, and resource management that allow users to submit parallel jobs to HTCondor.

4.2   Data Analysis

The hypervolume [10] has been used as performance metric, for which a representative set of the PF is required. For the DTLZ and WFG problems, we generate them using an open-source tool.[3] For the other problems, we use the representative sets available in jMetal. We also apply a Wilcoxon rank-sum test [8] to check if the hypervolume achieved by the original GWASF-GA is significantly different to that of its versions adjusting the weight vectors with each of the configurations considered. The null hypothesis is that the distribution of their hypervolume average values in the 30 runs differ by a value α, considering that the difference is significant if the obtained p-value is lower than $\alpha = 0.05$.[4]

4.3   Results

Table 2 reports the number of problems where each version of GWASF-GA with the weight vectors' adjustment performs significantly better than (▲) and worse than (▽) the original GWASF-GA. Note that the amount of problems in which the performance did not significantly differ can be known using the total number of problems. The column 'Configuration' shows the values of the parameters n_a, p, and N_a. The second, third, and fourth columns show the results obtained for the three-, five-, and six-objective problems, respectively ($k = 3, 5, 6$). In these columns, we highlight in gray color the configurations of the weight vectors' adjustment that perform better than the original algorithm in the highest number of cases, reaching an equal performance for the rest of them at the same time.

On average, the hypervolume achieved by GWASF-GA is higher when using the weight vectors' adjustment in 15/18 (81%), 11/14 (76%), and 10/14 (74%) of the cases, respectively, for three, five, and six objectives. In addition, for each group of problems, there always exists, at least, one configuration (or more)

---

[2] https://research.cs.wisc.edu/htcondor/index.html.

[3] https://github.com/rsain/Pareto-fronts-generation.

[4] We use the `wilcox.test` function from the R software available at https://stat.ethz.ch/R-manual/R-devel/library/stats/html/wilcox.test.html.

Configuration			$k = 3$ 18 prob.		$k = 5$ 14 prob.		$k = 6$ 14 prob.	
n_a	p	N_a	▲	▽	▲	▽	▲	▽
2	0.6	5	13	2	10	2	9	3
2	0.6	20	16	0	13	0	11	3
2	0.6	25	16	0	13	0	10	3
2	0.6	30	16	0	12	0	11	0
2	0.6	50	16	0	12	0	10	2
2	0.6	60	16	0	12	0	11	2
2	0.6	75	16	0	9	0	11	3
2	0.6	100	12	2	8	4	11	2
2	0.7	5	14	1	8	2	9	3
2	0.7	20	16	0	12	0	10	3
2	0.7	25	16	0	13	0	10	3
2	0.7	30	17	0	13	0	11	3
2	0.7	50	16	0	12	0	11	3
2	0.7	60	16	0	11	0	10	2
2	0.7	75	15	0	10	1	10	2
2	0.7	100	12	1	8	4	12	2
2	0.8	5	13	0	10	1	9	3
2	0.8	20	16	0	13	0	10	3
2	0.8	25	17	0	13	0	11	3
2	0.8	30	16	0	13	0	10	3
2	0.8	50	16	0	13	0	11	2
2	0.8	60	16	0	12	0	11	2
2	0.8	75	16	0	10	0	11	2
2	0.8	100	12	3	8	4	10	3
4	0.6	5	14	2	11	2	10	3
4	0.6	20	16	0	13	0	11	3
4	0.6	25	16	0	13	0	11	3
4	0.6	30	16	0	13	0	11	3
4	0.6	50	16	0	13	0	11	2
4	0.6	60	16	0	13	0	11	2
4	0.6	75	15	0	11	1	11	1
4	0.6	100	12	2	7	4	12	1
4	0.7	5	15	1	9	1	10	3
4	0.7	20	16	0	13	0	10	2
4	0.7	25	17	0	13	0	11	2
4	0.7	30	16	0	13	0	12	2
4	0.7	50	16	0	13	0	10	2
4	0.7	60	16	0	12	0	11	2
4	0.7	75	15	0	10	0	12	2
4	0.7	100	12	2	7	4	12	1
4	0.8	5	15	1	12	2	10	3
4	0.8	20	16	0	13	0	10	3
4	0.8	25	17	0	13	0	12	2
4	0.8	30	16	0	13	0	12	2
4	0.8	50	17	0	13	0	12	2
4	0.8	60	16	0	13	0	11	2
4	0.8	75	17	0	10	0	4	11
4	0.8	100	12	2	7	4	11	1

Configuration			$k = 3$ 18 prob.		$k = 5$ 14 prob.		$k = 6$ 14 prob.	
n_a	p	N_a	▲	▽	▲	▽	▲	▽
6	0.6	5	14	1	11	1	11	3
6	0.6	20	17	0	13	0	10	2
6	0.6	25	16	0	13	0	12	2
6	0.6	30	16	0	13	0	11	2
6	0.6	50	16	0	13	0	11	1
6	0.6	60	16	0	13	0	12	1
6	0.6	75	16	0	10	0	13	1
6	0.6	100	13	3	7	5	11	1
6	0.7	5	15	2	9	1	10	3
6	0.7	20	16	0	13	0	12	2
6	0.7	25	16	0	13	0	12	2
6	0.7	30	17	0	13	0	12	2
6	0.7	50	16	0	13	0	12	1
6	0.7	60	16	0	13	0	12	1
6	0.7	75	16	0	11	0	13	1
6	0.7	100	12	3	8	5	11	1
6	0.8	5	15	1	11	0	11	3
6	0.8	20	16	0	13	0	11	2
6	0.8	25	16	0	13	0	12	2
6	0.8	30	16	0	13	0	11	2
6	0.8	50	16	0	13	0	12	1
6	0.8	60	16	0	13	0	13	1
6	0.8	75	16	0	11	1	12	0
6	0.8	100	12	3	7	4	12	1

Fig. 2. Number of problems with three, five and six objectives ($k = 3, 5, 6$) where each version of GWASF-GA with the weight vectors' adjustment performs better (▲) and worse (▽) than the original GWASF-GA, for the hypervolume in the 30 runs.

that never performs worse than the original GWASF-GA. Therefore, we can conclude that adjusting the weight vectors enhance the outcome of GWASF-GA. In general, the best results in the test problems considered have been obtained when setting $n_a = 6$, $p = 0.7$, and $N_a \in \{30, 50, 60\}$ (for three-, five- and six-objective problems, respectively).

5   Conclusion

In this paper, we have proposed an improvement of GWASF-GA. This algorithm uses a set of weight vectors that determine the search directions for new non-dominated solutions and we have suggested to perform a dynamic adjustment of the weight vectors to enhance its convergence and diversity. Based on the population generated, we identify regions of the PF that are overcrowded and others that have been poorly approximated. Then, some weight vectors are replaced so that search directions pointing to overcrowded regions are re-directed towards parts with a lack of solutions that may be hard to be approximated.

The performance of our proposal depends on several parameters, such as the number of weigh vectors to be replaced, the generation at which the first adjustment is carried out, and the times the weight vectors are adjusted. In the experiments, we have tested different configurations for them to solve three-, five-, and six-objective optimization problems. According to the hypervolume, we have concluded that better results are provided when the weight vectors' adjustment is incorporated into GWASF-GA, in comparison to the original algorithm. In the future, we plan to compare our proposal to other EMO algorithms and using more test problems (such as ones with inverted PFs).

Acknowledgements. This research is funded by the Spanish government (ECO2017-88883-R and ECO2017-90573-REDT) and by the Andalusian regional government (SEJ-532). Sandra González-Gallardo has a technical research contract within "Sistema Nacional de Garantía Juvenil y del Programa Operativo de Empleo Juvenil 2014-2020 - Fondos FEDER", and thanks the University of Málaga PhD Programme in Economy and Business. Rubén Saborido is a post-doctoral fellow at Concordia University (Canada). Ana B. Ruiz thanks the post-doctoral fellowship "Captación de Talento para la Investigación" of the University of Málaga.

References

1. Coello, C.A.C., Lamont, G.B., Veldhuizen, D.A.V.: Evolutionary Algorithms for Solving Multi-Objective Problems, 2nd edn. Springer, New York (2007). https://doi.org/10.1007/978-1-4757-5184-0
2. Deb, K., Thiele, L., Laumanns, M., Zitzler, E.: Scalable multi-objective optimization test problems. In: Congress on Evolutionary Computation, pp. 825–830 (2002)
3. Huband, S., Hingston, P., Barone, L., While, L.: A review of multi-objective test problems and a scalable test problem toolkit. IEEE Trans. Evol. Comput. **10**(5), 477–506 (2007)
4. Li, H., Zhang, Q.: Multiobjective optimization problems with complicated Pareto sets, MOEA/D and NSGA-II. IEEE Trans. Evol. Comput. **12**(2), 284–302 (2009)
5. Nebro, A.J., Durillo, J.J., Vergne, M.: Redesigning the jMetal multi-objective optimization framework. In: Conference on Genetic and Evolutionary Computation, pp. 1093–1100 (2015)
6. Saborido, R., Ruiz, A.B., Luque, M.: Global WASF-GA: an evolutionary algorithm in multiobjective optimization to approximate the whole Pareto optimal front. Evol. Comput. **25**(2), 309–349 (2017)

7. Wierzbicki, A.P.: The use of reference objectives in multiobjective optimization. In: Fandel, G., Gal, T. (eds.) Multiple Criteria Decision Making, Theory and Applications, pp. 468–486. Springer, Heidelberg (1980). https://doi.org/10.1007/978-3-642-48782-8_32

8. Wilcoxon, F.: Individual comparisons by ranking methods. Biom. Bull. **1**(6), 80–83 (1945)

9. Zhang, Q., Zhou, A., Zhao, S., Suganthan, P.N., Liu, W., Tiwari, S.: Multiobjective optimization test instances for the CEC 2009 special session and competition. Technical report (CES-487, University of Essex and Nanyang Technological University) (2008)

10. Zitzler, E., Thiele, L.: Multiobjective evolutionary algorithms: a comparative case study and the strength Pareto approach. IEEE Trans. Evol. Comput. **3**(4), 257–271 (1999)

Reduction of the Size of Datasets by Using Evolutionary Feature Selection: The Case of Noise in a Modern City

Javier Luque, Jamal Toutouh[✉], and Enrique Alba

Departamento de Lenguajes y Ciencias de la Computación,
Universidad de Málaga, Málaga, Spain
javierluque@uma.es, {jamal,eat}@lcc.uma.es

Abstract. Smart city initiatives have emerged to mitigate the negative effects of a very fast growth of urban areas. Most of the population in our cities are exposed to high levels of noise that generate discomfort and different health problems. These issues may be mitigated by applying different smart cities solutions, some of them require high accurate noise information to provide the best quality of serve possible. In this study, we have designed a machine learning approach based on genetic algorithms to analyze noise data captured in the university campus. This method reduces the amount of data required to classify the noise by addressing a feature selection optimization problem. The experimental results have shown that our approach improved the accuracy in 20% (achieving an accuracy of 87% with a reduction of up to 85% on the original dataset).

Keywords: Smart city · Genetic algorithm · Feature selection · Noise

1 Introduction

Smart city has emerged as link of the different stakeholders of the cities to mitigate the negative effects of a very fast growth in urban areas [5]. With the raising of smart cities, countless new applications are appearing to improve our daily life [1,6,8,15,17]: smart parking, intelligent waste collection, intelligent traffic systems, efficient street lighting, etc. Most of these new applications use intelligent systems able to detect, predict, and efficiently manage different aspects of the city.

The high ambient noise level is an important problem in our present cities, because it causes discomfort, sleep disturbance, reduces cognitive performance, and is responsible of many disease [18]. As road traffic generates about 80% of the noise pollution [14], reducing its noise seems to be an efficient strategy to improve this aspect of the daily life. For this reason, there are many different approaches to measure and evaluate the ambient noise level in the roads [7,11, 16], e.g., people carrying a sound-meters to take temporal measures, installing wireless sensor networks (WSN) or, most recently, using smartphones of the

© Springer Nature Switzerland AG 2018
F. Herrera et al. (Eds.): CAEPIA 2018, LNAI 11160, pp. 230–239, 2018.
https://doi.org/10.1007/978-3-030-00374-6_22

inhabitants. Having a better knowledge about the sources of noise is vital to help city managers in taking better decisions.

In this work, we focus on the intelligent analysis of noise data captured by a cyber-physical system (WSN) installed in the University of Málaga's campus [16]. We want to characterize/label the noise level sensed during the day in order to identify the main noise source by using machine learning (i.e., *K-Means Clustering* classification method). In order to do so, we find the periods of time that efficiently classify the days in different groups (clusters) according to the noise levels for these periods. Thus, a feature selection method is applied to select these time periods, since this type of methods provide high competitive results in removing irrelevant data and improving the efficiency and accuracy of the application of machine learning methods [2].

When dividing the day is important to use time periods as small as possible to improve the accuracy of the evaluation of the sensed noise level. However, this critically increases the number of required periods to be evaluated. Therefore, it increases the number of studied features in our feature selection method. In order to deal with feature selection methods with large set of features, classic methods can not be applied. Thus, Genetic Algorithms (GA), as efficient tools to address *hard-to-solve* search and optimization problems, have been successfully applied to address feature selection achieving very competitive results [19,20].

In this study, we have divided the time line in blocks of 30 min. Thus, the size of the set of features is 48, that defines 2^{48} (2.81E14) possible subsets of features. In order to deal with this feature selection problem, we have applied a Genetic Algorithm (GA) to search for the best subset of features to efficiently classify the noise level information. Therefore, the main objective of this paper is to present this machine learning method based on GA and to use it to classify real noise data captured at the university campus.

The rest of this work is structured as follows: Sect. 2 introduces the noise analysis and presents the noise feature selection problem. Section 3 describes the evolutionary approach designed in this study to address the optimization problem. Section 4 presents the experimental analysis of our proposal. Finally, Sect. 5 outlines the conclusions and proposes the future work.

2 Smart Campus Efficient Noise Evaluation

In this section, we describe the process to capture noise data and we define the feature selection problem to efficiently classify the data.

2.1 Noise Measurement by the Smart Campus Sensing System

The noise level data analyzed in this study is gathered by the sensing system presented by Touotuh et al. [16]. These sensors measure the ambient noise and detect smart devices with a WiFi or Bluetooth connection. Their principal components are (see Fig. 1): two WiFi and a Bluetooth wireless interfaces, a Real

Time Clock (RTC), a noise sound meter, and a Raspberry Pi 3. The global architecture of the system is shown in Fig. 2, where the sensors send information to the data center via Internet by using a client/server model (see Fig. 3).

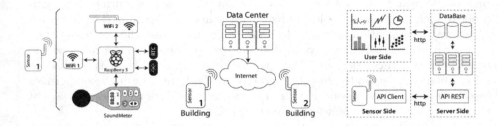

Fig. 1. Hardware scheme.   **Fig. 2.** Architecture.   **Fig. 3.** System model.

The noise sound meter captures the level of noise each second. The sound is evaluated in terms of *equivalent sound pressure level*, i.e. L_{eq}, which expresses the mean of the sound pressure perceived by an individual measured in decibels (dBa) [11] in an interval of time. In our case, L_{eq} is calculated for intervals of one minute. These measures are sent to the server to be stored in the database each hour. Finally, we generate the *daily noise curve* (see Fig. 4) by grouping all the noise level measures of a given day (from 0:00:00H to 23:59:59H).

2.2   The Noise Feature Selection Optimization Problem

The classic feature selection problem over a given dataset consists in selecting a subset of the most relevant data that allows us to characterize the original dataset without losing information [2]. In this work, we want to reduce the data required to accurately classify the daily noise level information.

As it is shown in Fig. 5, the noise levels follow three different patterns corresponding to: the *working days* (from Monday to Friday), *Saturdays*, and *holidays* (including Sundays). *K-Means Clustering* results with K = 3 confirmed that these daily noise levels may be grouped in the previously presented three different patterns. This is principally due to the road traffic patterns are mainly dependent on these three types of days, and as it has been stated before, road traffic is the main source of ambient noise in urban areas.

In order to characterize the noise level of a given day, we have calculated the *Area Under the Curve* (AUC) [9]. This method allows to simplify the analysis without loosing the information represented by the daily noise curve. AUC has been employed in different research areas for the same purposes, achieving high accurate results (e.g., medicine [10]).

AUC returns the area between a curve and the *x-axis*. We calculate the AUC by using a number of *trapezoidal approximations* (N), which are delimited by the noise curve, the x-axis and one of the N continuous periods of time. Equation 1 defines this approximation to compute the AUC, where the time line

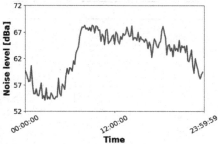
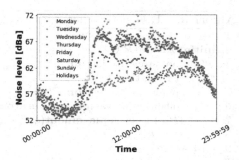

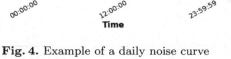

Fig. 4. Example of a daily noise curve of a given day.

Fig. 5. Level of noise sampled grouped by the weekdays and holidays.

(temporal space) is defined in the range between $a = 00{:}00{:}00$H and $b = 23{:}59{:}59$H ($x = [a, b]$), $f(x)$ is the daily noise curve, and each time period (x-$step$) is given by $\Delta x = \frac{b-a}{N}$.

$$\int_a^b f(x)\,dx = \frac{\Delta x}{2} \sum_{k=1}^N (f(x_{k-1}) + f(x_k)) \tag{1}$$

In the context of the *Noise Feature Selection* (NFS) optimization problem, we deal with the daily noise curve C, divided in an arbitrary daily rate DR (number of time blocks/samples/trapezoids per day). Thus, we define T the set of trapezoids or features that defines the original dataset of C ($\|T\| = DR$), and $ST \subseteq T$ to be a subset of the original set of features. The main target of this work is to find the most efficient (smallest) subset of trapezoids/features (ST) that allows us to maximize the accuracy ($ACC(ST)$) [4], i.e., minimize the error of a given machine learning classification approach (see Eq. 2). In this work, we have evaluated our feature selection method over *K-Means Clustering*.

$$Maximize\ ACC(ST)\ subject\ to\ ST \subseteq T \tag{2}$$

3 Noise Feature Selection by Using a Genetic Algorithm

In this section, we describe the approach applied to address NFS optimization problem by using a GA. Thus, we introduce the proposed algorithm, we show the solution representation, we describe the evolutionary operators, and we formulate the objective/fitness function used to evaluate the solutions.

3.1 The Genetic Algorithm

Algorithm 1 shows the pseudocode of the GA. It iteratively applies stochastic operators on a set of solutions (named *individuals*) that define a *population* (P) to improve their quality related to the objective of the problem. Each iteration is called *generation*.

Algorithm 1. Generic schema for a GA.
_____
1: $t \leftarrow 0$
2: **initialization**$(P(0))$
3: **while** not stopcriterion **do**
4:       **evaluation**$(P(t))$
5:       parents $\leftarrow$ **selection**$(P(t))$
6:       offspring $\leftarrow$ **variation operators**(parents)
7:       $P(t+1) \leftarrow$ **replacement**(offspring, $P(t)$)
8:       $t \leftarrow t + 1$
9: **end while**
10: **return** best solution ever found
_____

The first steps produce the initial population $P(0)$ of $\#p$ individuals (Line 2). An evaluation function associates a fitness value to every individual, indicating its suitability to the problem (Line 4). Then, the search process is guided by a probabilistic technique of selection of the best individuals (parents and generated offspring) according to their quality (Line 5). Iteratively, solutions evolve by the probabilistic application of *variation operators* (lines 6–7). The stopping criterion usually involves a fixed computational effort (e.g., number of generations, number of fitness evaluations or execution time), a threshold on the fitness values or the detection of a stagnation situation.

3.2   Representation

Solutions are encoded as a binary vector $S_o = <i_1, \ldots, i_N>$, where N is the number of features of the original set. Each index of the vector represents a given feature (period of time to compute the trapezoids), and the corresponding binary value $S_o(i)$ represents the selection i-th feature. Thus, if the i-th feature is selected, then $S_o(i) = 1$, otherwise $S_o(i) = 0$.

3.3   Operators

The main evolutionary operators are presented in this section.

Initialization. The population is initialized $(P(0))$ by applying a uniform random procedure.

Selection. Tournament selection is applied, with tournament size of two solutions (*individuals*). The tournament criteria is based on fitness value.

Evolutionary Operators. We analyze the application of three different **recombination operators:** the standard one point (1PX) and two points (2PX) crossover [13], and the recombination by *selecting randomly half* (SRH) applied with probability p_C. SRH, which was specifically designed in this study

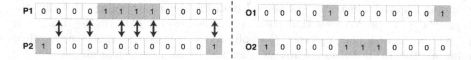

Fig. 6. Example of the SRH crossover operator.

to address NFS, randomly selects a set of features of size $N/2$ and it exchanges them between the two parents to generate two parent offspring solutions. Figure 6 shows an example of applying SRH to the parents P1 and P2, in order to get the offspring O1 and O2. Finally, we study the application of two different **mutation** operators: the flip bit (FB) and shuffle indexes (SI) mutation [13] used with probability p_M.

3.4 Fitness Function

The NFS optimization problem is addressed as a minimization problem. Thus, Eq. 3 presents the objective/fitness function ($fitness(S)$) used to evaluate the individuals during the evolutionary process, which should be minimized. Each solution S represents a subset of features ST_S and it is used to compute the accuracy measure ($ACC(ST_S) \in \mathbb{R} \subseteq [0, 1]$) of using $K\text{-}Means$ to classify/group the set of AUC daily values in the three day types (see Sect. 2.2). As it is a minimization problem, the fitness function takes into account the value of $1-ACC(ST_S)$. Besides, NFS is defined to minimize the size of the ST_s ($\|ST_s\| \leq N$). Thus, we use this metric in the objective function multiplied by an α factor. After some preliminary experiments α was set to 10^{-6}.

$$fitness(S) = (1 - ACC(ST_S)) + \alpha \times \|ST_s\| \in \mathbb{R} \subseteq [0, 1 + \alpha \times N] \qquad (3)$$

4   Experimental Analysis

This section presents the experiments carried out to evaluate the proposed optimization problem. The algorithms were implemented in Python, using the evolutionary computation framework DEAP [3] and the Anaconda framework.

The experiments conducted have been run in the cluster belonging to NEO (Networking and Emerging Optimization) group, which consists of 16 nodes (64 cores) equipped with an Intel Core2 Quad CPU (Q9400) @ 2.66 GHz and 4 GB of RAM.

In the following subsections, we define the problem instances, we present the results of configuring the applied GA, and we analyze the final results and discuss the performance of this proposal.

4.1   Problem Instances

In order to evaluate our proposal, we use the noise data sensed during seven weeks (from February 22nd, 2018) by one of the sensors installed in the university campus. Figure 5 illustrates the sensed data grouping (averaging) them by the type of day. It can be seen that there are three groups: *(a)* working days, that present a very similar pattern, *(b)* holidays and Sundays, that have a clear relation, and *(c)* noise on Saturdays, that has its own particular shape.

We defined two problem instances: the *1-h data blocks instance* (1-H) and the *30-min data blocks instance* (30-M). The 1-H instance is a low cost instance used to find the best configuration of the operators applied in the proposed GA (see Sect. 3.3). In this instance, the time line is divided in 24 features, i.e., trapezoids/data blocks of one hour ($N = 24$). The 30-M instance is applied to address the NFS optimization problem. In order to obtain more accurate results than in 1-H instance, the time line is divided in trapezoids/data blocks of 30 min, i.e., 48 features ($N = 48$).

4.2 Parameters Calibration

A set of parametric setting experiments were performed over the 1-H instance to determine the best operators and parameter values for the proposed GA. The analysis was carried out by setting the population size ($\#p$) to 50 and the maximum number of generations ($\#g$) to 100.

We analyzed the results of applying three crossover operators (1PX, 2PX, and SRH) with four different probabilities ($p_C \in \{0.2, 0.4, 0.6, 0.8\}$), and two mutation operators (BF and SI) with three different probabilities ($p_M \in \{1/10N, 1/N, 10/N\}$, where N is the number of features). Therefore, we evaluated 72 different parameterizations (three crossover operators, four values of p_C, two mutation operators, and three p_M). All these combinations were studied with the proposed GA, for a total number of 30 independent runs.

We applied Shapiro-Wilk statistical test to check if the results follow a normal distribution. As the test resulted negative, we analyzed them by applying non-parametric statistical tests [12]: first, we used Friedman rank test to rank the configurations, and second, we performed Wilcoxon tests to the three best ranked ones to determine the most competitive configuration.

According to Friedman ranking, the three best configurations were (2PX, 0.6, BF, 10/N), (1PX, 0.8, BS, 10/N), and (SRH, 0.8, BF, 10/N), with p-value $< 10^{-10}$. Wilcoxon test did not confirm the differences of comparing these three configurations with each other (p-value > 0.01). Therefore, we decided to select the parameterization that obtained the best (minimum) median value of these three best ranked ones (2PX, 0.6, BF, 10/N), i.e., the configuration defined by 2PX as crossover operator, $p_C = 0.6$, BF as mutation operator, and $p_M = 10/N$.

4.3 Numerical Results

This subsection reports the numerical results achieved in an exhaustive experimental evaluation of addressing NFS optimization problem over 30-M instance by applying the proposed GA and performing 100 independent executions. The GA configuration is defined by $\#p = 50$, $\#g = 200$, and the best configuration found in the previous section (2PX, 0.6, BF, 10/N).

Table 1. Optimization results: final fitness, number of features, and accuracy.

	Mean ± Std	Minimum	Median	Maximum	No NFS
fitness(S)	0.197 ± 13.8%	**0.125**	0.203	0.234	-
$\|ST_S\|$	-	**6**	7	10	48
$ACC(S)$	-	**0.875**	0.797	0.766	0.687

Table 1 reports the average, standard deviation, best (minimum), median, and worst (maximum) of the final fitness values, the number of selected features ($\|ST_S\|$), and the accuracy achieved by applying *K-Means* method ($ACC(S)$) over the 100 independent runs. In turn, it shows the result of the accuracy of no applying feature selection (No NFS), i.e., using 48 features.

Results in Table 1 clearly state that higher accurate classifications were achieved while the number of features was reduced. All the computed solutions obtained better accuracy and used critically lower number of features than the original dataset, i.e., *No NFS* obtained the worst accuracy (lower than 69%). The best solution found requires six features (data blocks) to achieve a high accuracy in classifying the noise (higher than 87%). The median fitness value was obtained by solutions that use just one feature more (seven). Thus, we can confirm that there are periods of the day in which the ambient noise is definitely different depending on type of the day, as can be infer from Fig. 5.

Finally, we identified the periods of time (features or data blocks) that best characterize the daily noise curve. Thus, we computed how often (the frequency) a given feature has been selected by the best solution found for each independent run, in order to observe the features selected with higher frequency. These results are shown in a histogram (see Fig. 7). This histogram illustrates the frequency of a given feature (*Feature ID*) and the corresponding time. The three most used features for classifying the noise are the ones numbered as 16, 18, and 19, which correspond to the noise level data captured during commercial and school opening hours (from 8:00:00H to 10:00:00H). This time period coincides with

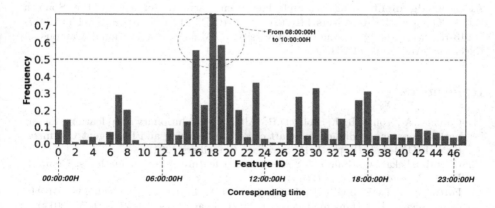

Fig. 7. Histogram of the features selected by the final solutions.

the morning peak traffic hours at the campus. Therefore, the main result of our work is that the daily noise level may be accurately classified by activating the noise sensors during just two hours (from 8:00:00H to 10:00:00H). This represents noticeable energy savings.

5   Conclusions and Future Work

The ambient noise is an important problem in the present cities, because it principally causes discomfort and it is responsible of many diseases. In this study, we have shown one interesting contribution in the intelligent data analysis of ambient noise level, which can be useful for city managers. The real data used are captured and processed by the cyber-physical system installed at University of Málaga by the authors.

In order to characterize the noise level captured during a day and use such data to classify/group the type of day, we have used the AUC metric as an input to *K-Means clustering* method. We have designed an intelligent and automatic feature selection method based on a GA to improve the accuracy of this classification approach.

As to the numerical findings, we have been able to drastically improve the accuracy of the classification method in about 20% (from less than 69% to higher than 87%), while the number of features were reduced from 48 to 6 (87%). These results have a direct benefit for city managers because we can save lots of time and energy by only measuring at some moments of the day. In this case, we have observed that the most characteristic period of time in the evaluated sensor is during the morning from 8:00:00H to 10:00:00H, which coincide with morning peak traffic hours.

As future work we plan to extend the proposed methodology by extending this approach by adding more noise data from other sensors, analyzing this approach over other datasets (road traffic data), and studying new evolutionary approaches as well as new machine learning methods.

Acknowledgements. This research has been partially funded by the Spanish MINECO and FEDER projects TIN2016-81766-REDT (http://cirti.es), and TIN2017-88213-R (http://6city.lcc.uma.es). University of Malaga. International Campus of Excellence Andalucia TECH.

References

1. Camero, A., Toutouh, J., Stolfi, D.H., Alba, E.: Evolutionary deep learning for car park occupancy prediction in smart cities. In: Learning and Intelligent OptimizatioN (LION), vol. 12, pp. 1–15. Springer, Heidelberg (2018)
2. Chandrashekar, G., Sahin, F.: A survey on feature selection methods. Comput. Electr. Eng. **40**(1), 16–28 (2014)
3. Fortin, F.A., De Rainville, F.M., Gardner, M.A., Parizeau, M., Gagné, C.: DEAP: evolutionary algorithms made easy. J. Mach. Learn. Res. **13**, 2171–2175 (2012)

4. García, S., Fernández, A., Luengo, J., Herrera, F.: A study of statistical techniques and performance measures for genetics-based machine learning: accuracy and interpretability. Soft Comput. **13**(10), 959 (2009)
5. McClellan, S., Jimenez, J.A., Koutitas, G.: Smart Cities: Applications, Technologies, Standards, and Driving Factors. Springer, Cham (2017). https://doi.org/10.1007/978-3-319-59381-4
6. Mir, Z.H., Toutouh, J., Filali, F., Alba, E.: QoS-aware radio access technology (RAT) selection in hybrid vehicular networks. In: Kassab, M., Berbineau, M., Vinel, A., Jonsson, M., Garcia, F., Soler, J. (eds.) Nets4Cars/Nets4Trains/Nets4Aircraft 2015. LNCS, vol. 9066, pp. 117–128. Springer, Cham (2015). https://doi.org/10.1007/978-3-319-17765-6_11
7. Murphy, E., King, E.A.: Testing the accuracy of smartphones and sound level meter applications for measuring environmental noise. Appl. Acoust. **106**, 16–22 (2016)
8. Nesmachnow, S., Rossit, D., Toutouth, J.: Comparison of multiobjective evolutionary algorithms for prioritized urban waste collection in Montevideo, Uruguay. Electron. Notes Discrete Math. **69**, 93–100 (2018). in Press
9. Nguyen, D.H., Rebello, N.S.: Students' understanding and application of the area under the curve concept in physics problems. Phys. Rev. Phys. Educ. Res. **7**(1), 010112 (2011)
10. Pruessner, J.C., Kirschbaum, C., Meinlschmid, G., Hellhammer, D.H.: Two formulas for computation of the area under the curve represent measures of total hormone concentration versus time-dependent change. Off. J. ISPNE **28**(7), 916–931 (2003)
11. Segura-Garcia, J., Felici-Castell, S., Perez-Solano, J.J., Cobos, M., Navarro, J.M.: Low-cost alternatives for urban noise nuisance monitoring using wireless sensor networks. IEEE Sens. J. **15**(2), 836–844 (2015)
12. Sheskin, D.J.: Handbook of Parametric and Nonparametric Statistical Procedures. CRC Press, Boca Raton (2003)
13. Spears, W.M.: Evolutionary Algorithms: The Role of Mutation and Recombination. Springer, Heidelberg (2000). https://doi.org/10.1007/978-3-662-04199-4
14. Steele, C.: A critical review of some traffic noise prediction models. Appl. Acoust. **62**(3), 271–287 (2001)
15. Stolfi, D.H., Alba, E.: Smart mobility policies with evolutionary algorithms: the adapting info panel case. In: Proceedings of the 2015 Annual Conference on Genetic and Evolutionary Computation. GECCO 2015, pp. 1287–1294. ACM, New York (2015)
16. Toutouh, J., Arellano-Verdejo, J., Alba, E.: Enabling low cost smart road traffic sensing. In: The 12th Edition of the Metaheuristics International Conference (MIC 2017), pp. 13–15 (2017)
17. Toutouh, J., Rossit, D., Nesmachnow, S.: Computational intelligence for locating garbage accumulation points in urban scenarios. In: Learning and Intelligent OptimizatioN (LION), vol. 12. pp. 1–15. Springer, Heidelberg (2018)
18. Van Kempen, E.E., Kruize, H., Boshuizen, H.C., Ameling, C.B., Staatsen, B.A., de Hollander, A.E.: The association between noise exposure and blood pressure and ischemic heart disease: a meta-analysis. Environ. Health Perspect. **110**(3), 307 (2002)
19. Xue, B., Zhang, M., Browne, W.N., Yao, X.: A survey on evolutionary computation approaches to feature selection. IEEE Trans. Evol. Comput. **20**(4), 606–626 (2016)
20. Yang, J., Honavar, V.: Feature subset selection using a genetic algorithm. In: Liu, H., Motoda, H. (eds.) Feature Extraction, Construction and Selection, vol. 453, pp. 117–136. Springer, Boston (1998). https://doi.org/10.1007/978-1-4615-5725-8_8

Pruning Dominated Policies in Multiobjective Pareto Q-Learning

Lawrence Mandow[(✉)] and José-Luis Pérez-de-la-Cruz

Andalucía Tech, Departamento de Lenguajes y Ciencias de la Computación,
Universidad de Málaga, Málaga, Spain
{lawrence,perez}@lcc.uma.es

Abstract. The solution for a Multi-Objetive Reinforcement Learning problem is a set of Pareto optimal policies. MPQ-learning is a recent algorithm that approximates the whole set of all Pareto-optimal deterministic policies by directly generalizing Q-learning to the multiobjective setting. In this paper we present a modification of MPQ-learning that avoids useless cyclical policies and thus improves the number of training steps required for convergence.

1 Introduction

The problem of solving Multi-Objective Markov Decision Processes (MOMDPs) and more concretely the application of Reinforcement Learning techniques to this problem (MORL) has raised some interest in recent literature [1]. MORL algorithms can be single-policy and multi-policy [2]. The later try to approximate part or the whole set of Pareto optimal policies. There are currently very few multi-policy MORL algorithms [6,7]; among them we will consider in this paper MPQ-learning [3], which offers a clear recovery procedure of optimal policies after learning is complete. It is an off-policy temporal-difference method. Off-policy methods are particularly interesting in multiobjective reinforcement learning, since they allow to learn a set of Pareto-optimal policies simultaneously. MPQ-learning was shown to solve standard problems from the benchmark proposed in [4]. However, there are efficiency issues due to the nature of the algorithm, that aims to learn *all* deterministic optimal policies (including non stationary ones). In this paper we propose a modification of MPQ-learning that controls the generation of cycles during the learning process and thus improves greatly its efficiency in terms of training steps.

The paper is structured as follows: in Sect. 2 MPQ-learning is summary described and some problematic features are identified and discussed. Then we describe the proposed modifications (Sect. 4) and present and discuss the results obtained in a set of experiments (Sect. 5). Finally some conclusions are drawn.

Supported by: the Spanish Government, Agencia Estatal de Investigación (AEI) and European Union, Fondo Europeo de Desarrollo Regional (FEDER), grant TIN2016-80774-R (AEI/FEDER, UE); and Plan Propio de Investigación de la Universidad de Málaga - Campus de Excelencia Internacional Andalucía Tech.

F. Herrera et al. (Eds.): CAEPIA 2018, LNAI 11160, pp. 240–250, 2018.
https://doi.org/10.1007/978-3-030-00374-6_23

2   MPQ-Learning

2.1   The Algorithm

MPQ-learning [3] is an extension of Q-learning to multiobjective problems. The goal is to obtain the set of all Pareto-optimal deterministic policies. These include both stationary and non-stationary policies, as nonstationarity is an essential feature of multiobjective reinforcement learning problems.

A detailed description of MPQ-learning can be found in [3]. Let us briefly describe here the rationale of the algorithm in an intuitive way. A basic under-lying idea of the algorithm (and also of the improvements presented later) is to reduce the set of candidate policies applying the principle of optimality. Just like Q-learning, MPQ-learning is an off-policy temporal-difference method. An agent interacts with the environment and learns the optimal expected long term rewards of its actions. Without loss of generality we shall assume that all rewards are to be maximized. Sets of labeled vectors $\mathbb{Q}(s, a)$ are learned for each state-action pair. Each labeled vector $\boldsymbol{q} \in \mathbb{Q}(s, a)$ estimates the expected vector reward when a particular Pareto-optimal policy is followed after choosing action a in state s. Additionally, set $\mathbb{V}(s')$ denotes the set of all Pareto-optimal vector esti-mates for all actions available to state s'. These sets are the multi-objective analogues of the scalar Q and V sets used in Q-learning.

An important feature of MPQ-learning is that it learns a partial model of the environment. This is in contrast with Q-learning, which is a model-free algo-rithm. The partial model is necessary for two main reasons. In the first place, it helps to reduce the number of candidate policies tracked by the algorithm. Additionally, once learning is over, the model can be used to recover the actions associated to a particular Pareto-optimal policy chosen by the agent. Therefore, in MPQ-learning each vector estimate $\boldsymbol{q} \in \mathbb{Q}(s, a)$ is labeled with a set of *indices* P, where each index $(s', i) \in P$ indicates that $\boldsymbol{q}$ is updated from the vector with label i in $\mathbb{V}(s')$.

Let us assume that at step n, transition $s, a, \boldsymbol{r}, s'$ is stochastically performed by the agent. The updating expression for MPQ-learning is,

$$\mathbb{Q}_n(s, a) = \begin{cases} \mathbb{N}_{n-1}(s, a) \cup \mathbb{U}_{n-1}(s, a) \cup \mathbb{E}_{n-1}(s, a) \\ \qquad \text{if } s = s_n \wedge a = a_n \\ \mathbb{Q}_{n-1}(s, a) \\ \qquad \text{otherwise} \end{cases} \tag{1}$$

The updated Q-set results from the union of three sets: $\mathbb{N}$, $\mathbb{U}$ and $\mathbb{E}$. These depend on the nature of the transition $s, a, \boldsymbol{r}, s'$. Let us consider the sample stochastic transition from some state s_1 through action a to states s_2 and s_3.

Let us assume that action a is selected *for the first time* from state s_1, and the agent transitions to state s_2 obtaining some vector reward $\boldsymbol{r}$. Let us further assume that currently there is a single optimal estimate for s_2, i.e. $\mathbb{V}(s_2) = \{\boldsymbol{q}_1^2\}$. According to the principle of optimality, optimal policies in s_1 can be obtained by combining optimal policies from s_2 and s_3. Since $\mathbb{Q}_{n-1}(s_1, a)$ is empty, a *new* estimate is added to $\mathbb{Q}(s_1, a)$ for each estimate in $\mathbb{V}(s_2)$. In this case $\mathbb{Q}_n(s_1, a) =$

$\{(\boldsymbol{q}_1^1, (s_2, 1))\}$, indicating that $\boldsymbol{q}_1^1$ is related to vector 1 in $\mathbb{V}(s_2)$ as indicated by the formula,

$$\mathbb{N}_{n-1}(s, a) = \{((1 - \alpha_n)\boldsymbol{q} + \alpha_n[\boldsymbol{r}_n + \gamma \boldsymbol{v}_j], P \cup \{(s', j)\}) \mid$$
$$(\boldsymbol{q}, P) \in \mathbb{Q}_{n-1}(s, a) \wedge \boldsymbol{v}_j \in \mathbb{V}_{n-1}(s')$$
$$\wedge s' \not\sqsubseteq \mathbb{Q}_{n-1}(s, a)\} \tag{2}$$

where $s' \not\sqsubseteq \mathbb{Q}_{n-1}(s, a)$ denotes that state s' does not participate in any index in $\mathbb{Q}_{n-1}(s, a)$, and α_n and γ are the usual *learning rate* and *discount factor* respectively.

Notice that if transition $s_1, a, \boldsymbol{r}, s_3$ is performed afterwards *for the first time*, and there is just one optimal estimate in $\mathbb{V}(s_3) = \{\boldsymbol{q}_1^3\}$, then $\boldsymbol{q}_1^1$ would be updated from $\boldsymbol{q}_1^3$ and its set of indices extended to $(s_2, 1), (s_3, 1)$, indicating that $\boldsymbol{q}_1^1$ is obtained combining estimate 1 from $\mathbb{V}(s_2)$, and estimate 1 from $\mathbb{V}(s_3)$.

When the transition $s_1, a, \boldsymbol{r}, s_2$ is repeated, vector $\boldsymbol{q}_1^1$ will be *updated* from $\boldsymbol{q}_1^2$, provided both are still part of $\mathbb{Q}(s_1, a)$ and $\mathbb{V}(s_2)$ respectively. Analogously for transition $s_1, a, \boldsymbol{r}, s_3$. The general expression for this situation is given by the formula,

$$\mathbb{U}_{n-1}(s, a) = \{((1 - \alpha_n)\boldsymbol{q} + \alpha_n[\boldsymbol{r}_n + \gamma \boldsymbol{v}_j], P) \mid$$
$$(\boldsymbol{q}, P) \in \mathbb{Q}_{n-1}(s, a) \wedge (s', j) \in P$$
$$\wedge \boldsymbol{v}_j \in \mathbb{V}_{n-1}(s')\} \tag{3}$$

Finally, let us consider that transition $s_1, a, \boldsymbol{r}, s_2$ is repeated one more time, but now $\mathbb{V}(s_2) = \{\boldsymbol{q}_1^2, \boldsymbol{q}_2^2\}$ is populated by two estimates: the previous one $\boldsymbol{q}_1^2$ (with possibly now a more accurate value), and a new, or *extra* one $\boldsymbol{q}_2^2$ that was not present the last time transition $s_1, a, \boldsymbol{r}, s_2$ was traversed. As described in formula 3, the value of estimate $\boldsymbol{q}_1^1$ will be rigthfully updated with that of $\boldsymbol{q}_1^2$, as indicated by the indices of the former. However, an additional estimate will be included in $\mathbb{Q}(s_1, a)$ reflecting a possibly new optimal policy that can be followed combining results obtained from the second vector in $\mathbb{V}(s_2)$, and the first one (and only known to date) in $\mathbb{V}(s_3)$. This operation is given in the general case by the formula,

$$\mathbb{E}_{n-1}(s, a) = \{(\alpha_n[\boldsymbol{r}_n + \gamma \boldsymbol{v}_j], (P \setminus s') \cup \{(s', j)\}) \mid$$
$$\boldsymbol{v}_j \in \mathbb{V}_{n-1}(s') \wedge s' \sqsubseteq \mathbb{Q}_{n-1}(s, a) \wedge \tag{4}$$
$$(s', j) \not\sqsubseteq \mathbb{Q}_{n-1}(s, a) \wedge \exists \boldsymbol{q}(\boldsymbol{q}, P) \in \mathbb{Q}_{n-1}(s, a)\}$$

where $(s', j) \not\sqsubseteq \mathbb{Q}_{n-1}(s, a)$ denotes that index (s, a) does not appear in the index set of any vector in $\mathbb{Q}_{n-1}(s, a)$.

Let us complete the description of the algorithm with the formal definition of the $\mathbb{V}(s)$ sets,

$$\mathbb{V}(s) = \mathrm{ND} \bigcup_{a \in A} \{\boldsymbol{q} \mid (\boldsymbol{q}, P) \in \mathbb{Q}(s, a)\} \tag{5}$$

where $ND(\mathcal{X})$ denotes the set of nondominated, (or Pareto-optimal) vectors in set $\mathcal{X}$.

In summary, optimal policies for each state-action pair are obtained combining optimal policies from reachable states. MPQ-learning incrementally builds a partial model of the environment through sets of indices, ensuring that each estimate is updated incrementally always from the same (optimal) policies. This reduces the number of combinations of policies tracked by the algorithm, increasing its efficiency.

2.2   Efficiency Issues with MPQ-Learning

By considering the algorithm at work, some issues can be detected. First of all, the "update" operation $\mathbb{U}$ is defined in a very conservative way respect to the use of memory. Notice that in the former definition of the $\mathbb{U}(s, a)$ set only optimal values in successors s' of s (i.e., values in $\mathbb{V}(s')$) are considered. In this way, new and potentially useful information in s' is not propagated to s.

Related to this point is the generation of useless cyclical (nonstationary) policies in intermediate steps of the algorithm. Indeed, since some of them can be optimal, the consideration of cyclical policies is unavoidable and, in fact, necessary. However, many cyclical policies appearing during the execution are mere artifacts due to the delayed propagation of true Q-values. These policies will be eventually discarded, since they will have –sooner or later– a dominated Q-value; but until that moment, they populate $\mathbb{Q}$ sets and slow down greatly the execution of the algorithm.

Finally, these spurious policies and Q-values difficult considerably an adequate finalization of the algorithm. At any given moment, it is difficult to identify them inside $\mathbb{Q}$ sets and hence it is difficult to output an useful approximation of the real set of optimal policies.

3   Example

Let us illustrate the concerns raised in Sect. 2.2. Let us consider a simple state space consisting of only two states, s_1 and s_2. Episodes start always at s_1 and terminate at s_2. There are two actions available at s_1. Action a_1 loops back deterministically to s_1, and the agent receives a vector reward of $(-1, 0)$. Action a_2 leads deterministically to s_2, and the agent receives a vector reward of $(-1, 0)$. At s_2 the episode ends, and the agent receives a reward of $(-1, 1)$. Let us assume all $\mathbb{Q}(s, a)$ sets are initialized with a zero vector, and $\mathbb{V}(s_2) = \{q_1^2 = (-1, 1)\}$. The agent maximizes all rewards.

There is just one optimal policy in this problem with value $(-2, 1)$, i.e. at s_1 the agent chooses action a_2 leading to s_2. However, there are infinitely many dominated nonstationary policies: looping with a_1 a number of times and then choosing a_2.

Let us assume that the process starts, and fixed values of $\alpha = 0'1$ and $\gamma = 1$ are used. We further assume that the agent chooses action a_2 leading to transition $s_1, a_2, \boldsymbol{r}, s_2$. Through application of the *new* rule in MPQ-learning (see formula 2), and since $\mathbb{V}(s_2) = \{q_1^2 = (-1, 1)\}$, a new value is calculated in $\mathbb{Q}_1(s_1, a_2) = \{(\boldsymbol{q}_1^1, (s_2, 1))\}$. The new vector $\boldsymbol{q}_1^1$ stands for the optimal policy, and is linked to the first (and only) estimate in s_2. With the provided data, the initial value for the estimate would be $\boldsymbol{q}_1^1 = (-0'2, 0'1)$.

Let us assume the process starts again and the agent performs transition $s_1, a_1, \boldsymbol{r}, s_1$. Now $\mathbb{Q}_2(s_1, a_2) = \mathbb{Q}_1(s_1, a_2)$, but through application of the *new* rule, and since $\mathbb{V}(s_1) = \{\boldsymbol{q}_1^1 = (-0'2, 0'1)\}$, a new value is calculated in $\mathbb{Q}_2(s_1, a_1) = \{(\boldsymbol{q}_2^1, (s_1, 1))\}$. The new estimate $\boldsymbol{q}_2^1$ is linked through its index to the first estimate in s_1, which in turn is linked to the first estimate in s_2. In other words, the new estimate stands for a nonstationary policy that loops once in s_1 and then proceeds to s_2. However, with the provided data, the initial value of this estimate would be $\boldsymbol{q}_2^1 = (-0'12, 0'01)$, which is nondominated with the value of $\boldsymbol{q}_1^1$.

Let us assume the process starts again and the agent performs transition $s_1, a_1, \boldsymbol{r}, s_1$ once again. Now $\mathbb{V}(s_1) = \{\boldsymbol{q}_1^1, \boldsymbol{q}_2^1\}$. Therefore, two different operations will be carried out to obtain $\mathbb{Q}_3(s_1, a_1)$. First, estimate $\boldsymbol{q}_2^1$ will be *updated* according to its index (formula 3). But additionally, an *extra* estimate $\boldsymbol{q}_3^1$ will be created (formula 4), since there is a new element in $\mathbb{V}(s_1)$. The new estimate will have an index $(s_1, 2)$, and stands for the nonstationary policy consisting of looping twice in s_1 and then proceeding to s_2. The calculated initial value for $\boldsymbol{q}_3^1$ would be $(-0'112, 0'001)$ which is again nondominated in $\mathbb{V}(s_1)$.

Now, each time the transition $s_1, a_1, \boldsymbol{r}, s_1$ is performed, an extra vector will be added to $\mathbb{Q}(s_1, a_1)$, and a long chain of dependencies between its estimates will be created ($\boldsymbol{q}_2^1 \leftarrow \boldsymbol{q}_3^1 \leftarrow \ldots \leftarrow \boldsymbol{q}_k^1$).

Let us assume that, after some experience in the environment is accumulated, the value of $\boldsymbol{q}_2^1$ eventually converges to its true value and is found to be dominated in $\mathbb{V}(s_1)$. The update rule in MPQ-learning would then remove $\boldsymbol{q}_3^1$ (i.e. the two loop policy) the next time transition $s_1, a_1, \boldsymbol{r}, s_1$ is performed. Two observations are in order here. In the first place, nothing prevents in the future the reconsideration of the two-loop policy provided an optimistic estimate happens to be nondominated again in $\mathbb{V}(s_1)$. Additionally, a set of *broken* estimates is left in $\mathbb{V}(s_1)$, since they not only depend on an estimate that was found to be dominated, but that was even removed from the Q-set.

In the next section we propose a new mechanism to tackle this kind of situation, noting that, once an estimate is found to be dominated, *all* estimates that depend on it through their indices cannot lead to nondominated policies, even if their current estimates happen to be locally nondominated.

4   Pruning MPQ-Learning

According to the aforementioned problems, the following modifications are proposed and implemented in the MPQ algorithm.

Nondiscriminating Update. Updating of Q-values depending of preexisting continuations are performed for *all* Q-values in $\mathbb{Q}(s', a)$. More concretely, the general expression for updating is not defined as in 3, but as,

$$\mathbb{U}_{n-1}(s,a) = \{((1-\alpha_n)\boldsymbol{q} + \alpha_n[\boldsymbol{r}_n + \gamma\boldsymbol{v}_j], P) \mid$$
$$(\boldsymbol{q}, P) \in \mathbb{Q}_{n-1}(s,a) \wedge (s',j) \in P$$
$$\wedge \boldsymbol{v}_j \in \cup_a \mathbb{Q}_{n-1}(s',a)\} \tag{6}$$

Suspended Q-Values. In order to avoid the generation of more and more cyclical policies, every Q-value can be labelled as "suspended". A suspended value in never considered for the operations "new" and "extra". In this way, the definitions are no more 2 and 4. They are now,

$$\mathbb{N}_{n-1}(s,a) = \{((1-\alpha_n)\boldsymbol{q} + \alpha_n[\boldsymbol{r}_n + \gamma\boldsymbol{v}_j], P \cup \{(s',j)\}) \mid$$
$$(\boldsymbol{q}, P) \in \mathbb{Q}_{n-1}(s,a) \wedge \boldsymbol{v}_j \in \mathbb{V}_{n-1}(s')$$
$$\wedge \neg\text{suspended}(\boldsymbol{v}_j) \wedge s' \not\sqsubset \mathbb{Q}_{n-1}(s,a)\} \tag{7}$$

and

$$\mathbb{E}_{n-1}(s,a) = \{(\alpha_n[\boldsymbol{r}_n + \gamma\boldsymbol{v}_j], (P \setminus s') \cup \{(s',j)\}) \mid$$
$$\boldsymbol{v}_j \in \mathbb{V}_{n-1}(s') \wedge s' \sqsubset \mathbb{Q}_{n-1}(s,a)$$
$$\wedge \neg\text{suspended}(\boldsymbol{v}_j) \wedge \tag{8}$$
$$(s',j) \not\sqsubset \mathbb{Q}_{n-1}(s,a) \wedge \exists \boldsymbol{q}(\boldsymbol{q}, P) \in \mathbb{Q}_{n-1}(s,a)\}$$

Suspension. Additional rules must be defined in order to (i) label and (ii) unlabel Q-values $\boldsymbol{q} \in Q(s,a)$ as "suspended",

(i.1) A Q-value is labelled as "suspended" when it is dominated by another Q-value in the state. Formally, the condition is as follows: $\forall s, a, \boldsymbol{q} \in Q(s,a)$, whenever $\mathbb{V}(s)$ is modified, if $\boldsymbol{q}$.suspended $= false \wedge \boldsymbol{q} \notin \mathbb{V}(s)$, then $\boldsymbol{q}$.suspended $\leftarrow$ true. (i.2) A Q-value is labelled as "suspended" whenever any of their indices is labelled as "suspended". This propagation is done recursively. In practice, this implements the idea that, if an estimate is found to be suspended, then any estimate depending on it is suspended as well.

(ii) A Q-value is unlabelled as "suspended" when it ceases to be dominated by another Q-value in the state and none of the estimates referenced by its indices is suspended. Formally, the condition is as follows: for every state s, whenever $\mathbb{V}(s)$ is modified, $\forall a, \boldsymbol{q} \in Q(s,a)$, if $\boldsymbol{q}$.suspended $= true \wedge \boldsymbol{q} \in \mathbb{V}(s) \wedge \forall (s',j)$ index of $\boldsymbol{q}, \boldsymbol{q}_j^{s'}$.**suspended** $= \textbf{false}$ then $\boldsymbol{q}$.suspended $\leftarrow$ false.

5 Experimental Results and Discussion

Both the MPQ-learning algorithm, and the alternative described in this paper (MPQ2), were implemented and tested over a set of sample problems. These are based on the standard Deep Sea Treasure (DST) problem proposed by [5].

More precisely, we consider a grid world like the one shown in Fig. 1. The agent controls a submarine that searches for treasures. These are found on the sea bed. Each grid cell represents a valid state of the agent. Grids are referenced according to their row (numbered top to bottom starting at 1), and column (numbered from left to right starting at 1). The agent is allowed to move in four directions (up, down, left or right) from its current cell to an adjacent one. Obstacles (represented as black cells) are unreachable for the agent. If an action tries to move the agent to an unreachable state or outside the grid, then the state remains unchanged. The start position is always $(1,1)$, and each episode ends whenever a treasure position is reached. The agent receives a vector reward with two components. The first one is a negative reward of -1 each time it moves. The second one is the value of the treasure found, or 0 if no treasure is available in the cell. The goal of the agent is to find all Pareto optimal policies that maximize both rewards.

Our experiments consider a number of subproblems of increasing difficulty. In problem P_i the state space is made up of all five rows, but only the first i columns of the grid shown in Fig. 1, i.e. a $5 \times i$ subgrid. This way we obtain a sequence of problems such that, for larger i we obtain a larger state space and a larger set of nondominated policies. Notice that in these problems all Pareto-optimal policies are stationary.

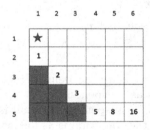

Fig. 1. Sample environment for deep sea treasure. Treasure values are indicated by values inside the cells.

In our implementation, the sets of optimal policies for each state $\mathbb{V}(s)$ are calculated in such a way that whenever two vectors are equal, we prefer the one that first entered $\mathbb{V}(s)$, i.e. the one with smaller index value. This way we try to ensure that, once values stabilize and several policies can propagate the same values, the algorithm consistently prefers the same one.

A limit of 700 actions per episode was set. We used the multiobjective ϵ-greedy behavior policy described by [3]. This basically calculates the ratio of nondominated vectors that each $\mathbb{Q}(s, a)$ contributes to $\mathbb{V}(s)$. With probability $(1 - \epsilon)$ each action a is selected with a probability proportional to the ratio of its $\mathbb{Q}(s, a)$ set. With probability ϵ each action is selected randomly with equal probability. The following parameters were used: Learning rate $\alpha = 0'1$; Discount rate $\gamma = 1$; Exploration probability $\epsilon = 0'4$.

Both algorithms were run until the estimates of the initial state $(\mathbb{V}(1,1))$ approximated the values of all optimal policies (which are readily known beforehand in this problem set) with a precision up to 1%. One hundred agents were run for each problem, and the number of training steps, episodes and maximum number of estimates were recorded for each run. All agents for both algorithms reached the termination criterion.

Table 1 summarizes the data gathered for MPQ (top) and MPQ2 (bottom). Figure 2 (left) compares the average value of overall training steps obtained by both algorithms for the different problems. Figure 2 (right) presents the analogous comparison for the maximum number of vector estimates stored in the $\mathbb{Q}(s, a)$ sets.

Table 1. No. of episodes, training steps and vector estimates for MPQ (top) and MPQ-2 (bottom).

		Trainig st.			Epi.	# estimates		
	Prob.	Avg.	Min	Max	Avg	Avg.	Min	Max
MPQ	1	188'99	128	258	44'00	133'56	51	255
	2	2310'64	1540	3049	229'64	502'18	329	721
	3	7756'59	6116	11553	507'13	1360'21	1012	2252
	4	22359'75	14838	30403	1046'93	2891'84	2364	3357
	5	35085'68	26367	45286	1371'28	5222'02	4517	5956
	6	50015'24	38679	78576	1740'66	7767'18	6844	9569
		Trainig st.			Epi.	# estimates		
	Prob.	Avg.	Min	Max	Avg	Avg.	Min	Max
MPQ2	1	136'86	118	169	44'00	95'95	27	186
	2	1031'34	886	1384	253'51	599'24	254	1324
	3	3519'36	2793	4665	633'51	2246'98	1352	3867
	4	8150'70	6448	9623	938'02	7433'49	4648	10652
	5	12110'93	9923	15281	968'05	15834'05	11689	21402
	6	18392'22	15719	25634	1205'79	30682'54	23763	38454

The results show an important trade-off between the number of training steps and the number of vector estimates required by both algorithms. This is to be expected, given the different design criteria for both algorithms. MPQ tries to reduce as much as possible the number of estimates. On the other hand, the main concern in MPQ2 is pruning dominated policies (and dominated nonstationary policies in particular), so that exploration can concentrate on the promising policies.

Figure 3 further illustrates the behavior and trade-off between both algorithms. The graphic displays the number of estimates stored by both algorithms on a particular run of problem P_6. The number of estimates kept by all $\mathbb{Q}(s, a)$ sets was recorded every 500 training steps. The growth in the number of policy estimates in MPQ2 is monotonic non-decreasing, since once an estimate is created, it is never discarded. If the estimate is suspended or found to be locally

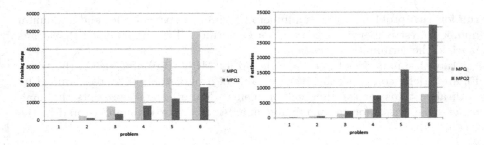

Fig. 2. Comparison of the average number of training steps (left) and maximum number of vector estimates (right) required by MPQ and MPQ2 for the deep sea treasure subproblems.

dominated, it is prevented from generating new or extra estimates in neighboring states. These values help MPQ2 limit the number of nonstationary policies. Therefore the exploration of the state space quickly concentrates on promising policies, achieving convergence in a small number of steps. In this particular instance, MPQ2 requires 16331 training steps distributed into 1052 episodes, and stores 28448 estimates.

In contrast, the number of policies tracked by MPQ can vary widely during the exploration of the state space. The only limit to the number of nonstationary policies considered is local dominance. Therefore, MPQ has to establish the dominance of each policy on an individual basis. However, once a policy is found dominated, it is forgotten, and is likely to reappear again and again. In practice, the number of local nonstationary nondominated estimates can be very high, making it hard for the behavior policy to concentrate on the interesting learned policies. This increases considerably the number of training steps required by MPQ. In this particular instance, MPQ requires 48650 training steps distributed into 1688 episodes, and stores a peak of 8291 estimates.

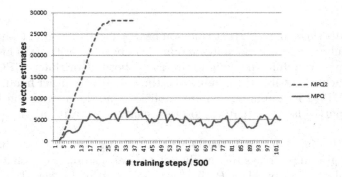

Fig. 3. Number of vector estimates kept by MPQ and MPQ2 agents on a sample run.

6   Conclusions and Future Work

The MPQ-learning algorithm is an interesting framework to investigate and ana-lyze the phenomena of multi-policy temporal-difference learning. In this paper we address one such penomenon, the presence of dominated policies, and par-ticularly the challenge of controlling the growth of dominated nonstationary policies. These frequently present nondominated estimates due to the delayed propagation of true Q-values, and their dominated nature can only be revealed after costly learning. This paper describes a variant of MPQ-learning aimed at pruning dominated policies. The pruning mechanism, called suspension, helps to reduce the number of nonstationary policies tracked by the algorithm. The per-formance of both algorithms is evaluated over sample problem instances based on a standard benchmark.

MPQ-learning does not incorporate a special treatment for nonstationary policies. These are left to grow and eventually disappear whenever learned to be dominated. MPQ is very conservative with the use of memory, deleting candi-date policy estimates as soon as they are found to be dominated. This results in an increase in the number of update steps required by the algorithm, since valu-able information can be lost due to local oscilations in dominance. Additionally, dominated nonstationary policies tend to reapear quickly after being discarded, and require learning their values each time, frequently to be dominated again and again.

The alternative contributed in this paper (MPQ2) is monotonic in the number of policy estimates considered, i.e. once an estimate is added to the $\mathbb{Q}(s, a)$ sets, it is never removed. The algorithm keeps updating its value even when it is locally dominated or suspended. In such cases, it will never be used to generate new or extra estimates, since new nondominated estimates can only be achieved through the combination of nondominated ones. Suspension is a recursive procedure, such that, if a particular estimate is suspended, all the estimates that depend on it will be suspended as well. However, if after some time the value is found to be locally nondominated again and suspension is lifted, it can immediatelly contribute to the determination of optimal policies. The experiments show that this strategy significantly reduces the number of training steps in the algorithm. The number of steps per episode is also considerably reduced, indicating that they are shorter and more focused on the interesting policies. This is the main advantage of MPQ2. The price to pay is an increase in space requirements.

Another important advantage of MPQ2 is that it clearly identifies the set of learned Pareto-optimal policies. Once the estimate of a policy converges to a dominated value, it is kept suspended. Through the recursive suspension mech-anism, all such policies are eventually identified and suspended. In consequence, the subsets of un-suspended estimates in the $\mathbb{V}(s)$ sets end up populated only with the estimates of Pareto optimal policies. In MPQ, on the contrary, even when the optimal policies have been found, temptative nonstationary policy estimates emerge in the $\mathbb{V}(s)$ sets to be discarded cyclically again and again.

Pruning dominated policies in general, and dominated nonstationary poli-cies in particular, remains an important challenge in multiobjective temporal-difference learning. The evaluation of the alternatives here presented over more ambitious and diverse benchmark problems is an interesting avenue of future research. It would also be interesting to reduce the memory requirements of MPQ2 in a cost-effective way.

References

1. Drugan, M., Wiering, M., Vamplew, P., Chetty, M.: Editorial: special issue on multi-objective reinforcement learning. Neurocomputing **263**, 1–2 (2017)
2. Roijers, D.M., Vamplew, P., Whiteson, S., Dazeley, R.: A survey of multi-objective sequential decision-making. J. Artif. Intell. Res. (JAIR) **48**, 67–113 (2013)
3. Ruiz-Montiel, M., Mandow, L., Perez-de-la Cruz, J.-L.: A temporal difference method for multi-objective reinforcement learning. Neurocomputing **263**, 15–25 (2017)
4. Vamplew, P., Dazeley, R., Berry, A., Issabekov, R., Dekker, E.: Empirical evaluation methods for multiobjective reinforcement learning algorithms. Mach. Learn. **84**(1–2), 51–80 (2011)
5. Vamplew, P., Yearwood, J., Dazeley, R., Berry, A.: On the limitations of scalarisation for multi-objective reinforcement learning of Pareto fronts. In: Wobcke, W., Zhang, M. (eds.) AI 2008. LNCS (LNAI), vol. 5360, pp. 372–378. Springer, Heidelberg (2008). https://doi.org/10.1007/978-3-540-89378-3_37
6. Van Moffaert, K., Nowé, A.: Multi-objective reinforcement learning using sets of Pareto dominating policies. J. Mach. Learn. Res. **15**, 3663–3692 (2014)
7. Wiering, M.A., Withagen, M., Drugan, M.M.: Model-based multi-objective reinforcement learning. In: 2014 IEEE Symposium on Adaptive Dynamic Programming and Reinforcement Learning (ADPRL) (2014)

Running Genetic Algorithms in the Edge: A First Analysis

José Á. Morell[(✉)] and Enrique Alba

Departamento de Lenguajes y Ciencias de la Computación, Andalucía Tech,
Universidad de Málaga, Málaga, Spain
{jamorell,eat}@lcc.uma.es

Abstract. Nowadays, the volume of data produced by different kinds of devices is continuously growing, making even more difficult to solve the many optimization problems that impact directly on our living quality. For instance, Cisco projected that by 2019 the volume of data will reach 507.5 zettabytes per year, and the cloud traffic will quadruple. This is not sustainable in the long term, so it is a need to move part of the intelligence from the cloud to a highly decentralized computing model. Considering this, we propose a ubiquitous intelligent system which is composed by different kinds of endpoint devices such as smartphones, tablets, routers, wearables, and any other CPU powered device. We want to use this to solve tasks useful for smart cities. In this paper, we analyze if these devices are suitable for this purpose and how we have to adapt the optimization algorithms to be efficient using heterogeneous hardware. To do this, we perform a set of experiments in which we measure the speed, memory usage, and battery consumption of these devices for a set of binary and combinatorial problems. Our conclusions reveal the strong and weak features of each device to run future algorihms in the border of the cyber-physical system.

Keywords: Edge computing · Fog computing
Evolutionary algorithms · Genetic algorithms · Metaheuristics
Smartphone · Tablet · Ubiquitous AI

1 Introduction

At present, cloud computing looks like a mature and stable domain. Processing all the data in the Cloud have been effective until now. However, when we analyze the near future in terms of volume of data produced, we find some issues for which we are not yet ready. However, we can say that cloud computing and even big data as we knew them until now are not sustainable in the long term. To

This research has been partially funded by the Spanish MINECO and FEDER projects TIN2014-57341-R (http://moveon.lcc.uma.es), TIN2016-81766-REDT (http://cirti.es), TIN2017-88213-R (http://6city.lcc.uma.es), the Ministry of Education of Spain (FPU16/02595), and Universidad de Málaga.

F. Herrera et al. (Eds.): CAEPIA 2018, LNAI 11160, pp. 251–261, 2018.
https://doi.org/10.1007/978-3-030-00374-6_24

understand the problem, we must realize the large amount of data that is created by new devices, not only laptops and smartphones, but also tablets, wearable devices, home and city sensors, and so on. According to Cisco, The "Internet of Everything", which means all of the people and things connected to Internet, will generate 507.5 zettabytes of data by 2019 [1]. That is a large amount of data that needs to be analyzed. Big Data analysis is currently performed in cloud or enterprise data centers. Consequently, it is necessary to find new approaches to solve problems at the edge without sending all the information to the cloud.

Several new approaches are now emerging to solve these problems. One of them is called **edge computing** (aka fog computing) [2], a paradigm where computing tasks move closer to the source of data. Currently, sensors are collecting so much data that it is necessary to perform some data analysis at the edge instead of sending all data to the cloud. To achieve this, edge computing uses endpoint processing capacity in a distributed way. Peter Levine, a partner at the Silicon Valley venture capital firm Andreessen Horowitz, has recently talked about edge computing as the next multibillion-dollar tech market [3]. He said that edge computing would overtake the cloud and that the cloud would still store important data, but as devices become more sophisticated, all the data curation and learning would take place at the edge. Furthermore, both research and industry are taking steps in this direction. For instance, some companies such as HTC and Samsung are developing apps to allow people to contribute with the unused computational power of their smartphones in mobile grids (HTC Power to Give[1] and Samsung Power Sleep[2]).

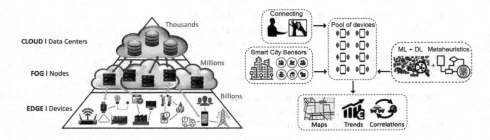

Fig. 1. Ubiquitous artificial intelligence at the edge.

Considering this, our main aim is to design a ubiquitous intelligent system (Fig. 1) for solving complex problems of smart cities using a highly decentralized computing model. However, this system will be composed of a wide variety of heterogeneous hardware. We have to deal with asynchronism, fault tolerance and the devices that join and leave the system. Therefore, we must analyze this hardware and test it to solve problems locally, before doing it in a distributed

---

[1] https://www.htc.com/es/go/power-to-give/.

[2] https://news.samsung.com/global/power-sleep-app-lets-you-be-a-part-of-an-advanced-scientific-research.

way. For this purpose, we perform a set of experiments in which we measure the speed, memory usage, and battery consumption of these devices solving a set of well-known optimization problems using **Genetic Algorithms** (GAs). This research is an extension of a previous work [4].

In summary, we will answer the following set of worth research questions:

- RQ1: Are these devices suitable to solve complex problems using smart algorithms?
- RQ2: What are the differences in resources usage between the different portable devices?
- RQ3: How can we develop efficient algorithms for so limited memory devices?.

This article is organized as follows: Sect. 2 describes the problems chosen for experimentation. Section 3 shows details about the design of our algorithm. Section 4 analyzes and discusses the experiments performed. Finally, Sect. 5 draws the main conclusions and the future work research.

2 Description of the Problems

In this section, we present the set of problems chosen for the experimentation, including different sizes, different types of underlying NP-Hard problems, different constraints, varied fitness functions, and considered the areas of research they represent. As a result, the benchmark contains binary and combinatorial problems with many different interesting features in optimization, such as epistasis, multimodality, and deceptiveness. The binary problems chosen are the Massively Multimodal Deceptive Problem (MMDP) [5], the Error Correcting Code Design (ECC) [6], the Minimum Tardy Task Problem (MTTP) [7], and finally, the One-Max problem [8] to ease comparisons to other domains and algorithms. Also, we have chosen a well-known combinatorial problem, the Capacitated Vehicle Routing Problem (CVRP) [9] that can scale and represent problems in smart mobility in cities. Let us now proceed to explain each one briefly.

MMDP is a problem in which bipolar deceptive functions with two global optima and with a number of deceptive optima are designed. In MMDP each subproblem s_i contributes to the total fitness value according to the number of ones it has. The number of local optma is quite large (n^k), while there are only 2^k global solutions. The degree of multimodality is regulated by the k parameter.

$$f_{MMDP(s)} = \sum_{i=1}^{k} fitness_{s_i} \qquad (1)$$

ECC consists of assigning codewords to an alphabet that minimizes the length of transmitted messages and that provides maximal correction of single uncorrelated bit errors, when the messages are transmitted over noisy channels. A code can be formally represented by a three-tuple (n, M, d), where n is the length of each codeword, M is the number of codewords and d is the minimum Hamming distance between any pair of codewords. An optimal code consists in constructing M binary codewords, each of length n, such that d, the minimum Hamming distance between each codeword and all other codewords, is maximized.

$$f_{ECC} = \frac{1}{\sum_{i=1}^{M} \sum_{j=1, i \neq j}^{M} \frac{1}{d_{ij}^2}} \quad (2)$$

$MTTP$ is a NP-hard task-scheduling problem wherein each task i from the set of tasks $T = 1, 2, \ldots, n$ has a length l_i (the time it takes for its execution), a deadline d_i (before which a task must be scheduled), and a weight w_i. The weight is a penalty that has to be added to the objective function in the event that the task remains unscheduled. The lengths, weights and deadlines of tasks are all positive integers. Scheduling the tasks of a subset S of T is to find the starting time of each task in S, such as at most one task at time is performed and such that each task finishes before its deadline. The objective function for this problem is to minimize the sum of the weights of the unscheduled tasks.

$$f_{MTTP}(x) = \sum_{i \in T-S} w_i \quad (3)$$

$OneMax$ (or $BitCounting$) is a simple problem consisting in maximizing the number of ones of a bitstring. Formally, this problem can be described as finding an string $x = \{x_1, x_2, \ldots, x_N\}$, with $x_i \in \{0, 1\}$, that maximizes the following equation:

$$f_{OneMax}(x) = \sum_{i=1}^{N} x_i \quad (4)$$

$CVRP$ is a problem in which a fixed fleet of delivery vehicles of the same capacity must service known customer demands for a single commodity from a common depot at minimum total transit costs. CVRP is defined as an undirected graph $G = (V, E)$ where $V = \{v_0, v_1, \ldots, v_n\}$ is a vertex set and $V = \{(v_i, v_j) | v_i, v_j \in V, i < j\}$ is an edge set. The depot is represented by v_0 and it is from where m identical vehicles of capacity Q must serve all the customers, represented by the set of n vertices $\{v_1, \ldots, v_n\}$. A solution for the CVRP is a partition $R_1, R_2, \ldots, R_m$ of V representing the routes of the vehicles. The cost of the problem solution is the sum of the costs of its routes $R_i = \{v_{i0}, v_{i1}, \ldots, v_{ik+1}\}$ where $v_{ij} \in V$ and $v_{i0} = v_{ik+1} = 0$, where 0 is the depot, satisfying $\sum v_{ij \in R_i}, q_j \leq Q$, which can be presented as:

$$Cost = \sum_{i=1}^{m} Cost(R_i) = \sum_{j=0}^{k} c_{ij+1} \quad (5)$$

where matrix $C = (c_{ij})$ is a non-negative cost between customers v_i and v_j defined on E.

3   Design of Our Algorithm

The algorithm used to test the performance of the portable devices is a Genetic Algorithm (GA). A GA is a population-based metaheuristic inspired by the process of natural selection. GAs are used to generate high-quality solutions to optimization and search problems. They are especially useful to solve NP-hard and even NP-complete problems. Exact algorithms are designed in such a way that it is guaranteed that they will find the optimal solution in a finite amount of time. However, when the problem is big, exact methods become useless while metaheuristic algorithms can still find good solutions in a reasonable amount of time.

The canonical GA uses three main types of rules at each step to create the next generation from the current population. First, selection rules select the individuals, called parents, that will be recombined. Second, recombination rules combine the parents to form a child for the next generation. Third, mutation rules apply random changes to the recombined individual to form the final child. Next, we replace the worst individual in the population with the new child. It is like natural selection. We repeat the process again and again until we achieve a good enough solution.

In this research we use a canonical steady state algorithm to solve our problems, adding an additional step in the case of the combinatorial problem (CVRP). In this case, we add a local search operator to the canonical GA due to its efficiency solving CVRP. We have implemented our algorithm similarly to that used by Dorronsoro and Alba [10].

4   Experimentation

For the experimentation we used five different portable devices of a wide variety of performances (Tables 1 and 2). These devices are a laptop, two smartphones, a tablet and a raspberry pi 3 (RP3). Their technical specifications are showed in Table 1. In experimentation we compare their performances solving known hard problems in literature in terms of speed, memory usage and battery consumption.

In this research, we focus on the behavior of these devices without using multithreading. In future work, we will check what happen when we use multithreading and compare it with these results.

The problems chosen for experimentation are detailed in Sect. 2. They are 4 binary problems (MTTP, ECC, MMDP, OneMax) and 1 combinatorial problem (CVRP). The algorithm used to solve these problems is a genetic algorithm and it is explained in Sect. 3.

We have chosen one instance of each problem (Table 3). The four binary instances have been chosen from JCell Library[3]. The CVRP instance is the CMT1 which is one of the set of instances proposed by Christofides [11] for this problem. The chosen instances are not large because they must be solved by devices with low processing capacity. For now, what interests us is to observe the performance in each platform and not solve very large problems. This is why we have chosen a varied set of problems that can be solved in an acceptable time by all the evaluated devices.

---

[3] http://neo.lcc.uma.es/software/jcell/.

Table 1. Hardware in the edge: features

	Laptop	RP3	mobileA	Tablet	mobileB
OS	Ubuntu 16.04 LTS (64-bit)	Raspbian 9 (64-bit)	Android 5.1.1 ARMv7 (32-bit)	Android 5.0.1 x86 (32-bit)	Android 6.0 AArch64 (64-bit)
Java VM	1.8.0_161	1.8.0_65	-	-	-
Model	Toshiba Satellite L50-B	Raspberry Pi 3	Zuk Z1	Lenovo TAB S8-50F	LeMobile LEX653
Memory	7893 MB 1600 MHz	745 MB 900 MHz	2871 MB 933 MHz	1870 MB 778 MHz	3759 MB 800 MHz
Factory battery	14.8 V 2800 mAh	3.7 V 3800 mAh	3.7 V 4000 mAh	3.8 V 4290 mAh	4.2 V 4000 mAh

Table 2. Processor specifications

	Laptop	RP3	mobileA	Tablet	mobileB
Name	Intel Core i7-4510U (64-bit)	ARM v7 BCM2709 (64-bit)	Qualcomm Krait 400 (32-bit)	Intel Atom Z3745 (32-bit)	ARM Cortex-A72 ARM Cortex-A53 MT6797D (64-bit)
Topology	1 Processor, 2 Cores, 4 Threads	1 Processor, 4 Cores	1 Processor, 4 Cores	1 Processor, 4 Cores	1 Processor, 10 Cores
Frequency	2.0–3.1 GHz	1.2 GHz	2.5 GHz	1.33–1.86 GHz	2.3 GHz × 2 1.85 GHz × 4 1.4 GHz × 4
L1 Inst. Cache	32.0 KB × 2	16.0 KB × 1	16.0 KB × 2	32.0 KB × 4	-
L1 Data Cache	32.0 KB × 2	16.0 KB × 1	16.0 KB × 2	24.0 KB × 4	-
L2 Cache	256 KB × 2	512 KB × 1	2.00 MB × 1	1.00 MB × 1	-
L3 Cache	4.00 MB × 1	-	-	-	-

Table 3. Problem instances

Problem	Chrom. size	Pop. size	Probabilities			Type
			Recomb.	Mutation	Local search	
OneMax	5000	100	0.8	$\frac{1}{chroSize}$	-	Binary
MTTP	200	100	0.8	$\frac{1}{chroSize}$	-	Binary
ECC	144	5000	0.8	$\frac{1}{chroSize}$	-	Binary
MMDP	240	5000	0.8	$\frac{1}{chroSize}$	-	Binary
CVRP	54	500	0.4	$\frac{1.2}{chroSize}$	1	Combinatorial

The parameters chosen to solve the binary problems are the most common in literature and the parameters chosen to solve CVRP can be found in [10], both are shown in Table 3. Regarding to population size, we tested with different sizes among 100, 200, 500, 1000, 2000, 5000 until we got 100% of success with 30 different seeds in each problem. For each experimental instance, we have carried out 30 independent runs with a different random seed in each repetition. Same seeds were used for each problem.

Our first experiment is to run some of the most well-known benchmarks on each platform to have an idea of their real performances. Most of these benchmarks are not multiplatform, however, we can obtain certain information using them in the platforms that we can.

Next, we evaluate the performance of each problem in each platform. Since all the devices use the same seeds, and since we are not using multithreading, all the devices solve the instances in the same number of evaluations. Therefore, what we compare is the time each device devotes to solve each instance.

Then, we analyze the use of the CPU and the memory of the different platforms when solving these instances. And finally, we compare the consumption of the battery.

4.1 Knowing Edge HW by Running Standard Benchmarks on It

First of all, we perform some benchmarks on these devices. These benchmarks are composed by the classical Whetstone [12], Dhrystone 2 [13], Linpack [14] and Livermore [15]. Also, we use Antutu benchmark[4] which is specific for Android OS, and we focus on their CPU (1 single core) and memory score. And finally we use one of the benchmarks most used today for multiplatforms devices, Geekbench 4 (1 single core)[5]. The results of the different benchmarks have been normalized getting a score. The score 1 represents the best result and the rest of scores are proportionals to this one (Table 4).

Table 4. Benchmarks

	Laptop	RP3	mobileA	Tablet	mobileB
GeekBench 4	1	-	2.15	2.43	1.38
Antutu CPU	-	-	1.12	1.26	1
Antutu maths	-	-	1.06	1.90	1
Antutu mem.	-	-	1	1.39	1.13
Whetstone	1	1.38	1.44	1.56	3.55
Dhrystone 2	1	4.42	2.65	4.13	2.18
Linpack	1	9.71	3.95	12.16	40.89
Livermore	1	13.53	6.89	17.90	5.91

The implementation of the classical benchmarks have been taken from Roy Longbottom's Benchmark Collection[6] which have implementations of many

---

[4] http://www.antutu.com/en/.
[5] https://www.geekbench.com/.
[6] http://www.roylongbottom.org.uk/.

benchmark for a wide variety of platforms. The results obtained in these benchmarks are not very accurate, apparently. This is because these benchmarks are very dependents of the compilers and the hardware architectures.

The results obtained in Antutu and GeekBench 4 are closer to what we expected. The problem is that Antutu is just for Android OS and GeekBench 4 does not have an implementation for Raspbian OS.

Based on the benchmark results and the features of the devices (Table 1 and 2) we can expect that the order from the faster to the slower device is as follow: *laptop, mobileB, mobileA, tablet, RP3.*

These benchmarks help us to have a first approach to these platforms as a first step before our experiments begin. Although, we should know that these benchmarks are just a number and does not have to be similar to the final results. For instance, Antutu is so common that hardware manufacturers have taken to cheating on the benchmark which makes the benchmark unreliable.

4.2   Time Results

In Table 5 we can see the experiment results per problem and per device. The most important thing here is the time performance because the number of evaluations is the same for all devices in each problem. We have normalized the times getting a score. The device with the fastest time in each problem gets a score of 1, and the rest of them get a score proportional to this in a incremental way.

We can see that the performance in Linux devices (laptop and RP3) is much better than in Android devices. It is interesting to see that RP3, a device of €30 cost, can obtain a result very similar than a laptop even when it has a much worse CPU. RP3 needs 10 times the time of the laptop to solve the same problem, that is not too much because of the cost of the RP3 compared to the laptop.

The bad news is the time Android devices need to solve these problems. It is clear that something is not working as well as expected in the optimization of the Java VM on Android. Android does not use Oracle Java VM, but uses Android Runtime (ART) (which replaced Dalvik Virtual Machine after Android KitKat). DVM was created to avoid JVM copyright and to better use memory and power in more limited devices. There are many studies [16] that shown that DVM was much slower than JVM, we have to say that some years later ART does not improve that results.

The tablet, and the smartphones have CPU much better than the RP3, however, they are getting a very bad performance solving these problems. Android OS is an obstacle to obtain our long-term goal and we will have to deal with it in future work. If benchmarks can obtain good results on these devices it means that it is possible and maybe we have to use native code to solve the restrictions of the Android OS.

Table 5. Problem results

Device	Prob.	Fitness avg	hit	Evaluations avg ± sd	max	min	Time (s) avg ± sd	max	min	score
laptop	OneMax	5×10^3	100 %	$(18 \pm 1) \times 10^4$	23×10^4	16×10^4	1.95 ± 0.14	2.38	1.76	1.00
RP3	OneMax	5×10^3	100 %	$(18 \pm 1) \times 10^4$	23×10^4	16×10^4	21.47 ± 1.46	26.28	19.43	11.01
mobileB	OneMax	5×10^3	100 %	$(18 \pm 1) \times 10^4$	23×10^4	16×10^4	98.79 ± 7.65	117.13	87.46	50.66
mobileA	OneMax	5×10^3	100 %	$(18 \pm 1) \times 10^4$	23×10^4	16×10^4	99.91 ± 6.85	119.85	90.33	51.23
tablet	OneMax	5×10^3	100 %	$(18 \pm 1) \times 10^4$	23×10^4	16×10^4	176.32 ± 12.34	215.39	158.40	90.41
laptop	MTTP	25×10^{-4}	100 %	$(48 \pm 48) \times 10^4$	24×10^4	39×10^3	0.52 ± 0.53	2.80	0.05	1.00
RP3	MTTP	25×10^{-4}	100 %	$(48 \pm 48) \times 10^4$	24×10^5	39×10^3	5.86 ± 5.82	29.65	0.50	11.23
mobileB	MTTP	25×10^{-4}	100 %	$(48 \pm 48) \times 10^4$	24×10^5	39×10^3	55.67 ± 54.97	278.69	4.42	106.67
tablet	MTTP	25×10^{-4}	100 %	$(48 \pm 48) \times 10^4$	24×10^5	39×10^3	66.01 ± 65.44	334.04	5.63	126.49
mobileA	MTTP	25×10^{-4}	100 %	$(48 \pm 48) \times 10^4$	24×10^5	39×10^3	71.75 ± 71.93	369.39	6.48	137.49
laptop	MMDP	40	100 %	$(37 \pm 46) \times 10^4$	22×10^5	19×10^4	2.12 ± 0.66	4.68	1.69	1.00
RP3	MMDP	40	100 %	$(37 \pm 46) \times 10^4$	22×10^5	19×10^4	17.94 ± 5.66	40.09	14.29	8.47
mobileB	MMDP	40	100 %	$(37 \pm 46) \times 10^4$	22×10^5	19×10^4	331.35 ± 195.25	1180.94	199.50	156.46
tablet	MMDP	40	100 %	$(37 \pm 46) \times 10^4$	22×10^5	19×10^4	416.22 ± 74.83	683.50	349.13	196.53
mobileA	MMDP	40	100 %	$(37 \pm 46) \times 10^4$	22×10^5	19×10^4	476.01 ± 122.47	983.44	379.61	224.76
laptop	ECC	674×10^{-4}	100 %	$(33 \pm 6) \times 10^4$	45×10^4	23×10^4	2.68 ± 0.34	3.41	1.99	1.00
RP3	ECC	674×10^{-4}	100 %	$(33 \pm 6) \times 10^4$	45×10^4	23×10^4	25.42 ± 3.45	32.85	18.94	9.48
mobileB	ECC	674×10^{-4}	100 %	$(33 \pm 6) \times 10^4$	45×10^4	23×10^4	470.22 ± 75.97	621.57	310.18	175.30
tablet	ECC	674×10^{-4}	100 %	$(33 \pm 6) \times 10^4$	45×10^4	23×10^4	533.14 ± 66.72	672.33	403.31	198.76
mobileA	ECC	674×10^{-4}	100 %	$(33 \pm 6) \times 10^4$	45×10^4	23×10^4	577.81 ± 71.82	725.69	438.77	215.41
laptop	CVRP	524.61	100 %	$(45 \pm 11) \times 10^6$	80×10^6	24×10^6	9.85 ± 2.45	17.29	5.09	1.00
RP3	CVRP	524.61	100 %	$(45 \pm 11) \times 10^6$	80×10^6	24×10^6	76.39 ± 19.54	136.23	39.89	7.76
mobileB	CVRP	524.61	100 %	$(45 \pm 11) \times 10^6$	80×10^6	24×10^6	1637.20 ± 553.04	3389.71	771.45	166.28
tablet	CVRP	524.61	100 %	$(45 \pm 11) \times 10^6$	80×10^6	24×10^6	2376.80 ± 601.78	4218.04	1247.25	241.40
mobileA	CVRP	524.61	100 %	$(45 \pm 11) \times 10^6$	80×10^6	24×10^6	2440.79 ± 589.31	4150.62	1332.99	247.90

4.3 Resources Usage

To measure memory we use the unique set size (USS) which is the portion of main memory (RAM) occupied by a process which is guaranteed to be private to that process. It was proposed by Matt Mackall because of the complications that arose when trying to count the "real memory" used by a process.

It is interesting to see in Fig. 2 how smartphones, the tablet and the RP3 manage the memory better than the laptop. It seems like the VM is optimized to use the memory better on these devices with memory restrictions than on the laptop. We have to highlight the use of memory in the tablet and in the RP3 that need much less memory than the other devices to solve the same problems.

Regarding to CPU, we have to say that to run the same java code on Android we needed a simple interface with a button to start the algorithm on background. That is why Android devices are using two cores when laptop and RP3 are using only one. However, we can see how the tablet uses less CPU than smartphones. It seems that Intel x86 is using the CPU in a smarter way than ARM/AArch on Android OS.

4.4 Battery Consumption

We have measured the battery consumption during 2 h with and without the algorithm running. As a result, we have obtained that all devices increment their battery consumption in a similar way from 1.5 to 1.97 depending on the device. The laptop increment their battery consumption 1.72 with respect to their normal consumption, RP3 did 1.8, mobileA did 1.97, tablet did 1.5 and mobileB did 1.76.

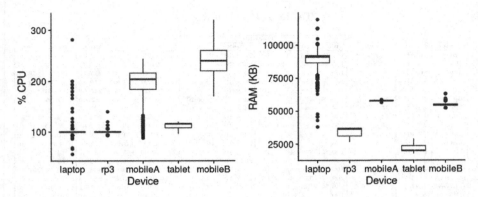

Fig. 2. Resources usage.

These are good results comparing it with making a call, web browsing or playing a video. These activities can increment battery drain in 5 or 6 times depending on the device.

5   Conclusions

In this article, we have analyzed the performance of five different devices while running smart algorithms in the edge. We have measured their performance, CPU and memory usage, and battery consumption.

First, we chose a representative set of problems with different interesting features and solved them using a GA. Second, we showed that benchmarks are not so useful to known the real performance of the devices although they can help to have a first impression about it. Also, we showed that OS could be more important than HW to get good results running these algorithms. Another remarkable result is that battery consumption is lower than many applications we commonly use, and much more lower than a call for instance.

Finally, we have got many unexpected results. On the one hand, RP3 arises as a perfectly suitable platform to run algorithms in the edge. RP3 is very cheap, and it uses less memory than laptop and an acceptable time solving the problems. RP3 is commonly used in smart city sensors, so it is very suitable device to perform edge computing. On the other hand, Android devices got very poor results in terms of performance. It seems like Android OS has many restrictions to run high-performance applications, Android OS seems more interested in maintaining battery consumption low and in using less memory (cleaning garbage collector more often). Anyway, this only encourages us to look for new approaches to solve the problem of performance in Android. This platform is crucial for our long-term goals, so in future work we will use native code (JNI).

We have shown how devices with low processor capacity as RP3 are perfectly useful solving optimization problems. In the case of smartphones and tablets, we still have to deal with the limitations of Android OS. In future work, we will check what happen when we use multithreading and we will research to new ways to improve the performance in Android devices like using native code.

References

1. Index, C.G.C.: Forecast and methodology, 2015–2020 white paper. Accessed 1 June 2016
2. Shi, W., Cao, J., Zhang, Q., Li, Y., Xu, L.: Edge computing: vision and challenges. IEEE Internet Things J. **3**(5), 637–646 (2016)
3. Levine, P.: Return to the edge and the end of cloud computing, December 2016. https://a16z.com/2016/12/16/the-end-of-cloud-computing/. Accessed 17 Feb 2018
4. Morell, J., Alba, E.: Distributed genetic algorithms on portable devices for smart cities. In: Alba, E., Chicano, F., Luque, G. (eds.) Smart-CT 2017. Lecture Notes in Computer Science, vol. 10268. Springer, Cham (2017). https://doi.org/10.1007/978-3-319-59513-9_6
5. Goldberg, D.E., Deb, K., Horn, J.: Massive multimodality, deception, and genetic algorithms. Urbana **51**, 61801 (1992)
6. MacWilliams, F.J., Sloane, N.J.A.: The Theory of Error-Correcting Codes. Elsevier, New York City (1977)
7. Stinson, D.: An introduction to the design and analysis of algorithms. The Charles Babbage Research Centre, St. Pierre (1985)
8. Eshelman, L.: On crossover as an evolutionarily viable strategy. In: Proceedings of the Fourth International Conference on Genetic Algorithms, pp. 61–68 (1991)
9. Dantzig, G.B., Ramser, J.H.: The truck dispatching problem. Manag. Sci. **6**(1), 80–91 (1959)
10. Alba, E., Dorronsoro, B.: Cellular Genetic Algorithms, vol. 42. Springer, Boston (2009). https://doi.org/10.1007/978-0-387-77610-1
11. Christofides, N., Mingozzi, A., Toth, P.: Loading problems. In: Christofides, N., et al. (eds.) Combinatorial Optimization, pp. 339–369 (1979)
12. Curnow, H.J., Wichmann, B.A.: A synthetic benchmark. Comput. J. **19**(1), 43–49 (1976)
13. Weicker, R.P.: Dhrystone: a synthetic systems programming benchmark. Commun. ACM **27**(10), 1013–1030 (1984)
14. Dongarra, J.J., Bunch, J.R., Moler, C.B., Stewart, G.W.: LINPACK Users' Guide. SIAM, Philadelphia (1979)
15. McMahon, F.H.: The livermore fortran kernels: a computer test of the numerical performance range. Technical report, Lawrence Livermore National Laboratory, CA (USA) (1986)
16. Batyuk, L., Schmidt, A.-D., Schmidt, H.-G., Camtepe, A., Albayrak, S.: Developing and benchmarking native linux applications on android. In: Bonnin, J.-M., Giannelli, C., Magedanz, T. (eds.) MOBILWARE 2009. LNICST, vol. 7, pp. 381–392. Springer, Heidelberg (2009). https://doi.org/10.1007/978-3-642-01802-2_28

Developing Genetic Algorithms Using Different MapReduce Frameworks: MPI vs. Hadoop

Carolina Salto[1,2](✉), Gabriela Minetti[1], Enrique Alba[3], and Gabriel Luque[3]

[1] Facultad de Ingeniería, Universidad Nacional de La Pampa, General Pico, Argentina
{saltoc,minettig}@ing.unlpam.edu.ar
[2] CONICET, Buenos Aires, Argentina
[3] Departamento de Lenguajes, Universidad de Málaga, Málaga, Spain
{eat,gabriel}@lcc.uma.es

Abstract. MapReduce is a quite popular paradigm, which allows to no specialized users to use large parallel computational platforms in a transparent way. Hadoop is the most used implementation of this paradigm, and in fact, for a large amount of users the word Hadoop and MapReduce are interchangeable. But, there are other frameworks that implement this programming paradigm, such as MapReduce-MPI. Since, optimization techniques can be greatly beneficiary of this kind of data-intensive computing modeling, in this paper, we analyze the performance effect of developing genetic algorithms (GA) using different frameworks of MapReduce (MRGA). In particular, we implement MRGA using Hadoop and MR-MPI frameworks. We analyze and compare both implementations considering relevant aspects such as efficiency and scalability to solve a large dimension problem. The results show a similar efficiency level between the algorithms but Hadoop presents a better scalability.

1 Introduction

MapReduce (MR) is a programming paradigm, created by Google researchers [3], to process and generate big data sets with a parallel and distributed algorithm on clusters and clouds. The programmer may abstract from the issues of distributed and parallel programming, because the MR manages the load balancing, the network performance, and the fault tolerance in some frameworks. This made MR popular for those not knowing on parallel programming but still wanting to use it, creating a new branch of parallel studies where the focus is on the application and not in exploiting the underlying hardware.

The Google implementation of MapReduce is a property C++ library with communication between networked machines via remote procedure calls. It allows for fault tolerance when large numbers of machines are employed, and can use disks as out-of-core memory to process petabyte-scale data sets [4]. Incorporating these characteristics, the Hadoop implementation of MR [1] is an

© Springer Nature Switzerland AG 2018
F. Herrera et al. (Eds.): CAEPIA 2018, LNAI 11160, pp. 262–272, 2018.
https://doi.org/10.1007/978-3-030-00374-6_25

open-source Java library that also supports stand-alone *map* and *reduce* kernels written as shell scripts or in various programming languages.

Since Hadoop allows parallelism of data and control, it is common science to ask ourselves on other software tools doing similar jobs. The MapReduce-MPI (MR-MPI) [11] is a library built on top of MPI with a somewhat similar goal. Since it is an open-source C++ library, it allows precise control over the memory and format of data allocated by each processor during a MapReduce operation. The *map* and *reduce* are user-specified functions that can be written in C, C++, and Python languages. Here you can have more control on the platform, allowing to improve the bandwidth performance and reduces the latency costs. Though their services (for Hadoop and MR-MPI) are not exactly the same, most applications can be written on top of them, and a comparative analysis is in order for any scientific curious mind, so as to offer evidences on their performance.

Given the scientific and industrial need for new evolutionary algorithms, particularly for scalable genetic algorithms (GAs), a data-intensive computing modeling can greatly benefit the field to build new optimization and machine learning models with them. For this reason, the objective of this work is to analyze different ways to introduce the MR elements into a Simple Genetic Algorithm (SGA) [8]. This fundamental research crystallizes in a technique named MRGA, coming out from a parallelization of each iteration of a SGA under the Iteration-Level Parallel Model [12].

In the literature, many researchers report GAs programmed on Hadoop's MapReduce [2,5,7,13,14] but no under MR-MPI, according to the authors knowledge. We then will implement MRGA using both open-source MapReduce frameworks, Hadoop and MR-MPI, to yield algorithms MRGA-H and MRGA-M, respectively. Later on, we analyze and compare both implementations considering relevant aspects such as efficiency and scalability, among others, to solve a large problem size of industrial interest as the knapsack problem [6].

This article is organized as follows: next section discusses the MR paradigm, since our position is to research in its meaning and workings, not that much to go on it as a black box and use it without any control of what it is doing. Then, Sect. 3 develops on the similitudes and differences in the two mentioned frameworks. Section 4 presents a brief state of the art in implementing GAs in MR, and contains our proposal. Sections 5 and 6 define meaningful experiments to reveal information on the two systems, perform them, and give some findings. Finally, Sect. 7 summarizes our conclusions and expected future work.

2 The MapReduce Paradigm

Let us first make a bit of history. The MapReduce paradigm was inspired by the *map* and *reduce* primitives present in Lisp and other functional languages, though it is not equivalent to them, just it was inspired by them. In functional programming, the *map* function is used when a same unary operation is to be performed over each element of a list, and the *reduce* operation combines all the elements of a sequence using a binary operation. In the MR paradigm, this functional idea allows to parallelize large computations easily by incorporating user-specified *map* and *reduce* functions.

The MR paradigm organizes an application into a pair (or a sequence of pairs) of *map* and *reduce* functions [3] The *map* functions are independent of each other and execute in parallel without connection among them. The second ones play the role of aggregating the values with the same key.

Each *map* function takes as input a set of key/value pairs (records) from data files and generates a set of intermediate key/value pairs. Then, MapReduce groups together all these intermediate values associated with a same intermediate key. A value group and its associated key is the input to the *reduce* function, which combines these values in order to produce a new and possibly smaller set of key/value pairs that are saved in data files. Furthermore, this function receives the intermediate values via an iterator, allowing the model to handle lists of values that are too large to fit into main memory.

The input data is automatically partitioned into a set of M splits when the *map* invocations are distributed across multiple machines, where each input split is processed by a *map* invocation. The intermediate key space is divided into R pieces, which are distributed into R *reduce* invocations.

Finally, we would like to mention that MR is not the only paradigm to analyze large amount of data. Frameworks, like Apache Spark, are currently very popular. Spark allows a more flexible organization of the processes, appropriate for iterative algorithms, and eases efficiency due to the use of in memory data structures (RDDs). But, MR approaches can be also useful in many cases. For example, Spark requires a quite large amount of RAM memory and it is quite greedy in the utilization of the resources of the cluster, while MR can be used in low-resource and non-dedicated platforms. Also, many problems in industrial domains are implemented in C/C++, which are only natively supported by MR implementations. Therefore, the research on efficient uses of MR approaches (as done in this article) is today an important domain [2,5].

3   Hadoop MapReduce and MR-MPI Frameworks

Our aim in this work is to perform a comparison of the Hadoop MapReduce framework [1] and the MR-MPI library [11] functionalities, with the goal of identifying the advantages and limitations of each one. In this process, the focus is put in the installation, use, and productivity characteristics of each framework.

On the one hand, the Hadoop framework consists of a single master *Resource-Manager*, one slave *NodeManager* per cluster-node, and a *MRAppMaster* per application, which is implemented using the Hadoop YARN framework [1]. The Hadoop client submits the job/configuration to the *ResourceManager*, which distributes the software/configuration to the slaves, schedules and monitors the tasks, providing status and diagnostic information to the client including fault tolerance management. In this sense, it is noticeable that the installation and the configuration of the Hadoop framework required a very specific and long sequence of steps, becoming difficult to adapt it to a particular cluster of machines. Moreover, at least one node (master) is dedicated to the system management. In contrast, MR-MPI is a small and portable C++ library that only uses MPI for inter-processor communication, thus the user writes a main program that runs on each processor of a cluster, making *map* and *reduce* calls to the MR-MPI library. As a consequence, no fault tolerance is provided by MR-MPI, but no extra installation and configuration tasks are needed.

On the other hand, the use of MPI within the MR library follows the traditional mode to call the *MPI_Send* and *MPI_Recv* primitives between pairs of processors, using large aggregated messages to improve bandwidth performance and reduced latency costs. But as MPI, for MR-MPI the non-detection of a dead processor or retrieval of the data is carried out. Instead, the Hadoop framework stores both the input and the output of the job in the Hadoop Distributed File System (HDFS), whereas MR-MPI allocates pages of memory. In this way, only Hadoop provides data redundancy. The price is that the data access time is greater than for MPI. Moreover, the HDFS creation and its configuration also require a careful setting of properties (time consuming), which defines the memory sizes for the *map* and *reduce* tasks, being related with the memory in each cluster's node. Instead, in MR-MPI the use of C++ allows precise control over the memory usage, and no specific configurations must be done by users.

4   GAs and MapReduce

In the literature, we can find some proposals to use the MapReduce to model GAs. The most representative works are described in the following paragraphs.

Verma et al. [13] proposed a SGA based on the selecto-recombinative GA, proposed by [8], which only uses two genetic operators: selection and recombination. SGA was developed using the Hadoop framework. The authors match the *map* function with the evaluation of the population fitness, whereas the *reduce* function performs the selection and recombination operations. They proposed the use of a custom-made Partitioner function, which splits the intermediate key/value pairs among the reducers by using a random shuffle. The same authors in [14] modeled the compact genetic algorithm (CGA) and the extended CGA (eCGA), which are an estimation distribution algorithm (EDA). The proposal consists of two MapReduce steps. The first one computes the fitness and the selection operation in the *map* and *reduce* functions, respectively. After this step, the model is built, which is followed by the second step in charge of the crossover operator according to the model that has just been built.

In [7], the authors proposed a similar model of GA than Verma et al. [13] for software testing. The main difference relays in the use of only one reducer that receives the entire population. Thus, reducer can perform the selection and apply on the generation the crossover and mutation operators to produce a new offspring to be evaluated in the next MapReduce job.

In the next subsections, we describe the SGA used in this work and how it is modeled using MapReduce (MRGA). The idea is to implement the same MRGA using the Hadoop (MRGA-H) and the MR-MPI frameworks (MRGA-M).

4.1 MRGA

MRGA is based on the SGA proposed by [8]. The selection of this SGA is based on their minimal set of operations, which contributes to focus the effort in the way to map the MapReduce elements to the GA model.

The SGA starts by generating an initial population. During the evolutionary cycle, the population is evaluated and then a set of parents are selected by tournament selection [9] without repetition. After that, the uniform recombination operator is applied to them. The recently created offspring conform the new population for the next generation (using the generational replacement). The evolutionary process ends when either the optimum solution to the problem at hand is found or the maximum number of iterations is reached.

Our proposed MRGA algorithm preserves the SGA behavior, but it resorts to parallelization for some parts: the evaluation and the application of genetic operators. Although, our technique performs several operations in parallel, its behaviour is the same to the sequential GA.

A key/value pairs has been used to represent individuals in the population. The representation of an individual for the knapsack problem is a sequence of bits, one for each item, indicating if the item is in the knapsack or not. To distinguish identical individuals (with the same genetic configuration), a random identifier (ID) is assigned in the *map* function to each one. The ID use prevents that identical individuals were assigned to the same reduce function, in the phase of shuffling when the intermediate key/multivalues space are generated. The sequence of bits together with the ID correspond to the key in the key/value pair. The value part, which is the fitness of an individual, is added by the *map* function in charge with evaluating that individual, which is their input. The sequence of bits is represented by an array of long ints (64 bits), in order to optimize the amount of memory to represent an individual. The ID is another long int, assigned in the last position of the array. Consequently, this optimization enables the use of efficient bit operations for recombination and evaluation.

For large problem sizes, the population initialization could be a consuming time process. The situation can get worse with large individual sizes, as the case in this work. According to this situation, this initialization is parallelized in a separate MapReduce phase. The *map* functions are only used to generate random individuals, which are written to permanent files in order to be read in the following MapReduce phase.

After that, the iterative evolutionary process begins, but the MapReduce model does not offer methods to implement iterative algorithms. Consequently, a chain of multiple MapReduce tasks are generated using the output of the last one as the input of the next one. An important consideration is that the termination condition must be computed within the main process of the MapReduce program. Each iteration consists of a *map* and *reduce* functions.

Considering large problem sizes, big individuals, and complex evaluation functions, the *map* functions compute the fitness of individuals. As each *map* has assigned different chunk of data, they evaluate a set of different individuals in parallel. This fitness is added as value in the key/value pair. Each *map* finds their best individual and writes it to a file. Using these files, the main process computes the best current individual to determine if the stop criterion is met.

The *reduce* functions carry out the genetic operations. The binary tournament selection is performed locally with the intermediate key space, which is distributed in the partitioning stage after *map* operation. The uniform crossover (UX) operator is applied over the selected individuals. The generational replacement is implemented by writing these new individuals into a file in order to become the input for the *map* in the next MapReduce task (a new iteration).

Regardless of the MapReduce framework used, the key/value pairs, generated at the end of the *map* phase, are shuffled and split to the *reducers* and converted in an intermediate key/multivalues space. The shuffle of individuals consists in a random assignment of individuals to reducers instead of using a traditional hash function over the key. This modification, as suggested in [13], responds to avoid that all values corresponding to a same key will be send to the same reduce function, as a consequence identical individuals, generating a biased partition and fix assigned of individuals to the same partition through evolution and an unbalance load of reduce functions at the end of evolution. Therefore, the intermediate key/value pairs are distributed into R partitions by computing a simple formula (rand()%R).

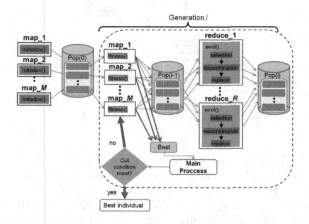

Fig. 1. MRGA-H scheme

4.2   Our Proposals

This section presents the MRGA implementation details under the two selected frameworks to solve the Knapsack problem [6]. The implementation of MRGA algorithm under the Hadoop framework (MRGA-H) is obtained from [13]. In the case of the MR-MPI implementation (MRGA-M) was developed from scratch.

The scheme of MRGA-H is plotted in Fig. 1. Chunks of data processed by each *map* are represented by shaded rectangles. Some modifications were introduced in the code developed in [13]. The most important ones are related to the changes imposed by passing from the old MRV1 to the new Java MapReduce API MRV2 [1], because they are not compatible with each other. These important differences involve the new package name, the context objects that allow the user code to communicate with the MapReduce system, the Job control that is performed through the Job class in the new API (instead of the old one JobClient), and the *reduce*() method that now passes values as an Iterable Object. Another modification is related with the generation of a random individual ID, which is added as another long into the array used to encode the solution. Finally, the individual evaluation was included in the method fitness of the GAMapper Class. The rest of the code with the functionality of the GA remains without important modifications.

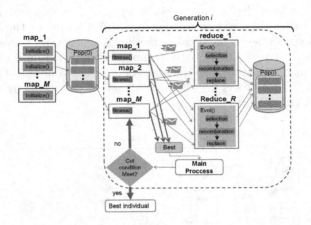

Fig. 2. MRGA-M scheme

The MRGA-M follows the scheme shown in Fig. 2 where boxes with solid outlines are files. Chunks of data processed by each *map* are represented by shaded rectangles. First, it creates the MPI environment for the parallel execution. Then, the sequence begins with the instantiation of an MR object and the setting of their parameters. The first MR phase consists of only one *map* function (which called to a serial *Initialize*()) to create the initial population. In our implementation the main process (process with MPI id equal to zero) generates a list of filenames. Our *Initialize*() function processes each file to build the initial population. The second and following MR phases have a sequence of *map* and *reduce* functions. These *map* functions receive a chunk of the large file passed back to our *fitness*() function as a string to process. This type of *map* function allows to concatenate all files in the input directory together and then split it in M chunks. The *fitness*() function processes each key (an individual) received, evaluates it obtaining the fitness value and emits a key/value pairs. After that, the MR-MPI *aggregate*() function shuffles the key/value pairs across processors by distributing the pairs randomly. Then, the MR-MPI *convert*() function transforms a key/value pairs into a key/multi-value pairs. The duplicate keys in a same processor are found and their values are concatenated into a list of values associated with the key. Finally, the $Evol()$ function (from the *reduce* method) will be called once for each key/multi-value pair assigned to a processor. This function selects a pair of individuals by tournament selection and performs the recombination. The new individuals generated are written into permanent storage to be read by the *map* methods in the following MR phase.

5   Experimentation Methodology

In order to analyze the algorithms, we use the Knapsack problem instances obtained with the generator described in [10]. To deal with large volumes of data, three different instances with huge dimensions were generated: kp50K, kp100K, and kp200K with 50000, 100000, and 200000 items, respectively.

The computational environment used in this work to carried out the experimentation consists of five nodes. Four of them are machines with INTEL I7 3770 K quad-core processors 3.5 GHz, 8 GB RAM, and the Slackware Linux with 3.2.29 kernel version. The fifth node is a machine with INTEL I5 4440 dual-core processors 3.1 GHz, 4 GB RAM, and the Slackware Linux with 3.10.17 kernel version. All the machines are connected by a Fast Ethernet.

Each execution of MRGA-H and MRGA-M launches 18 *mappers* and 18 *reducers* in each generation, where 50000 randomly initialized individuals are evolved during 1500 generations. Individuals are represented by arrays of long ints (64 bits) and their lengths depends on the instance dimension. For example, the individual length for the kp50K instance is 782 long ints (50000/64) requiring 6.3 KB, and the population needs 312.8 MB. For kp100K and kp200K instances, the required memory sizes are 78 MB and 1,25 GB, respectively. These memory requirements justify the use of parallel programming paradigms as MapReduce.

6   Result Analysis

In this section, we first present an analysis concerning the best fitness reached by MRGA-H and MRGA-M together with the generation that the best fitness was found and the total execution time. Then, additional analyses concerning the particular time involved by *map* and *reduce* functions are performed.

Table 1 presents the results obtained by MRGA-H and MRGA-M. The best fitness values (the Fit_{Best} column) found by both algorithms are similar, but MRGA-M presents a little advantage over MRGA-H. Furthermore, the iteration number that the best individual was found is analyzed. Both algorithms find their best solution in the last iterations ($Iter_{Best}$ column), indicating that they need more iterations to converge. Given we work with big data sets, for the smallest instance the mean execution time is approximately 7 h for MRGA-M and 16 h for MRGA-H (TT[min] column). As a consequence, the total execution time takes 98 h, when all instances are solved once by both algorithms.

Table 1. Results obtained by both MapReduce algorithms for the knapsack problem

Instance	MRGA-H			MRGA-M		
	Fit_{Best}	$Iter_{Best}$	TT[min]	Fit_{Best}	$Iter_{Best}$	TT[min]
kp50K	199024666	1499	919,45	199694097	1496	391,63
kp100K	381284661	1499	1054,80	384889578	1499	681,62
kp200K	696920716	1499	1324,95	710860713	1499	1444,29

Analyzing scalability according to the instance dimensionality, the mean total execution time (see TT[min] column) and the mean execution time of the *map* and *reduce* tasks (Fig. 3) are studied. In this sense, we observe the mean total execution times required to solve kp50K and kp100K are very different for each algorithm, i.e., MRGA-H is 2.35 times slower than MRGA-M to solve kp50K, and 1.55 times slower for the kp100K instance. However, for kp200K both algorithms spend a similar execution time.

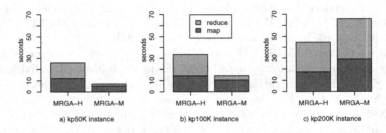

Fig. 3. Mean execution time of *map* and *reduce* tasks of our proposed algorithms.

Moreover, focusing on the scalability of each algorithm, the MRGA-M's time requirements increase proportionally to the increments of the instance dimensions, e.g. it needs twice as much time for kp100K as for kp50K. In contrast, MRGA-H does not present the previous behavior, it only needs 15% more of time when kp100K is solved with respect to kp50K. Similar situations are observed when the time requirements to solve kp100K and kp200K are compared.

The mentioned differences are also detected in the Fig. 3, where the mean execution time of the *map* and *reduce* tasks is shown. The differences arise because MRGA-M keeps and manages the population from memory while MRGA-H uses HDFS to manage it. Thus, for MRGA-M, the access time to each individual and the memory sizes required are different for each instance. However, for MRGA-H the Java Virtual Machine container and chunk sizes are equals for all instances, resulting in the same time to access to the individual. This suggests a better performance for MRGA-H than for MRGA-M, when bigger instances will be solved. Therefore, the MRGA-H system continues to perform adequately as its workload grows until the capacity of the containers allows it.

7   Conclusions

In this work, different ways to introduce the MapReduce elements into a Simple Genetic Algorithm are studied. As a consequence, two MRGA implementations arose using both open-source MapReduce frameworks, Hadoop and MR-MPI, to yield algorithms MRGA-H and MRGA-M respectively. These algorithms solve a large problem size of industrial interest as the knapsack problem. When the efficiency is analyzed, we detect that the two MRGA algorithms needs the same number of iterations to find their best solutions. Being quite similar the solution quality obtained by them with a little advantage in favor of MRGA-M. This is an expected outcome because both MRGA algorithms are based on the same selecto-recombinative genetic algorithm.

Considering the algorithmic scalability from the problem dimension point of view, we observe that MRGA-H is more scalable than MRGA-M. The reasons of this is the growth on the MRGA-H time requirement is not directly proportional to the instance dimension increment, i.e. the growth percentage of the execution time decreases as the number of items increases.

In future works, we will further conduct the experiments on much larger datasets in order to determine the maximum scalability level supported by these algorithms. We will also study the performance of these MRGA algorithms considering different numbers of *map* and *reduce* tasks.

Acknowledgments. This research has been partially funded by the Spanish MINECO and FEDER projects TIN2014-57341-R, TIN2016-81766-REDT, TIN2017-88213-R, and by Andalucía Tech, Universidad de Málaga. In addition, the research was funded by the National University of La Pampa, the Incentive Program from MINCyT, and CONICET.

References

1. Welcome to Apache Hadoop! Technical report, The Apache Software Foundation (2014). http://hadoop.apache.org/
2. Cano, A., García-Martínez, C., Ventura, S.: Extremely high-dimensional optimization with MapReduce: scaling functions and algorithm. Inf. Sci. **415–416**(Supplement C), 110–127 (2017)
3. Dean, J., Ghemawat, S.: MapReduce: simplified data processing on large clusters. In: OSDI04: Proceedings of the 6th Conference on Symposium on Operating Systems Design and Implementation. USENIX Association (2004)
4. Dean, J., Ghemawat, S.: MapReduce: simplified data processing on large clusters. Commun. ACM **51**(1), 107–113 (2008)
5. Ferrucci, F., Salza, P., Sarro, F.: Using Hadoop MapReduce for parallel genetic algorithms: a comparison of the global, grid and Island models. Evol. Comput. 1–33 (2017). https://doi.org/10.1162/evco_a_00213
6. Garey, M.R., Johnson, D.S.: Computers and Intractability: A Guide to the Theory of NP-Completeness. Freeman, New York (1979)

7. Di Geronimo, L., Ferrucci, F., Murolo, A., Sarro, F.: A parallel genetic algorithm based on Hadoop MapReduce for the automatic generation of JUnit test suites. In: 2012 IEEE Fifth International Conference on Software Testing, Verification and Validation, pp. 785–793, April 2012
8. Goldberg, D.E.: The Design of Innovation: Lessons from and for Competent Genetic Algorithms. Kluwer Academic Publishers, Dordrecht (2002)
9. Miller, B., Goldberg, D.: Genetic algorithms, tournament selection, and the effects of noise. Complex Syst. **9**, 193–212 (1995)
10. Pisinger, D.: Core problems in knapsack algorithms. Oper. Res. **47**, 570–575 (1999)
11. Plimpton, S., Devine, K.: MapReduce in MPI for large-scale graph algorithms. Parallel Comput. **37**(9), 610–632 (2011)
12. Talbi, E.: Metaheuristics: From Design to Implementation. Wiley, Hoboken (2009)
13. Verma, A., Llorà, X., Goldberg, D.E., Campbell, R.: Scaling genetic algorithms using MapReduce. In: ISDA 2009, pp. 13–18 (2009)
14. Verma, A., Llorà, X., Venkataraman, S., Goldberg, D.E., Campbell, R.: Scaling eCGA model building via data-intensive computing. In: IEEE Congress on Evolutionary Computation, pp. 1–8, July 2010

Strain Design as Multiobjective Network Interdiction Problem: A Preliminary Approach

Marina Torres[1](✉), Shouyong Jiang[2], David Pelta[1], Marcus Kaiser[2], and Natalio Krasnogor[2]

[1] Models of Decision and Optimisation Research Group, Department of Computer Science and A.I., Universidad de Granada, 18014 Granada, Spain
{torresm,dpelta}@decsai.ugr.es
[2] School of Computing, Newcastle University, Newcastle upon Tyne NE4 5TG, UK
math4neu@gmail.com, {marcus.kaiser,natalio.krasnogor}@ncl.ac.uk

Abstract. Computer-aided techniques have been widely applied to analyse the biological circuits of microorganisms and facilitate rational modification of metabolic networks for strain design in order to maximise the production of desired biochemicals for metabolic engineering. Most existing computational methods for strain design formulate the network redesign as a bilevel optimisation problem. While such methods have shown great promise for strain design, this paper employs the idea of network interdiction to fulfil the task. Strain design as a Multiobjective Network Interdiction Problem (MO-NIP) is proposed for which two objectives are optimised (biomass and bioengineering product) simultaneously in addition to the minimisation of the costs of genetic perturbations (design costs). An initial approach to solve the MO-NIP consists on a Nondominated Sorting Genetic Algorithm (NSGA-II). The shown examples demonstrate the usefulness of the proposed formulation for the MO-NIP and the feasibility of the NSGA-II as a problem solver.

Keywords: Strain design · Network interdiction
Metabolic networks · Multiobjective bilevel optimisation

1 Introduction

Microorganisms have been increasingly recognised as a very promising platform for the production of industrially relevant biochemicals, such as antibiotics and amino acids, as well as biomaterials. Whatever the desired product is, it is important to have a good understanding of metabolic networks of the host and then use the biological information to manipulate the corresponding biosynthesis pathway such that more flux can be directed toward the desired "target" product without rendering the cell inviable. With recent advances in genome sequencing technologies and high-throughput screening experiments, computational biologists are able to reconstruct genome-scale metabolic networks with high prediction accuracy for a range of well-characterised microorganisms, such as *E. coli*

© Springer Nature Switzerland AG 2018
F. Herrera et al. (Eds.): CAEPIA 2018, LNAI 11160, pp. 273–282, 2018.
https://doi.org/10.1007/978-3-030-00374-6_26

[7,15] and *B. subtilis* [9]. Such high-quality metabolic networks allow to predict phenotypes from genotypes in an efficient manner. This is particularly beneficial for strain design that is aimed at the identification of genetic perturbations needed for the production of biochemicals of interest.

Computational strain design has gained increasing attention in recent years due to the improvement on genome-scale reconstruction of metabolic networks. It involves the handling of two primary objectives, i.e. the bioengineering objective (mostly the target product) and the cellular objective (mainly biomass), which are often in conflict. It is widely accepted that cells' main function is to optimise the cellular objective. OptKnock is known to be the first computational strain design technique focusing on genetic deletions while maximising both objectives [3]. OptKnock mathematically formulates strain design as a mixed-integer bilevel optimisation problem with the leader maximising the bioengineering objective and the follower maximising the cellular objective while subject to mass balance constraints. The leader's decision on the state of some reactions (represented by binary variables) constrains the corresponding fluxes to zero in the follower's mass balance constraints. Based on this idea, a number of similar methods have been developed in the literature, and some of them differ mainly in the techniques used to solve the high-complexity bilevel problem [12] whereas others discuss the choice and suitability of the two objectives [11,17]. The concept of multiobjective optimisation has also be introduced to strain design in a few studies [4,10].

Network interdiction problem [18] involves competition of two network adversaries. One adversary optimises its network-associated objective (e.g. maximum flow) while the other destroys the network with limited budget in order to aggravate the first adversary's objective. Network interdiction has a number of success applications, including city-level drug enforcement [13] and cyber-physical security [1]. However, to best of the authors' knowledge, multiobjective network interdiction has not been applied to metabolic network analysis.

In metabolic networks, cellular growth and product yield are often in conflict as they have to compete for limited substrate resources. The aim of this paper is to interpret strain design as a Multiobjective Network Interdiction Problem (MO-NIP) and explore the use of NSGA-II as a solver to find the trade-off between cellular growth and product yield. The network owner is the cell who maximises its cellular biomass as it tries to remain viable and in homoeostasis, and the interdictor is the bioengineer who tries to disrupt the owner's network in order to have more substrate resources left for target chemical production.

The rest of the paper is organised as follows. Section 2 introduces the basics of metabolic networks and the proposed MO-NIP formulation for strain design. Section 3 presents the NSGA-II algorithm for the mix-integer program of MO-NIP. The MOP-NIP model is demonstrated through several design examples in Sect. 4. Section 5 concludes the paper.

2   MO-NIP Model Formulation

In this section, the model for the genes knockout discovery is explained and the formulation for the MO-NIP is detailed.

A metabolic network of an organism can be represented as a flow network where the arcs are the reactions between metabolites on the metabolic network and the metabolites are represented as nodes in the network. Each arc has a capacity that cannot be exceed by the flow/flux that goes through that arc. A fluxes distribution associated to the reactions will determine the production of the target product (the quantity of flux that can be directed toward the product). Due to all possible fluxes distributions between reactions of the network, usually, different distributions lead to the same production rate. In general, when a production rate is calculated it consists on an interval of the minimum and maximum achievable values for target product.

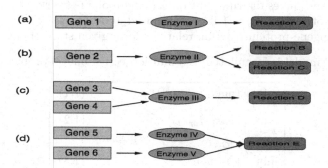

Fig. 1. The gene-protein-reaction relationship in four scenarios. (a) gene 1 has a unique function for reaction A. (b) gene 2 has multifunctions, catalysing both reaction B and C. (c) gene 3 and gene 4 encode an enzyme complex to catalyse the enzymatic reaction D. (d) gene 5 and gene 6 are isozyme genes, each of which can catalyse reaction E.

The reactions of a network are determined by the genes of the organism. The gene-protein-reaction (GPR) relationship describes the links between gene activities and reactions. Figure 1 presents four common gene-protein-reaction associations. The removal of a gene gives rise to the loss of the gene product, i.e., enzymatic proteins, and thus leads to the deactivation of the gene-encoding reactions, that is, the reactions are no longer active in the network. As a consequence, there exists no flux for the reactions associated with the removed genes. The idea in gene-knockout based metabolic engineering then, is that the target production (the bioengineering objective) can be increased by removing genes that catalyse reactions on competing pathways.

Given a metabolic network, the related GPR and a budget limiting the number of knockouts, a set of knockouts can be performed on genes to identify the genes that reduce the cellular objective (the biomass), increase the production of a desired product (the bioengineering objective) in the organism. While the cellular objective is reduced, a threshold on the growth rate is imposed to sustain cellular growth.

The objective is to identify gene knockouts that reduce the maximum cellular objective, increase the maximum value for the bioengineering objective and reduce the design costs simultaneously, where the maximum production value for the bioengineering objective is obtained within the flux distributions that maximises the cellular objective first (due to the underdetermination of metabolic networks, many flux distributions exist for the maximum cellular objective). Then, denoting

j, G index and set of genes with $j \in \{1, 2, \cdots |G|\}$,
K maximum allowable number of knockouts, $K \in (0, |G|)$,
$f_{cellular}$ the cellular objective,
$\delta_{cellular}$ a threshold for minimum $f_{cellular}$,
f_{prod} the production rates for the bioengineering objective,
i, v indexes and fluxes distributions on the metabolic network,
y_j the state of gene g_j, $y_j = 1$ if g_j is removed, 0 otherwise,
$GPR(y)$ the gene-protein-reaction relations for a given state of genes y,

the MO-NIP model is formulated as a multiobjective bilevel problem as follows:

$$
\begin{aligned}
&\min_{y \in \Gamma} \quad (\max_{v}(f_{cellular})) && (1.1) \\
&\max_{y \in \Gamma} \quad (\max_{v}(f_{prod})) && (1.2) \\
&\min_{y \in \Gamma} \quad (\sum_{j=1}^{|G|} y_j) && (1.3) \\
&s.t. \quad f_{cellular} \geq \delta_{cellular} \\
&\qquad \boxed{\begin{aligned} &v \in \underset{v_i}{\operatorname{argmax}} \; f_{cellular} \\ &\quad s.t. \quad \text{mass-balance} \\ &\qquad\qquad GPR(y) \end{aligned}}
\end{aligned}
\tag{1}
$$

where $\Gamma = \{y| \sum_{j=1}^{|G|} y_j \leq K, y_j \in \{0, 1\}\}$ are the feasible knockout strategies. From engineering point of point, it is desirable to have as few gene knockouts as possible, since each knockout is a design cost and too many knockouts likely induce cell death. Growth rate is a widely used cellular objective for $f_{cellular}$ [14].

The MO-NIP formulation consists on a bilevel problem where the outer problem is a multiobjective optimisation problem following the $min - max$ schema of the interdiction problem on the first objective (Eq. (1.1)). The outer problem (the outer box in Eq. (1)) finds the knockout strategies y that minimises the maximum production rate for $f_{cellular}$, maximises the maximum production rate for f_{prod}, and minimises the number of knockouts (less gene knockouts generally means lower design costs), while the inner problem (the inner box in Eq. (1)) solves a Flux Balance Analysis (FBA) problem [14] searching for feasible flux distributions v that maximise the $f_{cellular}$ subject to standard mass balance constraints while considering the knockout strategy y imposed by the outer level.

Algorithm 1. Generation

1: *absents* ← random K genes from the list of genes
2: *individual* ← initialisation with 0 for all genes
3: **for** each absent in *absents* **do**
4:      *individual* ← 1 at the position of *absent*
5: **end for**

Algorithm 2. Evaluation

1: m ← metabolic model (Cobra model)
2: **for** each *absent* gene in individual **do**
3:      perform knockouts of *absent* in m
4: **end for**
5: f_1 ← maximum production of $f_{cellular}$ in m (using a Cobra solver)
6: **if** $\delta_{cellular} \leq f_1$ AND number of absent genes $\leq K$ **then**
7:      f_2 ← maximum of the fluxes f_{prod} in m (using flux balance analysis)
8:      f_3 ← number of absent genes
9: **else**
10:      f_1, f_2, f_3 ← worst solution
11: **end if**
12: return f_1, f_2, f_3

---

3   MO-NIP Algorithm

This section details the Genetic Algorithm implementation to solve the formulated MO-NIP.

The proposed implementation to solve the MO-NIP for strain design is an adaptation of the Nondominated Sorting Genetic Algorithm (NSGA-II) [5] which has previously shown satisfactory results on solving the Multi-Objective Deterministic Network Interdiction Problem (MO-DNIP) [16]. The NSGA-II implementation is focused on multiobjective oriented methods: nondominated sorting with crowding for replacement, binary tournament selection to produce the new population and a Pareto archival strategy to store non-dominated solutions. The NSGA-II implementation available in Inspyred [8], a python framework for bio-inspired algorithms, is used. The rest of the algorithmic components, i.e., initialisation, fitness evaluation and mutation, are detailed next.

Representation

An individual representation consists on a ordered list of present/absent genes in the organism indicated as binary values: 1 if the gene is absent, 0 otherwise. This representation allows to reduce the considered set of genes to interdict in case some genes knockouts are not desired.

Initialisation

Individuals are randomly generated using the strategy shown in Algorithm 1. The absents genes for the individual are obtained using a random selection of non-unique genes being possible to obtain duplicated genes. This implies that each individual starts with K or less absent genes.

Fitness evaluation

The evaluation of an individual requires the calculation of all objective functions from the multiobjective formulation detailed in Eq. (1). It uses Cobrapy [6], a metabolic constraint-based reconstruction and analysis python package.

The pseudocode for an individual's evaluation is shown in Algorithm 2. The evaluation method first performs the knockouts for the absent genes in the individual. Then, the maximum production of $f_{cellular}$ is calculated. Only when this value is greater than the threshold, the maximum value of f_{prod} is obtained. In case the threshold requirement is not met, the solution is set as the worst possible solution: ∞ for objectives 1 and 3 and $-\infty$ for objective 2. The evaluation's average running time from 1500 runs is 0.1180 s, tested with an i7-5700HQ CPU at 2.70 GHz.

Mutation

The generated individuals are mutated using different methods explained next.

The individual is first modified using a *n-point crossover* (NPX) method: the individual is 'cut' at n unique random points. Then, the cuts are later recombined as a new individual. An individual is also mutated with a probability of 0.1 using a scramble mutation. This mutation method randomly chooses two locations along the individual and scrambles the values within that slice.

Two additional mutation methods are implemented to obtain more variation in the number of absent genes. The first one randomly swaps a present gene into absent (changes 0 to 1), if the budget K allows it, and the second one randomly swaps an absent gene into a present gene (changes 1 to 0), only if there is at least one more absent gene in the individual.

4   Examples

To illustrate the proposal, two examples producing different desired biochemicals are run with different GA parameters. In both examples the *E. coli* model iJO1366 [15] is used. It has 1805 metabolites, 2583 reactions and 1367 genes. The cellular objective is the biomass flux in the iJO1366 model. Therefore, the resulting MO-NIP has 1367 binary variables (representing on/off state for each gene) and 3 objectives as mentioned in Eq. (1). The fitness functions are evaluated after each generation to show the evolution of the quality of the solutions in the Pareto fronts. The evaluation is made calculating the hypervolume indicator

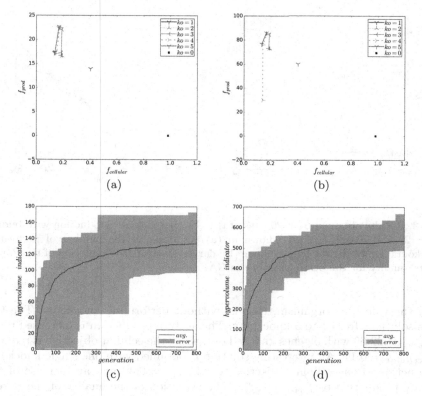

Fig. 2. Solutions obtained from 30 runs of the GA for products (a) acetate and (b) formate with budget $K = 5$ and $\delta_{cellular} = 0.1$. The solutions are shown in terms of the number of performed knockouts, ko. The average and error of the hypervolume [19] curves illustrate the Pareto fronts' quality for all 30 runs of the GA for (c) acetate and (d) formate.

[19] between the Pareto front and a reference point (a point dominated by the initial solution) using the scientific library Pygmo [2].

The GA is run multiple times on both examples. That means that some of the non-dominated solutions could be later dominated by solutions from other runs. The solutions shown in next examples are non-dominated by any other final solution.

Two products are considered in the first example as the bioengineering objective: acetate and formate. The knockout budget is $K = 5$, as suggested in other studies [12, 17], and the threshold on the cellular objective is $\delta_{cellular} = 0.1$. The GA is run 30 times with 800 generations with a population size of 15 individuals. Different population sizes ($N \in \{30, 45, 60\}$) show similar results. The results for all runs are depicted in Fig. 2 together with the GA's fitness functions evaluation.

For the acetate product, Fig. 2(a), the initial solution (shown with a black square) consists on $f_{cellular} = 0.98$ and $f_{prod} = 0$. This initial solution is obtained

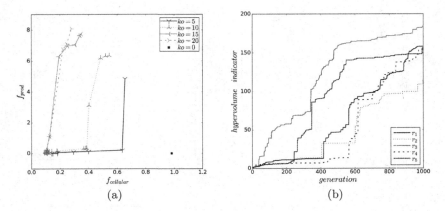

Fig. 3. Results obtained from 20 runs of the GA for succinate production with budget $K = 20$ and $\delta_{cellular} = 0.1$, shown in (a) according to the number of performed knockouts, ko. In (b) the hypervolume [19] curves illustrate the evolution of the Pareto fronts' quality for the best 5 runs of the GA.

from the wild-type organism, that is, without performing knockouts. The GA finds solutions from 1 to 5 knockouts. The acetate production is increased up to $f_{product} = 22.69$ with 5 genes knocked out, while the cellular objective is reduced to $f_{cellular} = 0.17$. An interesting solution is obtained with one single knockout: a reduction of the cellular objective to $f_{cellular} = 0.40$ and an increase of the acetate production to $f_{prod} = 13.94$. The evolution of the quality of the Pareto front obtained for each run is illustrated in Fig. 2(c). The algorithm is able to quickly improve the solutions within 300 generation showing a typical convergence curve. However, further improvement is required on the robustness of the algorithm such that the best Pareto trade-off can be obtained without multiple runs, which would reduce the duration of strain design and also give the decision maker high confidence in the identified strain design strategies.

The initial solution for the formate product is also $f_{cellular} = 0.98$ and $f_{prod} = 0$. It is obtained with no knockouts and shown as a black square in Fig. 2(b). The GA finds solutions from 1 to 5 genes knockouts. The greatest value for the formate production is achieved with 5 knockouts $f_{prod} = 86.01$ where the interdicted $f_{cellular}$ is reduced to 0.17. In the formate case, the Pareto front quality is improved drastically within the first 300 generations, as shown in Fig. 2(d).

It can be noticed that in both cases, solutions with more knocked out genes are better in general, as they obtain a greater increment on the bioengineering objective and a greater reduction on the cellular objective simultaneously.

On the second example, the product is succinate. The GA is run 20 times considering 1000 generations with a population of 30 individuals. In this case, the budget is increased to 20 due to known difficulty in increasing succinate production. The threshold on biomass is considered as $\delta_{cellular} = 0.1$.

Results are shown in Fig. 3(a), where the number of gene knockouts in the obtained solutions goes from 2 to 20. The initial solution (obtained with no knockout) is $f_{cellular} = 0.98$ and $f_{prod} = 0$. It is shown as a black square. The greatest succinate production f_{prod} is achieved with 20 knockouts where $f_{cellular} = 0.28$ and $f_{prod} = 8.08$ although other solutions with 19 and 18 knockouts are also very close. The evaluations of the quality of the best Pareto fronts are illustrated in Fig. 3(b). Even though the differences among different runs show that there is a lack of robust performance on the case of succinate production, the algorithm is able to find interesting knockout genes strategies within 20 runs.

5   Conclusions

Exploring genome-scale metabolic networks is challenging yet important for elucidating biological circuits. Computational strain design methods based on metabolic models have been increasingly popular for metabolic engineering in recent years. While most existing methods consider stain design as a single-objective bilevel problem, this work borrows the idea of network interdiction from network theory and proposes to apply it to strain design in a multiobjective manner. As a result, a Multiobjective Network Interdiction Problem (MO-NIP) model is derived. It formulates a Network Interdiction Problem with three objectives where the first one is interdicted. A modified NSGA-II is proposed to solve the resulting mix-integer program. The proposed MO-NIP formulation has been shown by two examples and has demonstrated its ability to help identify high-quality design strategies for the production of desired biochemicals.

With *in silico* strain design solutions at hand, further analysis are needed to investigate the feasibility of the solutions *in vivo*. However, this is not the focus of this paper but will be done in future studies.

As a proof of concept, the current MO-NIP model is relatively simple. It is interesting to further investigate the capacity of network interdiction in the context of metabolic engineering. For example, the formulation of the model is easily expanded to more objectives and can be used to interdict any bioengineering objective. This is left for future work.

Acknowledgements. DP acknowledges support through projects TIN2014-55024-P and TIN2017-86647-P from the Spanish Ministry of Economy and Competitiveness (including European Regional Development Funds). MT enjoys a Ph.D. research training staff grant associated with the project TIN2014-55024-P and co-funded by the European Social Fund.

SJ, MK, and NK acknowledge the EPSRC for funding project "Synthetic Portabolomics: Leading the way at the crossroads of the Digital and the Bio Economies (EP/N031962/1)".

282   M. Torres et al.

References

1. Amin, S.: Network interdiction and inspection models for cyber-physical security, 25 January 2017
2. Biscani, F., Izzo, D., Mrtens, M.: esa/pagmo2: pagmo 2.7 (Version v2.7). Zenodo, 13 April 2018. https://doi.org/10.5281/zenodo.1217831
3. Burgard, A.P., Maranas, C.D.: Optimization-based framework for inferring and testing hypothesized metabolic objective functions. Biotechnol. Bioeng. **82**(6), 670–677 (2003)
4. Costanza, J., Carapezza, G., Angione, C., Li, P., Nicosia, G.: Robust design of microbial strains. Bioinformatics **28**(23), 3097–3104 (2012)
5. Deb, K., Pratap, A., Agarwal, S., Meyarivan, T.: A fast and elitist multiobjective genetic algorithm: NSGA-II. IEEE Trans. Evol. Comput. **6**(2), 182–197 (2002)
6. Ebrahim, A., Lerman, J.A., Palsson, B.O., Hyduke, D.R.: COBRApy: constraints-based reconstruction and analysis for python. BMC Syst. Biol. **7**(1), 74 (2013)
7. Feist, A.M., et al.: A genome-scale metabolic reconstruction for Escherichia coli K-12 MG1655 that accounts for 1260 ORFs and thermodynamic information. Mol. Syst. Biol. **3**(1), 121 (2007)
8. Garrett, A.: Inspyred: a framework for creating bio-inspired computational intelligence algorithms in python. Software (2017). https://aarongarrett.github.io/inspyred
9. Henry, C.S., Zinner, J.F., Cohoon, M.P., Stevens, R.L.: i Bsu1103: a new genome-scale metabolic model of Bacillus subtilis based on SEED annotations. Genome Biol. **10**(6), R69 (2009)
10. Jiang, S., Torres, M., Pelta, D., Krabben, P., Kaiser, M., Krasnogor, N.: Improving microbial strain design via multiobjective optimisation and decision making. In: AI for Synthetic Biology 2 (2018)
11. Kim, J., Reed, J.L.: OptORF: optimal metabolic and regulatory perturbations for metabolic engineering of microbial strains. BMC Syst. Biol. **4**(1), 53 (2010)
12. Lun, D.S., et al.: Largescale identification of genetic design strategies using local search. Mol. Syst. Biol. **5**(1), 296 (2009)
13. Malaviya, A., Rainwater, C., Sharkey, T.: Multi-period network interdiction problems with applications to city-level drug enforcement. IIE Trans. **44**(5), 368–380 (2009)
14. Orth, J.D., Thiele, I., Palsson, B.Ø.: What is flux balance analysis? Nat. Biotechnol. **28**(3), 245 (2010)
15. Orth, J.D., et al.: A comprehensive genome scale reconstruction of Escherichia coli metabolism—2011. Mol. Syst. Biol. **7**(1), 535 (2011)
16. Rocco, C.M.S., Salazar, D.E.A., Ramirez-Marquez, J.E.: Multi-objective network interdiction using evolutionary algorithms. In: 2009 Annual Reliability and Maintainability Symposium (2009)
17. Tepper, N., Shlomi, T.: Predicting metabolic engineering knockout strategies for chemical production: accounting for competing pathways. Bioinformatics **26**(4), 536–543 (2009)
18. Wood, R.K.: Deterministic network interdiction. Math. Comput. Model. **17**(2), 1–18 (1993)
19. Zitzler, E., Thiele, L., Laumanns, M., Fonseca, C.M., Da Fonseca, V.G.: Performance assessment of multiobjective optimizers: an analysis and review. IEEE Trans. Evol. Comput. **7**(2), 117–132 (2003)

Data Mining

Risk Factors for Development of Antibiotic Resistance of *Enterococcus Faecium* to Vancomycin. A Subgroup Discovery Approach

Andrej Fajfar[1], Manuel Campos[2(✉)], Francisco Palacios[3],
Bernardo Canovas-Segura[2], Gregor Stiglic[1], and Roque Marin[2]

[1] Faculty of Health Sciences, University of Maribor, Maribor, Slovenia
[2] Faculty of Computer Science, University of Murcia, Murcia, Spain
manuelcampos@um.es
[3] University Hospital of Getafe, Madrid, Spain

Abstract. Health-care associated infections (HAI) are infections that are not present or incubated at the time of admission to hospital. HAI are one of the major causes of morbidity and mortality among immuno-compromised patients and have an important economic impact. The bacteria isolated in microbiology cultures can be treated with a limited combination of antibiotics owing to their resistance to many groups of antibiotics, which represents a major challenge. This paper focuses on the problem of vancomycin resistant *Enterococcus faecium* (VREfm) due to its high prevalence in HAI, multidrug resistance and ability to survive under intense selective pressure. We use the subgroup discovery technique to identify target populations with high risk clinical factors for VREfm infections that we shall be able to incorporate into a clinical decision support system for antimicrobial stewardship program. The dataset used contained 201 susceptibility tests with *Enterococcus faecium* from a University Hospital in years 2014 and 2015. The clinician evaluated and discussed the knowledge reported by the most interesting subgroups based on their positive predictive value and sensitivity.

1 Introduction

Health-care associated infections (HAI) caused by multidrug resistant (MDR) microorganisms are increasing the levels of morbidity and mortality in hospital settings, particularly among immunocompromised patients [3]. Data shows that 1 out of 20 hospitalized patients developed HAI in the United States (US) [18] with the *Enterococcus* species being among second or third cause of infections in US and Europe [17]. Reports show that in US over 2 millions of patients are infected with MDR bacteria and 23,000 patients die annually. The statistic for the European Union is even higher, as there is an annual mortality of 25,000 people, leading to costs of €1,5 billion per year [4].

F. Herrera et al. (Eds.): CAEPIA 2018, LNAI 11160, pp. 285–295, 2018.
https://doi.org/10.1007/978-3-030-00374-6_27

HAI (or nosocomial infections) are infections that are not present or incubated at the time of admission to hospital. This definition means that any infection developed in 48–72 h after admission or in the first 10 days after discharge from a clinical environment is classified as HAI [3]. The sources of infections are endogenous (i.e., patient microflora and antibiotic driven changes in normal flora) or exogenous (i.e. cross-contamination with healthcare personnel, invasive medical devices or the hospital environment) [18]. *Enterococcus faecium* is gram positive bacteria that is a major problem in HAI owing to its MDR [10] and rapid spread among patients [18]. *Enterococcus faecium* has developed resistance to almost all antienterococcal antibiotics currently available. Vancomycin is a basic first line treatment when infection by gram positive bacteria under a patient's certain clinical condition is suspected. The two main alternatives are linezolid and daptomycin, but both of them are starting to show resistances [10]. In this paper we face the problem of vancomycin resistant *Enterococcus faecium* (VREfm).

The clinical authorities have proposed the implementation of antimicrobial stewardship program (ASP) in hospitals as part of the fight against the global antimicrobial resistance. Such programs are designed for a better use of antimicrobial drugs and enable the physicians to prescribe treatments more accurately. The introduction of clinical decision support systems (CDSS) into the daily work of an ASP is an effective tool for the management of antimicrobial resistance, improves the safety of the patient being treated and decreases costs of antimicrobials [14].

The motivation of this research is to obtain insights into patient subpopulations by finding determinants or variables that may be risk factors for VREfm infections and that can be acted upon. Our objective is to obtain descriptions of target populations that could be easily integrated into the CDSS for ASP. To that end, we opt for using a subgroup discovery approach. This technique describes subgroups that are easily interpreted by the clinician in order to evaluate the results.

The remainder of the paper is organized as follows: Section 2 presents the background of *Enterococcus* genus and CDSS for infection management; Sect. 3 describes the data and methods applied; Sect. 4 shows our results along with a discussion; and Sect. 5 presents important findings.

2   Background

2.1   *Enterococcus* Infection and Treatment

Enterococci are gram positive nonsporulating facultatively anaerobic bacteria. They are ubiquitous in environments outside hosts and the gastrointestinal track (GIT) of animals and humans. The *Enterococcus* genus presents 1% of the normal microflora in human GIT and is a well known opportunist in hospital environments [11]. Their long evolutionary history has given them the ability to adapt and survive in harsh environmental conditions such as UV radiation, hypertonic and hypotonic conditions, pH and temperature changes, starvation, inhibitory substances (i.e. antibiotics, immune–response), etc.

VREfm developed resistance with four general mechanisms of resistance: repelling of antibiotic (i.e. daptomycin), changing the target (i.e. vancomycin, linezolid, tigecycline), modification of the antibiotics or efflux pumps (i.e. quinupristin–dalfopristin) [10]. Miller *et al.* [10] proposed treating complicated VREfm infections with daptomycin, over linezolid, because of its bacteriolitic activity. They described treatment based on minimal inhibitory concentration and the type of infection with one or combinations of antibiotics (i.e. daptomycin plus β–lactam).

2.2   CDSS in Infection Management

Clinical decision support systems (CDSS) are intelligent medical informatic support tools designed to provide "computer–generated clinical knowledge and patient–related information intelligently filtered or presented at appropriate times, to enhance patient care" [16]. Our focus will be on the use of CDSS for infection prevention and associated antimicrobial stewardship.

There has been a long history of intelligent systems working on infection diagnosis and treatment [12]. We found different approaches for the detection of HAI such as production systems [7] or fuzzy rules [1]. A number of infection control intelligent systems have dealt with the search for patterns such as association rules [9] or the selection of antibiotics with probabilistic networks in the TREAT system [8].

Palacios *et al.* [14] are developing a CDSS called WASPSS (Wise Antimicrobial Stewardship Program Support System) at a hospital in Madrid. WASPSS proposes an integral approach for the development of CDSS for infection management, including guidelines for patient infection treatment, the support of ASP actions that are not directly related to antimicrobial supervision and visual knowledge representation, and the integration of both expert knowledge and knowledge extracted from the database as a result of data mining techniques.

2.3   Subgroup Discovery

Subgroup discovery (SD) is a supervised technique for descriptive and exploratory data mining. It can be used to obtain general relations in database, automatic hypothesis generation and data exploration. Prominent results of SD application were obtained in technical and medical domains [2]. Its main goal is to extract the "most interesting" subgroups of individuals according to the property of interest or the target variable. The most interesting subgroups are as large as possible, and have an unusual statistical distribution (characteristic) with regard to the property of interest. The subgroups do not have to be complete, and partial relations suffice. We can describe a typical SD process with three components: search strategy, quality measure and subgroup representation.

The search strategies are algorithms to explore the space of all possible solutions. The searching time is in exponential relationship with input data dimensions (the larger the dimensions, the longer the execution time). Various search strategies exist, i.e., beam and exhausting search or evolutionary algorithms, and its selection also depends on the target value (binary, nominal or numeric), because not all algorithms can work with all those classes [2,6].

The quality measure (QM) evaluates subgroups at each iteration of the algorithm and only subgroups above a selected threshold continue the search process. There are many QMs that focus on different features of importance and interestingness of the subgroups extracted. There is no gold standard for QM and is depended on research goals.

Two representative measures are WRACC and binomial test (BT). WRACC is a hybrid measure that trades generality (coverage, $\frac{n}{N}$) for accuracy ($p - p_0$) of the subgroup [2,6], as presented with the following equation:

$$WRACC(R) = \frac{n}{N} \cdot (p - p_0),$$

where n is the size of the subgroup, N is the size of the initial population, p is the fraction of the target variable samples in the subgroup and p_0 is the fraction of the target variable samples in the initial population. WRACC is an unbiased accuracy measure and is equivalent to maximizing area under curve (AUC) in predictive data mining [15].

The BT measure is presented in the following equation:

$$BT(R) = \frac{p - p_0}{\sqrt{p_0 \cdot (1 - p_0)}} \sqrt{n} \sqrt{\frac{N}{N - n}}$$

Differences in unusualness ($p - p_0$) and size, coverage and generality (differences between n and N) provides us with small and very accurate subgroups with a low sensitivity.

3   Data and Methods

3.1   Data

The data employed originated from 201 susceptibility tests with E. Faecium isolates during 2014–2015. The specimens were obtained from the first culture taken at any location of a patient (there are no two cultures from the same patient at the same location). The target variable was vancomycin resistance, with 20 positive and 181 negative values. A total number of attributes is 20 without missing values. Attributes contain information about the patient, the episode, the culture, antibiotic treatment received (a selection of indicators of antibiotics related to GIT flora) and antecedents, and are summarized in Table 1. Apart from these variables, the database contains admission date, admission department, culture date, culture department, and culture location.

Table 1. Summary of variables

Section	Attribute	Summary
Demographic	Sex (MF)	106/95
	Age (Avg ± sd)	69.12 (±13.55)
Episode	Indicator for admission ER (T/F)	166/35
	Indicator for surgery dept (T/F)	46/155
	Indicator for ICU (T/F)	69/132
	Indicator for burns unit (T/F)	2/199
	Indicator for ER (T/F)	15/186
Antibiotics	Standard SDD (T/F)	65/136
	Mixed SDD (T/F)	31/170
	Colistine (T/F)	5/196
	Tobramycin (T/F)	28/173
	Amphotericin B (T/F)	1/200
	Cefotaxime (T/F)	35/166
	Vancomycin (T/F)	64/137
Antecedents	Had previous bacteria (T/F)	160/41
	Had previous gram+ bacteria (T/F)	103/98
	Readmission in last 30 days (T/F)	92/109
	Antibiotics in last 30 days (T/F)	76/125

Regarding the preprocessing, we included binary indicators and discretized numeric variables. We included binary indicators for some labels of interest that, according to the clinician, could prove relevant: admission through any of the different emergency departments, or stay in any of the departments with a higher risk of infection (surgery, emergency, ICU, or burns unit). These same departments were also include in an indicator $JOINED\_DEPARTMENT$, where the SEIBC-dep value represents any of the above mentioned departments.

We discretized the age according to clinical criteria as infants (<16 yo), adults (<66 yo) and elderly. Susceptibility testing was performed on 61 adult (28 male and 33 female) and 140 elderly patients (62 male and 78 female). We also discretized the variable for the time of the culture according to the definition of HAI by merging them into one variable, $CULTURE\_GAIN\_TIME$, with 3 categories: first 72 h (F72H), first 10 days (F10D) and more than 10 days (M10D). A last variable included was the days from ICU into 6 categories: no icu, at icu, less than 15 days, less than 30 days, less than 60 days and more than 60 days. This variable was used to identify whether the fact of being at the ICU was relevant despite the culture having been taken at another department.

3.2 Methods

In this paper, we have applied exhaustive search using the SD–MAP [2] algorithm owing to the limited search space in our case (see Subsect. 3.1).

For the selection of the QM we considered that our target variable is imbalanced (see Subsect. 3.1). Therefore, we describe the subgroups from two points of view:

1. Subgroups that capture as many resistance occurrences as possible. The QM chosen was Weighted relative accuracy (WRACC) because it allows us to focus on subgroups that are large and highly sensitive and that have low positive predictive value (PPV).
2. Subgroups with PPV as high as possible. We use BT because it obtains very specific subgroups with a low FP rate and a high true positive (TP) rate.

In our case, the subgroups are presented as conjunction of attribute-value pairs. A more general representation by means of relational operators or disjunctive normal form is also possible [6].

An advantage of SD discovery as opposed to classification models such as decision trees or rules with exceptions is that the latter are not able to select overlapped regions of the search space. What is more, the rules extracted from decision trees contain the same root path and provide a less intuitive interpretation than the subgroups. In [5] the authors propose an algorithm for fuzzy rules that also improves the interpretability of the subgroups with numerical variables.

Data processing and analysis was performed in R version 3.3.1 and package rsubgroup version 0.6.

4 Results and Discussion

We should first like to make some general comments about the results. In the first place, the number of subgroups found in the experiments is low (around 20 subgroups) in both quality functions. This facilitates the clinician's task of evaluating the models. In the second place, SD is a non–parametric method that learns from data, and the subgroups may, therefore, overfit (i.e. PRIM avoid this with cross–validation [13]). In our case, it is an imbalanced problem since the data shows low incidence of resistance, and we sought a balance between more sensitive subgroups, which are more general and capture most of the resistances, and subgroups with a high PPV and a low FP. To that end, we studied only subgroups with a PPV above 75%, and a sensitivity above 30% (6 resistances).

The process followed to obtain insights into the main factors consisted of several steps. In a first experiment, we used the original variables and an additional indicator that considers high risk medical services that a priori are consider to be relevant (SEICB: surgery, emergency, ICU, burns). This variable turned out to be included in almost all the subgroups (see Table 2).

Table 2. Subgroups with SEICB indicator using WRACC QM

Subgroup	Subgroup description	PPV	Sensitivity
4	CULTURE_GAIN_TIME=M10D, JOINED_DEPARTMENT=SEICB-dep, admission_in_ER=t	0.23	0.80
1	CULTURE_DEPARTMENT=ICU, CULTURE_GAIN_TIME=M10D	0.30	0.70
2	AGE=adult, CULTURE_GAIN_TIME=M10D, HAD_AMFOTERICINE=f	0.35	0.65
3	AGE=adult, CULTURE_GAIN_TIME=M10D, admission_in_ER=t, HAD_COLISTINE=f	0.43	0.60

The results depicted in Table 2 show that there are three main factors. Firstly, since the M10D variable appears in all the subgroups, it is clear that the resistance is associated with healthcare (it is not a community problem). Secondly, it is more prevalent in the services contained in the SEIBC variable than in the rest. Thirdly, admission through the emergency room (ER) appears to be relevant, but its clinical value was not explained by the clinician and merits further research. It is possible that the patients in question had an acute pathology that forced them to be moved to the ICU. The appearance of other variables in the subgroups but with false value does not provide any information. However, the absence of information may also be relevant, i.e., we cannot find the site of infection or the location at which the sample was taken. This may reveal that the resistance is not related to any specific pathology.

In the second experiment (see Table 3) which did not include the SEIBC indicator, some relevant rules formerly present no longer appeared. Moreover, this experiment concerned the profile of adult (as opposed to elderly patients), without previous bacteria or antibiotic treatment in last 30 days, who were admitted through the emergency department, that had vancomycin and that showed the resistance while staying at the ICU.

One conclusion drawn from this experiment is that the problem it seems to be associated with the ICU. The therapeutic peculiarity of the ICU that is not performed in other departments is SDD (the mixed SDD contains topic vancomycin), while vancomycin treatment is shared with other services. The fact of being adults may be related to a different length of stay, a different diagnostic and a different length of the use of SDD. The clinician found this interesting for and deserving of further research.

In a third experiment (see Table 4), we explored the subpopulation that includes the main factors found in the first experiment (SEIBC and M10D). There were 83 susceptibility tests in this subset, with 20 resistances. In this case, we used the original department name variable to see if any of them stood out. As a conclusion, we discarded the departments other than the ICU in the SEIBC as the main factor. Apart from confirming the knowledge obtained in

Table 3. Subgroups without SEICB indicator using BT QM

Subgroup	Subgroup description	PPV	Sensitivity
2	AGE=adult, HAD_VANCOMYCIN=t, ADMISSION_DEPT=EMERGENCY, ANTIBIOTICS_LAST_30D=f, HAD_CEFOTAXIME=f, CULTURE_GAIN_TIME=M10D	0.88	0.35
3	had_other_bacteria_before=f, AGE=adult, HAD_CEFOTAXIME=f, CULTURE_GAIN_TIME=M10D, ANTIBIOTICS_LAST_30D=f	0.88	0.35
6	AGE=adult, HAD_VANCOMYCIN=t, admission_in_ER=t, HAD_TOBRAMICYN=f, CULTURE_GAIN_TIME=M10D, HAD_CEFOTAXIME=f	0.78	0.35
7	had_other_bacteria_before=f, CULTURE_DEPARTMENT=ICU, HAD_VANCOMYCIN=t, SEX=F, CULTURE_GAIN_TIME=M10D	0.78	0.35
8	had_other_bacteria_before=f, AGE=adult, DAYS_FROM_ICU=AT_ICU, CULTURE_GAIN_TIME=M10D, ANTIBIOTICS_LAST_30D=f	0.78	0.35
1	AGE=adult, HAD_VANCOMYCIN=t, admission_in_ER=t, HAD_TOBRAMICYN=f, ANTIBIOTICS_LAST_30D=f, CULTURE_GAIN_TIME=M10D, HAD_CEFOTAXIME=f	1.00	0.30
15	had_other_bacteria_before=f, CULTURE_DEPARTMENT=ICU, HAD_VANCOMYCIN=t, SEX=F, HAD_SD_STANDARD=t, CULTURE_GAIN_TIME=M10D	0.86	0.30
16	had_other_bacteria_before=f, AGE=adult, HAD_VANCOMYCIN=t, ANTIBIOTICS_LAST_30D=f	0.86	0.30

the previous experiments, it is interesting to note that these patients had not been previously admitted, contrary to what happens with other bacteria such as the *Pseudomonas aeruginosa*, which tends to be recurrent in patients who have been re–admitted.

In the fourth experiment, we selected information regarding only: age, admission service, readmission in last 30 days, SDD, vancomycin, infections with other or gram positive bacteria, and removed the spatial information, that is, the

Table 4. Subgroups on filtered population using BT QM

Subgroup	Subgroup description	PPV	Sensitivity
10	AGE=adult, ADMISSION_DEPT=EMERGENCY, ANTIBIOTICS_LAST_30D=f, HAD_TOBRAMICYN=f, HAD_CEFOTAXIME=f	0.73	0.40
11	AGE=adult, HAD_VANCOMYCIN=t, ADMISSION_DEPT=EMERGENCY, HAD_CEFOTAXIME=f	0.73	0.40
2	AGE=adult, HAD_VANCOMYCIN=t, ADMISSION_DEPT=EMERGENCY, ANTIBIOTICS_LAST_30D=f, HAD_CEFOTAXIME=f	0.88	0.35
3	had_other_bacteria_before=f, AGE=adult, HAD_CEFOTAXIME=f, ANTIBIOTICS_LAST_30D=f	0.88	0.35
12	AGE=adult, HAD_VANCOMYCIN=t, admission_in_ER=t, HAD_TOBRAMICYN=f, HAD_CEFOTAXIME=f	0.78	0.35
13	had_other_bacteria_before=f, DAYS_FROM_ICU=AT_ICU, HAD_VANCOMYCIN=t, SEX=F	0.78	0.35
14	had_other_bacteria_before=f, AGE=adult, DAYS_FROM_ICU=AT_ICU, READMISSION_LAST_30D=f	0.78	0.35
1	AGE=adult, HAD_VANCOMYCIN=t, admission_in_ER=t, HAD_TOBRAMICYN=f, ANTIBIOTICS_LAST_30D=f, HAD_CEFOTAXIME=f	1.00	0.30
8	had_other_bacteria_before=f, DAYS_FROM_ICU=AT_ICU, HAD_VANCOMYCIN=t, SEX=F, HAD_SD_STANDARD=t	0.86	0.30
9	had_other_bacteria_before=f, AGE=adult, HAD_VANCOMYCIN=t, ANTIBIOTICS_LAST_30D=f	0.86	0.30

department where the sample was taken. We were able to verify that the use of vancomycin (either systemic or topic in SDD) proved to be the main factor. Another two ideas appear here. Upon comparing the quality of the subgroups it will be noted that the quality of the subgroup decreases without the ICU factor (PPV from 0.30 to 24 and sensitivity from 0.70 to 0.60). That is, the clinical condition of the patient at the ICU would appear to be more relevant than the mere use of SDD.

5   Conclusion

In this paper we have tackled the problem of finding risk factors that could help the clinicians of an antimicrobial stewardship program to consider the actions that may be taken with on patients suffering from Enterococcus faecium resistant to Vancomycin. We highlight several conclusions.

In the first place, we used a subgroup discovery technique. The subgroups found are easy to read and interpret and they provide few subgroups that can easily be examined by the clinician. This technique allows us to consider simultaneously a set of factors or markers and permit an overlap of the subgroups that increases the opportunity to find sound explanations. Apart from the quality measure used in the subgroup discovery algorithm, the use of PPV and sensitivity as measures to select the subgroups to be studied represents a trade-off between more sensitive subgroups, that are more general and captures most of the resistances, and subgroups with a high PPV with a low FP, for an imbalanced problem with a low incidence of resistance.

It is important to stress that the covariates in the subgroups are markers, not causal factors of resistance. On the one hand, the markers indirectly represent particular masked causal factors. We used only clinical and demographical covariates, but none of them was an etiological factor for a disease. For example, being attended at the ICU implies that the patient's condition is different from other departments. On the other hand, some of the factors found have no simple explanation. For example, the age of the patient (much higher prevalence of VREfm in adults than in elderly patients at the ICU) or admission through the emergency department. More clinical research is therefore necessary in order to show the causal relations.

From the clinical point of view, these findings allow the decision makers to take measures during the treatment of infections, such as, the use of digestive selective decontamination with only some subgroups of patients. One advantage of the representation chosen is that the membership in the subgroups can easily be included as alarms in the CDSS for antimicrobial management where this study has taken place.

Finally, regarding our next steps in this research, since we have seen a pattern of adult patients without previous admission or previous gram positive bacteria, we now need to carry out a spatio–temporal study in order to analyze the possibility of a horizontal transmission of the resistance.

Acknowledgments. This work was partially funded by the Spanish Ministry of Economy and Competitiveness under project TIN2013-45491-R, and by the European Fund for Regional Development (EFRD).

References

1. Adlassnig, K.P., Blacky, A., Koller, W.: Artificial-intelligence-based hospital-acquired infection control. Stud. Health Technol. Inform. **149**, 103–110 (2009)

2. Atzmueller, M.: Subgroup discovery. Wiley Interdiscip. Rev.: Data Min. Knowl. Discov. **5**(1), 35–49 (2015)
3. Babady, N.: Hospital-associated infections. Microbiol. Spectr. **4**(3), DMIH2-0003-2015 (2016). https://doi.org/10.1128/microbiolspec.DMIH2-0003-2015
4. Blair, J.M.A., Webber, M.A., Baylay, A.J., Ogbolu, D.O., Piddock, L.J.V.: Molecular mechanisms of antibiotic resistance. Nat. Rev. Microbiol. **13**(1), 42–51 (2015)
5. Carmona, C.J., González, P., del Jesus, M.J., Herrera, F.: NMEEF-SD: non-dominated multi-objective evolutionary algorithm for extracting fuzzy rules in subgroup discovery. IEEE Trans. Fuzzy Syst. **18**(5), 958–970 (2010). TIN-2008-06681-C06-01, TIN-2008-06681-C06-02, TIC-3928
6. Herrera, F., Carmona, C.J., González, P., Del Jesus, M.J.: An overview on subgroup discovery: foundations and applications. Knowl. Inf. Syst. **29**(3), 495–525 (2011)
7. Landers, T., Apte, M., Hyman, S., Furuya, Y., Glied, S., Larson, E.: A comparison of methods to detect urinary tract infections using electronic data. Jt. Comm. J. Qual. Patient Saf. **36**(9), 411–417 (2010)
8. Leibovici, L., Paul, M., Nielsen, A.D., Tacconelli, E., Andreassen, S.: The TREAT project: decision support and prediction using causal probabilistic networks. Int. J. Antimicrob. Agents **30**(Supplement 1), 93–102 (2007)
9. Ma, L., Tsui, F.C., Hogan, W.R., Wagner, M.M., Ma, H.: A framework for infection control surveillance using association rules. In: AMIA Annual Symposium Proceedings, pp. 410–414 (2003)
10. Miller, W.R., Murray, B.E., Rice, L.B., Arias, C.A.: Vancomycin-resistant enterococci: therapeutic challenges in the 21st century. Infect. Dis. Clin. N. Am. **30**(2), 415–439 (2016)
11. Montealegre, M.C., Singh, K.V., Murray, B.E.: Gastrointestinal tract colonization dynamics by different Enterococcus faecium clades. J. Infect. Dis. **213**(12), 1914–1922 (2016)
12. Nachtigall, I., Tafelski, S., Deja, M., Halle, E., Grebe, M.C., Tamarkin, A., et al.: Long term effect of computer-assisted decision support for antibiotic treatment in critically ill patients: a prospective beforeafter cohort study. BMJ Open **4**(12), e005370 (2014)
13. Nannings, B., Abu-Hanna, A., de Jonge, E.: Applying PRIM (patient rule induction method) and logistic regression for selecting high-risk subgroups in very elderly ICU patients. Int. J. Med. Inform. **77**(4), 272–279 (2008)
14. Palacios, F., Campos, M., Juarez, J.M., Cosgrove, S., Avdic, E., et al.: A clinical decision support system for an antimicrobial stewardship program. In: Proceedings of the 9th International Joint Conference on Biomedical Engineering Systems and Technologies (BIOSTEC 2016), vol. 5, pp. 496–501 (2016)
15. Powers, D.M.W.: Evaluation: from precision, recall and F-factor to ROC, informedness, markedness & correlation. Int. J. Mach. Learn. Technol. **2**(1), 37–63 (2007)
16. Sittig, D.F., et al.: Grand challenges in clinical decision support. J. Biomed. Inform. **41**(2), 387–392 (2008)
17. Tedim, A.P., Ruíz-Garbajosa, P., Rodríguez, M.C.: Long-term clonal dynamics of Enterococcus faecium strains causing bloodstream infections (1995–2015) in Spain. J. Antimicrob. Chemother. **72**(1), 48–55 (2017)
18. Weber, D.J., Anderson, D., Rutala, W.A.: The role of the surface environment in healthcare-associated infections. Curr. Opin. Infect. Dis. **26**(4), 338–344 (2013)

Intelligent Management of Measurement Units Equivalences in Food Databases

Beatriz Sevilla-Villanueva[1,3]([✉]), Karina Gibert[1,3],
and Miquel Sànchez-Marrè[2,3]

[1] Department of Statistics and Operations Research, Universitat Politècnica de
Catalunya-BarcelonaTech (UPC), Barcelona, Spain
bea.sevilla@gmail.com
[2] Department of Computer Science, Universitat Politècnica de
Catalunya-BarcelonaTech (UPC), Barcelona, Spain
[3] Knowledge Engineering and Machine Learning Group at Intelligent Data Science
and Artificial Intelligence Research Center, Universitat Politècnica de
Catalunya-BarcelonaTech (UPC), Barcelona, Spain

Abstract. It is currently well-known that diet plays an important role
in the promotion of healthy lifestyle and the prevention of chronic diseases. The Diet4You project is conceived to support the creation of an
intelligent decision support system that provides personalized menus fitting a nutritional plan and taking into account the characteristics, needs
and preferences of the person. The system involves a background food
database, recording a collection of foods and prepared dishes with their
standard portions as well as their nutritional decomposition in different food families. This DB is used to search the best combination of
dishes approaching the total intake of different nutrients specified in the
prescribed nutritional plan. The available background databases, specify the quantities of standard portions of several foods based on different measurement units which are not standardized, and it happens that
the weight specified by one cup of melon is different from that of one
cup of berries, among others. This arises the need of applying variable
conversion factors to the dish description, before assessing whereas the
total quantities of a certain menu fit well to the prescription. In this
paper, a knowledge based approach is presented to the automatically
management. An annotated reference food ontology is built on the basis
of additional documentation. However the granularity of the information provided is heterogeneous and non exhaustive. The ontology-based
missing values imputation is presented to overcome this limitations.

Keywords: Ontology · Missing imputation · Database

1 Introduction

Adhering a healthy lifestyle constitutes a major goal of the most recent health
paradigm, in accordance with its strategic role in the prevention of chronic diseases and for maintaining the quality of life. One of the most relevant factors

© Springer Nature Switzerland AG 2018
F. Herrera et al. (Eds.): CAEPIA 2018, LNAI 11160, pp. 296–306, 2018.
https://doi.org/10.1007/978-3-030-00374-6_28

for a healthy lifestyle is to follow a proper diet. Currently, nutritionists design diets based on the specific needs of each person, by using their expertise and some basic tools such as food composition tables. There are limited support tools available for this task and the existent ones consist in assisting the global counting of quantities of nutrients.

The Diet4You project arises from a need of providing intelligent assessment to the composition of personalized menus and proposes an intelligent decision support system (IDSS) oriented to the adaptive and dynamic preparation of personalized diets that satisfy nutritional requirements and fit with the specific situation of each person, including their health status, genomics, medication, personal habits, preferences and/or allergies among others. IDSS are applications domain knowledge dependent. Diet4You is a complex hybrid system that integrates artificial intelligence components with data mining, being able to compose personalized and complete menus for a certain period of time. This system is composed by two main components: the Nutritional Plan Generator (NPG) that is a complex component generating a nutritional plan adapted to the person's needs and balanced in terms of the different families of nutrients; and the Personal Menu Planner (PMP), the specific component that composes a final menu by searching from a background database containing foods and dishes. The final proposal must fulfil the specifications of the nutritional plan, in terms of total quantity of food and distribution among the several food families. The composition of the menu is found through a case-based reasoning methodology.

The foods database used to support the dishes search in Diet4You system is called Food Proportions database (FPBD). It contains a set of foods decomposed in quantities of a set of attributes (food families, nutrients or similar). In the project, the *Food Patterns Equivalents Database* (FPED) and *Food and Nutrient Database for Dietary Studies* (FNDDS) provided by the USDA (U.S Department of Agriculture) are used with its additional documentation.

The nutritional plan received by PMP specifies a total number of Kcal to be consumed along a certain period, together with a specified distribution of food families. To evaluate the goodness of fitting of a certain menu, both the Kcal of every selected dish and the percentage of Kcal per food family are required. The database provides the dish composition in units of food families, which must be transformed into a vector of proportions of Kcal per food family. Typically, the measurement units change from one food family to another (litres for liquids, ounces for meat or cups for milk or fruits). Thus, some preprocessing is needed to make suitable the computation of total quantities. This transformation requires the conversion factor between the original measurement units and the common reference (usually grams or Kcal). Apparently, the documentation should provide the needed conversion factors but it has two important limitations:

- Same measurement unit can change meaning depending on the food, or cooking process of a given food. Although some foods, like meat, have a fixed conversion factor (1 ounces correspond to 28.35 g for all meats in the database), others have a variable quantity, like fruits, where 1 cup of melon corresponds

to 170 g, but 1 cup of berries corresponds to 145 g. Also, 1 cup of dried apple is assumed to be 45 gr whereas 1 cup of baked apple is 110 gr.).
- The information provided about these equivalences is incomplete and multi-granularity appears. For some families equivalences at a level of some specific ingredients (like baked apple) is available, whereas for others, only information about a general food family is found (meat).

The additional documentation is also giving information to build a reference food ontology that constitutes an essential component of Diet4You system. This provides the necessary knowledge structure to perform food decomposition transformation from a vector of quantities of heterogeneous measurement units to a vector of Kcal per food family. As the textual process do not guarantee the extraction of a conversion factor for all nodes in the reference food ontology, the *ontology-based missing imputation method* is proposed to manage the incompleteness and heterogeneity of the primary information source. The solution proposed is general for every situation where a total quantity must be decomposed in items that can be measured with heterogeneous magnitudes.

The structure of this paper is the following: In Sect. 2, the background is explained. Section 3 contains the proposed methodology and its application is presented in Sect. 4. Finally, the conclusions are outlined in Sect. 5.

2   Background

Nowadays, there is no well-established methodology to elaborate the design of an integral diet, by simultaneously taking into account the five dimensions: nutrition, genetics, metabolism, health and context. Traditional data modelling and analysis techniques have shown little attention to detect relationships as complex as those that concern us [5]. However, databases with sufficient information are available allowing to study the patterns of response to diets based on the whole characteristics of people genetics, health status and habits or context [7].

Currently, some available resources help experts on tailoring diets, but none of them take into account the special characteristics of the person. USDA offers an application for health professionals to compute the Dairy Recommended Intakes (DRI) based on the person's gender, age, height, weight, and physical activity [3]. Other systems allow the composition of menus interactively, such as the application of NHLBIN [1] where dishes can be manually combined. This system provides indications but does not control the balance of the final menu. "Eats This Much" [2] is an automatic menu planner that, given a number of Kcal, prepares a menu trying to maintain the proportions of macronutrients.

Diet4You is an IDSS based on reasoning by analogy that considers nutrigenomic knowledge and provides personalized nutritional advice. The PMP component is based on a Case-Based Reasoning (CBR) system. CBR is an stable and efficient methodology that has been used in a wide variety of applications and domains. It has shown good results in the preparation of recipes from a culinary the point of view [8]. CBR has been used in the construction of menus such as in MIKAS [9] or CAMPER [10] that use rules as well that have a case base of complete menus and modifies them to fit the requirements. The system DietPal [11] offers a menu generator that uses an exchange table by food groups and meals created by nutritionists. The menus can be designed by the nutritionist or they can be retrieved from a pre-existing Case Base. After that, the nutritionist can adjust the menu with the exchange table.

3   Methodology

As mentioned in Sect. 1, Diet4You recommends personalized menu. The recommendation must fit previously prescribed nutritional plan. In [12], the concept of nutritional plan ($\mathcal{N}$) was defined as the triplet ($\mathcal{N} = < F, T, Q >$) where:

- $F = \{\pi_1, \ldots, \pi_f, \ldots, \pi_N\}$ provides the balance between food families $f = \{1 : N\}$ (i.e., fruits, proteins, etc.) to be fulfilled by the diet and being π_f, the proportion of food family F_f prescribed ($\sum_{f=1:N} \pi_f = 1$). The food families referred in F correspond to a subset of nodes ($\mathcal{F}$) of the reference food ontology from a fixed level of granularity in the food hierarchy.
- $T = [t_o, t_f]$ is the period of time where the diet must be followed.
- Q: total quantity of food in Kcal to be consumed along the whole diet.

With this prescription, Diet4You recommends a personalized menu $\mathcal{M}$ as a list of dishes globally fulfilling $\mathcal{N}$. The menu is composed by a set of dishes (eventually simple foods) $\mathcal{M} = \{\delta_1, \ldots, \delta_D\}$ where $\delta = (p_\delta, q_\delta)$:

- $p_\delta = (p_{\delta_1}, \ldots, p_{\delta_f}, \ldots, p_{\delta_N})$, where p_{δ_f} is the proportion of Kcal of the food family F_f contained in one standard portion of dish δ ($f = 1 : N$).
- q_δ is the total quantity of energy associated to one standard portion of δ.

Thus, the personalized menu must hold

- $\sum_{\forall \delta \in \mathcal{M}} q_\delta = Q$
- $\sum_{\forall \delta \in \mathcal{M}} q_\delta * p_{\delta_f} = Q * \pi_f, \ \forall f \in \{1 : N\}$

The specification of foods in the original food database is $p_\delta^\bullet$ (see Table 1), where n_{δ_f} is the number of standard units of food family F_f contained in the dish δ. As said before, the measurement units can change along the food families (some available units are grams, ounces, cups, teaspoons, etc.). Being U the set of available measurement units, a second vector U_δ provides the information to properly interpret the quantities expressed in $p_\delta^\bullet$. The original definition of a dish $p_\delta^\bullet$ must be transformed from a vector of non standardized units per food family to a vector of homogeneous quantities $q_{\delta f}$ all using a single measurement unit. The DB documentation provides the equivalences between u_f and the reference measurement unit (grams or Kcal), which we will refer as β.

To transform $p_\delta^\bullet$, the conversion factor β_δ of each food family is used to find the equivalent quantity of each food family of the dish $q_{\delta f} = n_{\delta f} \beta_{\delta f}$, the total quantity of food of the dish $q_\delta = \sum_{\forall f} q_{\delta f}$ and its distribution per food families $p_{\delta f} = \frac{q_{\delta f}}{q_\delta}$. As said before, β_{δ_f} is fixed or variable depending on the food families:

Table 1. Decomposition o dish δ

Food family	vector	F_1	...	F_f	...	F_N
Decomposition in DB	$p_\delta^\bullet$	n_{δ_1}	...	n_{δ_f}	...	n_{δ_N}
Original measurement unit	U_f	u_1	...	u_f	...	u_N
Conversion factor	β_f	β_1	...	β_f	...	β_N
Conversion factor type		Fixed	...	Variable	...	Fixed
Decomposition in quantities	q_δ	q_{δ_1}	...	q_{δ_f}	...	q_{δ_N}

- Fixed conversion factors: Some food families use a constant conversion factor, such as the protein family measured in ounces where 1 ounce = 28.35 g. In this case $\beta_{\delta f} = \beta_f, \forall \delta$.
- Variable conversion factor: Some food families are defined with measurement units that follow different conversion factors. Here, $\beta_{\delta f}$ is not constant for all δ depending on the type of food, the food itself or even, the cooking process. In this case, $\beta_{\delta f} = g(\delta, u_f)$.

In this work, the reference food ontology, with food families in the nodes, is annotated with the conversion factor given in the additional DB documentation. The resulting annotation might be incomplete. Thus an strategy is required to determine the conversion factor of those food families that are not specified. The proposal is to use the inheritance properties provided by the structure of the ontology itself to impute the missing conversion factors.

Ontology-Based Missing Values Imputation. The reference food ontology includes all foods classified by food family and subfamilies in a hierarchical way. Nodes are foods and main relationship is *is-a*. In this work, two additional attributes are associated to each node f: the standard measurement units (u_f) used to express quantities of food δ in the DB and the conversion factor (β_f) used to transform the standard units to the equivalent quantity in the reference measurement unit. The documentation associated with the food database is used to extend the reference ontology of food families used in [12] with more subfamilies and the associated conversion factors.

For all components (f) of a given food δ, β_f is retrieved from the f node of the reference ontology. However, as the DB documentation is incomplete, some nodes might have missing values for β_f. In that case, the list of ancestors of f is retrieved and the β_f is estimated by taking the $\beta_{f'}$ of the deepest ancestor f' in the ontology that has a non-missing conversion factor. This determines how to use the taxonomical relationship between foods to find a complete system of conversion factors that permits menu composition and assessment of goodness of fit with prescribed nutritional plan.

However, when a dish is searched from the DB, determining its corresponding β_δ vector, requires to identify which nodes in the reference food ontology are involved in the decomposition of δ. The dishes are described in the DB by a textual short description like *roasted chicken with potatoes* or *baked apple*. A natural language processing parser has been built to automatically identify the relevant keywords of this description and map these keywords with the nodes of the reference food ontology. From this point on, corresponding β coefficients are directly retrieved, or found on the ancestor nodes, as explained before.

4 Application

Diet4You involves a Food Proportions Database (FPDB) as a search space. To run the system, the following informations should be provided by FPBD:

- Food: Identifier, description, food type/family.
- Quantity of one standard portion of the dish.
- Total energy in Kcal.
- Nutrient List: name, quantity, measurement unit and nutrient/food family.
- Eventually a hierarchy of nutrients.
- Eventually the season when the food/dish is available.

Although some available food DBs contain the nutritional composition of foods, none of them contain all the information listed before. In previous works [12], it was determined that a combination of two databases supports the Diet4You system: the FNDDS and FPED databases, both offered by USDA [13]. FNDDS [14] contains the nutritional composition of foods and drinks that have been reported in *What We Eat in America (WWEIA)*. FNDDS is published every two years and it can publicly be downloaded and provides for each dish, the standard portion (in grams) and corresponding energy in Kcal. FPED (Food Patterns Equivalents Database) [4] maps the food and drinks of the FNDDS into 38 food patterns indicating the corresponding recommended quantity of each pattern per 100 g of food. In these 38 food patterns, there are 8 main patterns and the rest are subdivisions. This classification into food families (see Fig. 1) is used as the reference food ontology mentioned in methodology. The $\mathcal{F}$ set of food families, is determined from the reference ontology, by selecting the appropriate horizontal cut of the concepts network. According to [12], the considered food families are: $\mathcal{F} = \{$ *Fruit, Vegetables, Grain, Protein, Dairy, Oils, Added Sugars, Alcohol*$\}$ (see Fig. 1). In FPED Fruits, Vegetables, and Dairy are measured in cups; Grains and Protein Foods in ounces; Alcoholic Drinks in number; Added Sugars in teaspoons; Solid Fats and Oils in grams. A preprocessing step is needed in order to obtain the Food Proportion Database (FPDB), as a combination of FNDDS and FPED. One of the critical steps of this preprocessing is to build p_δ vectors associated to each dish δ and the total Kcal associated to one standard portion of each dish. The following steps has been required to create the FPBD:

1. Selecting only the foods of FPED that have non missing energy (and the weight of a portion) in FNDDS.
2. Omitting all foods with all null values.
3. Detecting the quasi-equals foods and minimize redundancies.
4. Unifying measurement units.

Steps 1 to 3 refer the selection of foods to be included into the database and step 4 refers the transformation of the food families decomposition into a vector of proportions of Kcal per each food through conversion factors.

Selecting Quasi-Equal Foods. Preliminary tests produced menus including quasi-repeated foods. As more variety is desired in the recommended menus, the foods that are considered quasi-equals are grouped and only one of each family is kept in FPBD. This prevents redundancies like "roasted chicken" and "roasted chicken with rosemary" or "pumpkin, pie" and "pumpkin, pie, individual portion". Quasi-equality is determined from the description name, quasi-equal dishes are those with a equal $\lambda\%$ of their name. Two cases are tackled: for dished described with less than 15 characters, considered *short-names*, a $\lambda = 0.7$ is used; for *Long names*, those with larger descriptions $\lambda = 60\%$ is considered.

Unifying Units. The FPED includes the amounts of the 38 Food Families per 100 g of each of the FNDDS foods. Most of the foods included in this database are multi-ingredient foods like pizza, salads, etc.

The food families ontology is created and extended using the documentation provided by the USDA of the FPED database since the USDA does not offer any tables or databases with this information. For doing this, the documentation is scanned in order to find keywords associated with grams equivalences both in subfamilies and foods. From text and tables included in the documentation, the information is taken in order to build the food family ontology with annotation for measurement units and conversion factors. This documentation provides the information for transforming the measurements into grams.

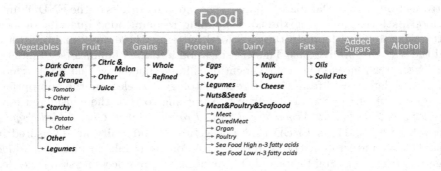

Fig. 1. Food families ontology

The resulting ontology shows that Fruit, Vegetables and Grain are measured with a measurement unit with variable associated conversion factors. The remaining food families can be transformed using a fixed conversion factors. Table 2 contains a synthesis of the original units per food family and conversion factors.

Figure 1 shows the upper levels of the reference food ontology. In 2 partial view of the food ontology is shown with deeper levels and conversion factors

Table 2. Measurement unit with the corresponding equivalence in grams per food family

Food family	Fruit	Vegetables	Grain	Protein	Dairy	Oils	Added sugars	Alcohol
Quantity	q_F	q_V	q_G	q_P	q_D	q_O	q_S	q_A
Unit	cup	cup	grain unit	ounce	cup	grams	tsp	unit
Conversion factor	Variable	Variable	Variable	28.35	Variable	1	7	14

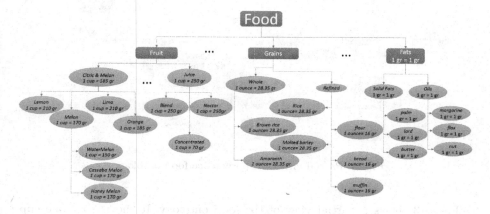

Fig. 2. Extended food families ontology

annotation. Three main food families are shown with three different situations. The oils family presents an easy scenario where the measurement unit is grams with fixed conversion factor ($1gr = 1gr$) for the whole family. The Fruit family is measured in cups with variable weight. The Grain family is measured in ounces, but for *flour* $1ounce = 16gr$ whereas for other *refined grains* like *Rice* $1ounce = 28.35gr$. Using constant conversion factors (see Table 2) and the annotated food ontology, the food descriptions can be transformed by unifying the original measurement units to percentages of Kcal. Since the documentation provides the conversion factors into grams, an extra step is needed to transform grams into Kcal by using the constant conversion factors described in Table 3. Then, for each food δ:

1. For each food family f.
2. If β_f is constant, use for all foods in the family to compute $q_{\delta f}$.
3. If *legumes* $\in f = Protein$, also copy into the subfamily of *vegetables*.
4. If $\beta_f = g(\delta, u_f)$, use the reference food ontology to compute $\beta_{\delta u_f}$:
   (a) Identify the ontology nodes containing the keywords of the description of the food (note that the name of food families do not need to be in the food name).
   (b) Find the deepest nodes in each involved branch that have a non-missing conversion factor and use it to transform the several components in $q_{\delta f}$.
5. Compute q_f and transform into Kcal using Table 3.
6. Assess the proportions in Kcal and build p_δ.

Table 3. Equivalence of 1 g to Kcal per food family

Food family	Fruit	Vegetables	Grain	Protein	Dairy	Oils	Added sugars	Alcohol
Kcal/1gr	4	4	4	4	4	9	4	7

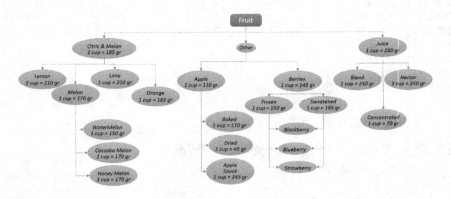

Fig. 3. Partial view of the reference food ontology

Figure 3 shows a partial view of the food ontology. It shows that one cup of all types of berries such as raw blackberries, blueberries, or strawberries corresponds to 145 gr, unless they are frozen, where 1cup = 150gr or sweetened (1cup = 165gr). Also, melons are given 170 gr per cup except for the watermelon. Another example is the different weights for apples depending on its presentation (raw, baked, dried or as sauce). The ontology also contains the vegetables and dairy family, treated similarly than fruit family, with the exception of beans and peas (legumes) which are considered as a subfamily of both protein foods and vegetables. The Legume component is computed both ways - one measured in cups and the other in ounces. As the ounces can be directly transformed, the legumes are assessed using the subfamily legumes descending from Proteins food family and copied to the subfamily legumes in Vegetables. Ounce equivalences for grains also depends on the grain itself as well as the presentation (flour or not) (see Fig. 2). The ontology-based missing value imputation method helps to get conversion factors for food not explicitly mentioned in the DB documentation. Indeed, as 1 ounce of flour is 16 g, grain products such as breads, pancakes, waffles, muffins, and grain-based snacks made of flour, the correspondence is estimated to be 1gr = 16 g of flour, as flour is the deepest ancestor in the food ontology with conversion factor annotation. For cooked grains such rice, pasta, hot breakfast cereals are converted to uncooked form, and the proposed method suggests that conversion factor for this group of foods is 1 ounce = 28.35 g, as it is for ready-to-eat cereals. The systematic application of the ontology-based missing value imputation method along all dishes in the database, produces a new food database in the form specified in FPDB. This guarantees a proper conversion factor for all dishes in the FPDB and the composed menu can be assessed with regards to the prescribed nutritional plan.

5 Conclusions

This work copes with a crucial step in all data analysis or data mining process, which is the preprocessing [6]. A hard preprocessing is required to prepare the Diet4You system to work with homogeneous measurement units along the food families considered in the food descriptions. A particular complex situation is faced, as the underlying food database uses different measurement units for different kind of foods (protein, dairy, etc.) and what was more unexpected, the same unit corresponds to a different quantity of grams depending on the foods. Apart from this intrinsic polysemy of the measurement units used in the food databases, which is not restricted to FPED, there are also some additional challenges: the conversion factors information and equivalences between measurement units are only available from the additional documentation provided with the database and have not been conceived to be automatically processed. As a consequence, not all foods have conversion factors specified. So, we are in front of incomplete information and polysemy in a non-structured support (a pdf document). The *ontology-based missing value imputation* method is proposed based on the construction of an annotated ontology, built from the additional documentation. The proposal permits to impute missing conversion factors using the closest ancestors, so guaranteeing the semantic proximity with the food providing the conversion factor estimation. The development of this method opens the door to combine several food databases using different metric systems and opens the door to integrate food databases coming from different countries or food style (vegan, Asian food, etc.) into a single polysemic Diet4You.

A particular kind of preprocessing emerges here, far from the classical pretreatment of a variable in an standard dataset. Several food databases are preprocessed to build a knowledge representation formalism, in this case an annotated ontology. The preprocessing is not on data, but on a conceptual level and it is tackling the issue of standardization of concepts (in this case the concept of *measurement units*). In turn, the preprocessed knowledge can be used as a basis to a knowledge-based imputation method of missing values (ontology-based missing values imputation method), which happens in the space of available dishes, when transforming decomposition of dishes into a standardized measurement units, to enable intelligent menus elaboration.

The scan of the additional documentation provided with the food database has been the unique source of knowledge to build the annotated food database with the food taxonomy annotated with available conversion factors for each food family. By using regular expressions, the foods in the database can be mapped into the corresponding set of nodes of the ontology and ontology-based imputation method is used to recover the missing conversion factors. This permits computation of p_δ as a combination of the ontology-based imputation method and the conversion factors expressions, and permits to use a complete reference food database to compose the personalized method. In [12] some successful menus recommended by Diet4You, which include the preprocessing steps presented in this paper, can be found.

Acknowledgements. Work supported by projects Diet4You (TIN2014-60557-R, Spanish Government) and IDEAI (SGR2017-574, Catalan Government).

References

1. Menu planner, june 2015. http://hp2010.nhlbihin.net/menuplanner/menu.cgi
2. Eat this much (2017). http://www.eatthismuch.com
3. Interactive DRI for healthcare professionals, April 2017. http://fnic.nal.usda.gov/fnic/interactiveDRI/dri_results.ph
4. Bowman, S., Clemens, J., et al.: Food patterns equivalents database 2011–12: methodology and user guide (2014)
5. Chavez, A., de Chávez, M.M.: Nutrigenomics in public health nutrition: short-term perspectives. Eur. J. Clin. Nutr. **57**, S97–S100 (2003)
6. Gibert, K., Sànchez-Marrè, M., Izquierdo, J.: A survey on pre-processing techniques in the context of environmental data mining. AICOM (2016, in press)
7. Gibert, K., Horsburgh, J.S., Athanasiadis, I.N., Holmes, G.: Environmental data science. Environ. Model. Softw. **106**, 4–12 (2018)
8. Hammond, K.: CHEF: a model of case-based planning. In: AAAI, pp. 267–271 (1986)
9. Khan, A., Hoffmann, A.: An advanced artificial intelligence tool for menu design. Nutr. Health **17**(1), 43–53 (2003)
10. Marling, C., Petot, G., Sterling, L.: Integrating case-based and rule-based reasoning to meet multiple design constraints. Comp. Intell. **15**(3), 308–332 (1999)
11. Noah, S.A., Abdullah, S.N., et al.: DietPal: a web-based dietary menu-generating and management system. J. Med. Internet Res. **6**(1), e4 (2004)
12. Sevilla-Villanueva, B., Gibert, K., Sànchez-Marrè, M.: Generating complete menus from nutritional prescriptions by using advanced CBR and real food databases. In: Recent advances in artificial intelligence research and development, pp. 166–175. IOSPress (2017)
13. USDA: USDA department of agriculture, agricultural research service, nutrient data laboratory (2017). http://www.ars.usda.gov/ba/bhnrc/ndl
14. USDA, Agricultural Research Service: USDA food and nutrient database for dietary studies 2013–2014 (2016). http://www.ars.usda.gov/nea/bhnrc/fsrg

Modeling the Navigation on Enrolment Web Information Area of a University Using Machine Learning Techniques

Ainhoa Yera[✉], Iñigo Perona, Olatz Arbelaitz, and Javier Muguerza

Faculty of Informatics, University of the Basque Country, 20018 Donostia, Spain
{ainhoa.yera, inigo.perona,
olatz.arbelaitz, j.muguerza}@ehu.eus

Abstract. This work analyses the navigation in the enrolment web information area of the University of the Basque Country. A complete data mining process shows that successful and failure navigation behaviors can be modeled using machine learning techniques. Unsupervised learning algorithms have been applied on two different domains: URLs visited by the users in each session (navigation sequence) and some interaction parameters extracted from the recorded click-stream (navigation style). Both domains have been used satisfactorily to model the behavior of success and failure navigation sessions achieving more than 78% of accuracy predicting success or failure sessions. Furthermore, the clustering based on the navigation style was able to identify the main characteristics of each type of session and to build a subsystem that enables to detect failure type sessions with high precision.

Keywords: Web usage mining · Navigation models
Web interaction characterization

1 Introduction

During the last years an information revolution has happened which has been important in eGovernment and eSociety. Almost all the governments in the world have been part of it [11] and they are including eGovernment in the government processes [1]. The development of the eGovernment is becoming an integral part of the strategies of the governments [14] and is quickly changing the way the institutions provide information and services to the citizen.

To this regard in 2016 the European Parliament and the Council of the EU established a Directive to make the websites and mobile apps of the public sector bodies more accessible [5] recommending to follow the Web Content Accessibility Guidelines (WCAG) 2.0 [13]. However, experience has shown that the standards do not ensure an effective accessibility and more efforts should be done to adapt the web content to the user needs and its usage context. Furthermore, it is now acknowledged that more than just usability is needed in the design and evaluation of e-Systems and, in general, the way the information of the institutions is presented on the Web. In fact,

© Springer Nature Switzerland AG 2018
F. Herrera et al. (Eds.): CAEPIA 2018, LNAI 11160, pp. 307–316, 2018.
https://doi.org/10.1007/978-3-030-00374-6_29

there has been an apparent shift in research on human-computer interaction from cognitive-task performance to user experience [10].

In this context, to improve the accessibility and user experience data mining and modeling techniques can be used to discover user navigation patterns. These techniques allow discovering and then predicting the interactive behavior of the users and also provide the opportunity to discover the problems arising from the structure or content of a site [12].

Therefore, the advantages of applying machine learning techniques in this context and then introducing their outcome can be useful [6] to make Web services and information adaptable to the diversity of users and to improve their digital inclusion.

As a step to these contributions this paper presents a research resulting from our collaboration with the University of the Basque Country (UPV/EHU). The university has provided us access to the navigation logs of its website from the end of February to the middle of April 2016. We used data mining processes to model the university enrolment web information area (www.ehu.eus/web/sarrera-acceso) to provide clues for future website improvements, and to extract as much knowledge as possible from it.

We analyzed the usage of the site accessing to the enrolment area and labeled the sessions as success or failure based on the type of information accessed at the end of the navigation. The user navigation in the enrolment area was analyzed from two points of view: visited URLs and navigation style. Unsupervised learning techniques showed that both aspects affect to the success or failure of a user session but however, a comparison revealed that both perspectives were independent. We used the two systems to classify new sessions and the navigation style showed to be more effective than the visited URLs for this aim. Experiments showed that both systems could be complementary since the system performing the best to classify new sessions combined both approaches. Being failure sessions the ones with higher urgency to react, we designed a heuristic that detects sessions with high failure probability. Finally, we characterize the failure and success type sessions according to the navigation style.

The paper proceeds with the description of the research in Sect. 2. The next section, Sect. 3, describes the results achieved and finally in Sect. 4 we present the main conclusions of the study.

2 Preprocessing, Session Classification and Generation of the Navigation Sequence and Navigation Style Databases

In this section we first describe the preprocessing and the session classification carried out with the navigation data of the University of the Basque Country and then we explain how we generated the navigation sequence and navigation style databases (DBs) for the analysis.

2.1 Preprocessing

The UPV/EHU provided us access to their logs of 53 days, from the 23/02/2016 to the 16/04/2016, and we focused the analysis on the navigation in the area that supplies enrolment information. These dates partly cover the pre-enrolment dates for the secondary school students, from the middle of January to the middle of March (2016).

In particular, we only considered the user sessions with a meaningful navigation in the area of interest: minimum length of three clicks and at least one URL within the enrolment area (www.ehu.eus/web/sarrera-acceso).

The preprocessing carried out produced a DB of 35,467 sessions, around 8.4% of the 422,448 sessions available. After ordering this data on time, the first 25,467 sessions, first 49 days, were used to model the web area (modeling DB), whereas the last 1,000 sessions, last four days, were used to validate the system (validation DB).

2.2 Session Classification

In our research we related the users' interest to the end of their navigation and we accordingly classified user sessions, concretely, based on the type of the last webpage visited. Specifically, the URLs requested in the sessions were characterized considering two criteria: the content of the URL (whether the text or the links were dominant) and the area of the URL (whether it corresponded to the enrolment area or not). Regarding the content, those URLs with text format (.pdf/.doc/.docx) were classified as of content type. The remaining URLs were classified as content or scatter type (when links were dominant) using the LCIndex (Link Content index) [3] described in expression (1):

$$LCIndex = \frac{Nwords - NwordsLinks}{Nlinks}; URLtype = \begin{cases} scatter, LCIndex \leq 10 \\ content, LCIndex > 10 \end{cases} \quad (1)$$

Where, *Nwords* is the number of words appearing in the webpage, *NwordsLinks* is the number of words used in the links of the page and *Nlinks* is the number of links in the webpage.

Finally, we considered to be successful (classified as success) the user sessions finished in a URL with enrolment information (content type and within the enrolment area) because this will probably help in the enrolment process. On the contrary, the user sessions ended in web pages with little information (scatter type and any area) were considered to be of failure type. Since we are focusing in users that want to enroll, the sessions ended in a content type URL outside the enrolment area were considered out of our scope (possible success in another area). Table 1 shows the session classification and the number of examples obtained in the modeling and validation DBs for each type of session.

Table 1. User session classification of the enrolment area of the UPV/EHU.

Type of session	Last URL of the session		No. sessions	
	Area	Type	Modeling DB	Validation DB
Success	Enrolment	Content	10,734 (42.1%)	379 (37.9%)
Failure	Enrolment	Scatter	14,733 (57.9%)	621 (62.1%)
	Not enrolment	Scatter		

2.3   Generation of the Navigation Sequence and Navigation Style Databases

After the log was preprocessed and the user sessions were classified, data was prepared for analysis. As stated in the introduction, user sessions were analyzed from two points of view: navigation sequence and navigation style. Accordingly, we created two databases with the selected sessions: the first one containing the sequence of URLs visited in each user session, and the second one, representing the user sessions with a vector of interaction parameters calculated from the information contained in the log files combined with the content and the structure of the site. The parameters that represent the user sessions in the second DB were computed according to the time, the URL classification (content/area) or the number of clicks. Table 2 shows the list of parameters mentioned and their description.

Table 2. Interaction parameters used to represent the user sessions in the navigation style DB.

Parameter	Description
No. click	Number of clicks (length of the session)
No. scat. %	Number of scatter type URLs/length of the session
No. cont. %	Number of content type URLs/length of the session
No. enr. %	Number URLs of the enrolment area/length of the session
No. not-enr. %	Number URLs from outside the enrolment area/length of the session
No. ind. %	Number of times the start page of the enrolment area (index) is visited/length of the session
No. ref-sear. %	Number of URLs that have a web search engine as reference
T-ses.	Duration of the session (s)
T-click_avg	Average duration of a click (s)
T-scat._avg	Average duration of a click on a scatter type URL (s)
T-cont._avg	Average duration of a click on a content type URL (s)
T-enr._avg	Average duration of a click on a URL of the enrolment area (s)
T-not-enr._avg	Average duration of a click on a URL outside the enrolment area (s)
T-ind.-avg	Average duration of a click on the start page of the enrolment area (s)
No. cont.-scat.	Number of transitions: content type URL - scatter type URL
No. scat.-cont.	Number of transitions: scatter type URL - content type URL
No. enr.-not-enr.	Number of transitions: enrolment area URL - outside enrolment area URL
No. not-enr.-enr.	Number of transitions: outside enrolment area URL - enrolment area URL

3   Modeling of the Enrolment Web Information Area

The goal of this work was to model the enrolment web information area so that it could be improved in the future. We mainly focused on characterizing successful and failure sessions and detecting failure sessions so that actions can be taken. With this aim, we first analyzed whether the two aspects studied, navigation sequence and navigation style were meaningful to decide if a user session will be of success or failure type, whether they can be used to foresee the type of new sessions, and if they are complementary or not.

In order to analyze the interrelationship between the two navigation representations with the sessions' type, unsupervised learning procedures were used. These techniques allow on the one hand, characterizing success and failure patterns. On the other hand, the centroids we computed in the clustering procedures used in the two DBs, provide stable patterns that enable to tune the success and failure session classification system in order to control its level of precision. The next subsections describe the analysis carried out.

3.1   Session Modeling

Based on Navigation Sequence

The DB built with sequences of URLs was used to analyze how the navigation sequence is interrelated with the type of sessions defined. We used PAM (K-medoids) [8] clustering algorithm that allows to group sequences into high quality clusters [4] with edit distance. Broadly, in a clustering procedure the number of clusters selected is ideally small, as it contributes to create clusters with as many cases as possible of the same type. We selected the K value according to the Silhouette Cluster Validity Index [2] which indicated that K = 50 was the most appropriate cluster configuration when compared to K = 25, K = 75 and K = 100. The scores for each K in ascending order were 0.046, 0.069, 0.055 and 0.047 respectively.

To evaluate the discernment power of the approach for the two types of navigation sessions, we focused on the clusters where the superiority of success or failure cases was over 74%. In the selected clusters, the total number of sessions grouped was significant (the 42% of the whole DB), and a suitable representation of each type of session can be found (12% of success and 29% of failure). Table 3 summarizes the types of examples grouped in the eight success type clusters and the 17 clusters of failure type.

Table 3 shows that half of the clusters (25) have a proportion of one class or the other one above 74%. The whole DB contains 14,733 user sessions of failure type and the clusters labeled as failure, 6,669. Consequently, although the percentage of failure type sessions in the DB is 58%, within the failure type clusters that probability raises to 89%. Similarly, being the success type user sessions 42% in the whole DB, in the selected success type clusters this percentage increases up to 82%.

Table 3. Results of PAM (K = 50) in the DB built with sequences of URLs.

Parameter	Clusters with a no. success-sessions $\geq 74\%$	Clusters with a no. failure-sessions $\geq 74\%$
No. clusters	8	17
No. success-sessions	2,551 (81.9%)	842 (11.2%)
No. failure-sessions	564 (18.1%)	6,669 (88.8%)
No. sessions-clusters	3,115	7,511
No. sessions-DB (%)	12.2%	29.5%

These results suggest that the navigation sequence (URL sequence visited) and the success/failure of a user session are connected. Hence, we can gather that there is a chance to automatically classify the navigation of new users of the UPV/EHU enrolment eService, based on the navigation sequence.

Based on Navigation Style

The database with the interaction parameters of the sessions (Table 2) was used to determine whether the navigation style is tightly interrelated with the success/failure of a session or not. We ran the K-means algorithm [9] with Euclidean distance using K = 50 in the previously normalized (normal distribution) database. Then, we selected the clusters with a superiority of success or failure cases over 74%, which grouped 43% of the total number of sessions of the DB (14% of the success and 29% of failure). Results are shown in Table 4.

Table 4. Results of K-means (K = 50) in the interaction parameters DB.

Parameter	Clusters with a no. success-sessions $\geq 74\%$	Clusters with a no. failure-sessions $\geq 74\%$
No. clusters	6	21
No. success-sessions	3,212 (87.7%)	544 (7.4%)
No. failure-sessions	451 (12.3%)	6,842 (92.6%)
No. sessions-clusters	3,663	7,386
No. sessions-DB (%)	14.4%	29%

As it happened with the navigation sequence DB, Table 4 shows that in more than half of the clusters (27/50), the proportion of one of the types of sessions defined or the other is higher than 74%. In the case of the failure clusters there are 6,842 sessions of failure type what raises form being 58% of the sessions in the complete DB, to 93% within those clusters. Likewise, being the success type sessions 42% of the DB, in the success clusters this number raises to 88%. Thus, the results show that the navigation style is discriminant for the two types of navigations defined, success and failure.

Comparison of the Navigation Sequence and Navigation Style

According to the results shown in the previous sections, we can state that whether a session will be of success or failure type depends on both the navigation sequence and the navigation style. But it would be interesting to know if both perspectives are closely related or not. With this aim we compared the partitions of the two clustering procedures using the Jaccard index [7] which provided a very low value (0.04) in the comparison, showing that both results are quite different. This suggests that in the navigation of the UPV/EHU enrolment area, the navigation sequences (URLs visited) and the navigation style described by the interaction parameters are independent, and thus the design of each concrete URL does not affect much to how the user navigates.

Hence, in principle to classify the navigation of new users in the enrolment area of the website, both view points, could be useful and might be complementary. To this regard, in the next section we describe the validation process performed to test these hypotheses.

3.2   Classification of New Sessions

As mentioned before, a total of 1,000 sessions of the initial database were kept for validating the system (validation DB). These sessions were also represented as sequences of URLs visited (navigation sequence perspective) and as vectors of the interaction parameters described in Table 2 (navigation style perspective).

Classification Based on Navigation Sequence

In this system we first computed the medoids of the 25 clusters with a wide percentage of sessions ($\geq 74\%$) of success or failure type (Table 3) and labeled them with the majority class of their corresponding cluster. Then, for each new session of the navigation sequence DB, we calculated the 10 nearest medoids (10-NM) using the edit distance and finally labeled each new session with the most voted type according to the simple voting performed using different numbers (km) of nearest medoids: km-NM; $km \in \{1, 3, 5, 7, 9, 10\}$. We measured the accuracy (%) in terms of number of examples where the type of sessions is guessed among the total number of new sessions (see Table 5). The voting approach that involved the 10 nearest medoids was found to be the best one, reaching an accuracy of 61.90% thus the results were not very good.

Table 5. Accuracy (%) of the system built with the km-NM of the 25 clusters obtained with PAM (K = 50) procedure used in the navigation sequence DB.

	Voting approaches performed with different number of medoids (km-NM)					
	1-NM	3-NM	5-NM	7-NM	9-NM	10-NM
Accuracy (%)	55.7%	56.1%	55.9%	61.8%	59.9%	61.9%

Classification Based on the Navigation Style

Similarly, in this system we computed the centroids (average) of the 27 clusters with a wide majority of sessions ($\geq 74\%$) of success or failure type (Table 4) and labeled them with the majority class of their corresponding cluster. Then, each session of the new navigation style DB was labeled with the most voted type according to the different numbers (kc) of nearest centroids (NC) computed with the Euclidean distance: kc-NC; $kc \in \{1, 3, 5, 7, 9, 10\}$. The results of this system (see Table 6) were better than the previous ones. In this case, the voting that involved the five nearest centroids (5-NC) was found to be the best, reaching an accuracy of 78.2%.

Table 6. Accuracy (%) of the system built with the kc-NC of the 27 clusters obtained with K-means (K = 50) procedure used in the navigation style DB.

	Voting approaches performed with different number of centroids (kc-NC)					
	1-NC	3-NC	5-NC	7-NC	9-NC	10-NC
Accuracy (%)	75.50%	76.70%	78.20%	77.50%	76.90%	74.80%

Classification Based on the Navigation Sequence and the Navigation Style
In order to analyze if both systems, the navigation sequence and the navigation style, were complementary or not, we built a system that combined the votes of the system based on the navigation sequence DB, km-NM; $km \in \{2, 3, 4, 5, 6\}$, and the votes of the best approach of the system based on the navigation style DB (5-NC). Accordingly, each new session was classified using the most voted type among the km nearest medoids and kc nearest centroids involved. The best voting was the one that involved the nine nearest neighbors (9-NN), using the four nearest medoids (4-NM) for the navigation sequence DB and the five nearest centroids (5-NC) for the navigation style DB. Table 7 shows the accuracy obtained for different configurations. Although no weighting of other more complex strategies where used in this first approach, it seems that both options could be complementary.

Table 7. Accuracy of the system built combining the votes of the systems built with the km-NM and the kc-NC of the navigation sequence and navigation style DBs respectively.

	Voting approaches performed with different number of medoids and centroids (k-NN = km-NM + kc-NC)					
	7-NN	8-NN	9-NN	10-NN	11-NN	7-NN
	2-NM	3-NM	4-NM	5-NM	6-NM	2-NM
	5-NC	5-NC	5-NC	5-NC	5-NC	5-NC
Accuracy (%)	78.20%	77.90%	78.70%	76.90%	77.40%	78.20%

Failure Detection Subsystem Based on the Navigation Style
We built a subsystem to detect users that were having problems (failure type sessions) enabling to adapt the restriction level (minimizing the false positives).

In this subsystem new sessions were classified based exclusively on the 21 clusters of the navigation style DB that detected failure type sessions (Table 4). Specifically, for each new session we computed the nearest centroid (1-NC) of the selected clusters and then, we reordered the validation DB based according to the distance. The smaller the distance is, the higher probability will the pattern have to be of failure type.

This allowed us to segment the new users and only work with those who were more similar to the failure patterns detected with the following precision values: 100%, 99% and 91% for the 10%, 15% and 25% nearest new sessions to the computed centroid respectively. This method also enabled us to define a distance threshold (1.657) to classify new users individually with a high certainty to be of failure type.

Characterization of the Types of Session Based on the Navigation Style
Additionally, we extracted the main characteristics of the six success type clusters and the 21 failure type clusters to model both types of navigation.

These are the main characteristics for the failure type sessions compared to the success type sessions: the click-streams on average tend to be larger (*No. click* = 13.1 vs. 6.2); they are more focused on scatter type URLs (*No. scat. %* = 80% vs. 30%); the internal and external navigations are more balanced (*No. enr. %* = 48% vs. 92%); the total duration of these sessions is higher (*T-ses* = 184.6 s vs. 106.3 s); the duration on average of a click on a scatter type URL is longer ($T\text{-}scat.\_avg$ = 27.8 s vs. 11.7 s); the duration on average of a click on a content type URL is shorter ($T\text{-}cont.\_avg$ = 7.4 s vs. 20.9 s); there is at least one transition from enrolment area to outside this area ($No.\ enr.\text{-}not\text{-}enr.$ = 0.8 vs. 0); there is almost no transition from outside the enrolment area to the enrolment area ($No.\ not\text{-}enr.\text{-}enr$ = 0.4 vs. 1.1).

4   Conclusions and Further Work

This research presents the modeling of the enrolment web information area of the UPV/EHU using web mining techniques. The navigation sessions were represented based on the navigation sequence (sequences of URLs) and based on the navigation style (interaction parameters) and two types of sessions were identified: success and failure.

Our experiments showed that the two perspectives used give rise to automatically detect the two navigation types defined. The application of PAM and K-means clustering algorithms in the databases built with sequences of URLs and with the interaction parameters respectively came out with partitions where half of the clusters had a high proportion (>74%) of one of the navigation types defined. However, the low value provided by Jaccard index (0.04) suggested that the navigation sequence and the navigation style were not connected.

Two voting systems built using different numbers of nearest medios and nearest centroids of the clusters with a high proportion of success or failure type sessions previously obtained, showed that the navigation style perspective was more effective than the navigation sequence perspective to classify new sessions, achieving a maximum accuracy of 78.2% when using the five nearest centroids (5-NC).

Besides, the two perspectives showed to be complementary. A third voting system that combined different numbers of medoids and centroids together showed to perform slightly better than the previous, reaching an accuracy of 78.7%.

On the other hand, a failure detection subsystem based on the navigation style and using the nearest centroid (1-NC) for new sessions, achieved a 100% of precision using the nearest 10% of the new sessions and it allowed defining a distance threshold of 1.567 to classify new sessions as failure type with high probability.

Finally, the characterization of each type of session, showed that success and failure type sessions have many differences is their interaction parameters.

This work is the first step to model the enrolment web information area of the UPV/EHU and considering the results, we can state that unsupervised learning techniques are useful in that process.

As future work, the complementary system should be more thoroughly studied, considering more complex voting criteria where for instance the distance to the centroids or medoids could be taken into account. We also want to review the parameters selected to represent the sessions in the navigation style DB and find out if the use of a subset of these parameters leads to better results. Additionally, we wish to do an indepth analysis of the models created in order to anticipate to the future users, and to identify and improve those elements having a negative influence on the usability.

To conclude, we would like to use the conclusions of this analysis as a kind of guidelines to improve other kinds of websites.

Acknowledgment. This work has been funded by the following units: Firstly, by the University of the Basque Country UPV/EHU (PIF15/143 grant). Secondly, by the research group ADIAN that is supported by the Department of Education, Universities and Research of the Basque Government, (grant IT980-16). Finally, by the Ministry of Economy and Competitiveness of the Spanish Government, co-founded by the ERDF (eGovernAbility, TIN2014-52665-C2-1-R).

References

1. Alshehri, M., Drew, S.: Challenges of eGovernment services adoption in Saudi Arabia from an e-ready citizen perspective. World Acad. Sci., Eng. Technol. **66**, 1053–1059 (2010)
2. Arbelaitz, O., Gurrutxaga, I., Muguerza, J., Pérez, J.M., Perona, I.: An extensive comparative study of cluster validity indices. Pattern Recognit. **46**(1), 243–256 (2013). https://doi.org/10.1016/j.patcog.2012.07.021
3. Arbelaitz, O., Lojo, A., Muguerza, J., Perona, I.: Web mining for navigation problem detection and diagnosis in Discapnet: a website aimed at disabled people. J. Assoc. Inf. Sci. Technol. **67**(8), 1916–1927 (2016)
4. Barioni, M.C.N., Razente, H.L., Traina, A.J.M., Traina, C.: Accelerating k-medoid-based algorithms through metric access methods. J. Syst. Softw. **81**(3), 343–355 (2008). https://doi.org/10.1016/j.jss.2007.06.019
5. European Parliament and Council of the European Union: Directive (EU) 2016/2102 of the European parliament and of the council of 26 October 2016 on the accessibility of the websites and mobile applications of public sector. Off. J. Eur. Union, L 327, 2 December 2016, pp. 1–15 (2016). https://eur-lex.europa.eu/legal-content/EN/TXT/?uri=CELEX%3A32016L2102
6. Gugliotta, A., et al.: Semantic web service-based architecture for the interoperability of e-Government services. In: WISM 2005, p. 21 (2005)
7. Jaccard, P.: Nouvelles recherches sur la distribution florale. Bull. de la Soc. Vaudoise de Sci. Nat. **44**, 223–370 (1908)
8. Kaufman, L., Rousseeuw, P.J.: Finding Groups in Data: An Introduction to Cluster Analysis. Wiley, London (1990)
9. Lloyd, S.: Least squares quantization in PCM. IEEE Trans. Inf. Theory **28**(2), 129–137 (1982)
10. Standen, P.J., Karsandas, R.B., Anderton, N., Battersby, S., Brown, D.J.: An evaluation of the use of a computer game in improving the choice reaction time of adults with intellectual disabilities. J. Assist. Technol. **3**(4), 4–11 (2009)
11. Taylor, J., Lips, M., Organ, J.: Information-intensive government and the layering and sorting of citizenship. Public Money Manag. **27**(2), 161–164 (2007)
12. Vigo, M., Harper, S.: Challenging information foraging theory: screen reader users are not always driven by information scent. In: Proceedings of the 24th ACM Conference on Hypertext and Social Media, pp. 60–68. ACM (2013)
13. W3C. Web Content Accessibility Guidelines (WCAG) 2.0 (2008). http://www.w3.org/WAI/intro/wcag.php
14. Zhang, H., Xiaolin, X., Jianying, X.: Diffusion of eGovernment: a literature review and directions for future directions. Gov. Inf. Quart. **31**(4), 631–636 (2014)

Applications

A GRASP Algorithm to Optimize Operational Costs and Regularity of Production in Mixed-Model Sequencing Problems with Forced Interruption of Operations

Joaquín Bautista-Valhondo[1]([⊠]) [iD] and Rocío Alfaro-Pozo[2] [iD]

[1] IOC ETSEIB Universidad Politécnica de Catalunya,
Diagonal 647, 08028 Barcelona, Spain
joaquin.bautista@upc.edu
[2] EAE Business School, C/Aragó, 55, 08015 Barcelona, Spain
ralfaro@eae.es

Abstract. We present a GRASP algorithm to solve a problem that involves the sequencing of mixed products in an assembly line. The objective of the problem is to obtain a manufacturing sequence of models that generates a minimum operational cost with a forced interruption of operations and that is regular in production. The implemented GRASP is compared with other procedures using instances of a case study of the Nissan engine manufacturing plant in Barcelona.

Keywords: GRASP · Bounded Dynamic Programming · Work overload
Operational cost · Mixed-model sequencing problems

1 Introduction

The automotive industry has undergone a constant evolution. This evolution has based on production methods and management ideologies. The mass production, the increase of flexibility and the synchronous manufacturing are some of the innovations that the sector has incorporated during these last decades. The main objective of these changes has been to offer a wide range of products while reducing costs and increasing productivity. Today, this objective continues to govern any improvement in production and management systems where flexibility is an essential requirement.

Among the sequencing problems in assembly lines [1], the Mixed-Model Sequencing Problem (MMSP) is a clear example of the purpose of improving and innovating product-oriented manufacturing systems. The objective of MMSP is to determine the best sequence of products in terms of some productive criterion, such as the amount of completed work, the idle time of the line or even the operational costs.

The Mixed Model Sequencing Problem with Operational Costs Minimization (MMSP-Cost: [2, 3]) is a family of sequence problems in assembly lines that consists of establishing a bijection between the elements of a set T of manufacturing cycles $(t = 1, \ldots, T)$ and those of a set Ψ of products (T elements). The elements of the set Ψ can be grouped into exclusive classes ψ_i that satisfy the following: $\Psi = \bigcup_{i \in I} \psi_i$ and

© Springer Nature Switzerland AG 2018
F. Herrera et al. (Eds.): CAEPIA 2018, LNAI 11160, pp. 319–329, 2018.
https://doi.org/10.1007/978-3-030-00374-6_30

$\psi_i \cap \psi_{i'} = \emptyset, \forall \{i, i'\} \in I$, where I is the set of product types $(i = 1, \ldots, n)$. The units of Ψ pass, one after the other, through a set of workstations, K, that are arranged in series (a production line). A unit of type $i \in I$ requires a processing time $p_{i,k}$ for normal activity $(\alpha^N = 1)$ at workstation $k \in K$ $(k = 1, \ldots, m)$.

Obviously, the difference between the classes ψ_i (SUVs, Vans, Trucks) makes the times $p_{i,k}$ heterogeneous, while the cycle time c is identical at every workstation $k \in K$; therefore, the operators and robots (processors) have the same time to complete an operation regardless of the product and the station. This discrepancy between the cycle time and the processing times places the processors between two undesirable situations: (i) delays with idle time U and (ii) blockages due to work overload W.

To alleviate the blockages due to work overload, the processors are eventually granted a time to work that is longer than the cycle time, called the temporary window $l_k(l_k > c)$, which depends on the station $k \in K$ [4, 5]. This concession can reduce the operator's available time to work on the next product and can also reduce the available time for the next station $(k + 1)$ to work on that product when it is released by station k. When the temporary window is not sufficient to complete the work required by the product, it will be incomplete and will pass to the next station, which results in an interruption of the operation that generates an inefficiency that implies that we call work overload or have the production fall.

An operation can be interrupted in a *forced* or *free* way. The *forced interruption* occurs when the operator reaches the limit of the temporary window l_k without completing the operation [6, 7]. The *free interruption* occurs when the incomplete product passes to the next station before reaching the limit of the temporary window of the emitting station [8, 9]. It is clear that the operation is interrupted whenever the limit established by the temporary window l_k $(\forall k \in K)$ is reached.

The purpose of the MMSP-Cost (whether the interruption of the operations is forced or free) is to obtain a sequence of models $\pi(T)$ that minimizes the costs by production losses in the line, what we call Total Operational Cost $\Gamma(\pi(T))$ and which is determined from two addends: the total work overload cost $\Gamma_W(\pi(T))$ and the total idle time cost $\Gamma_U(\pi(T))$.

Minimize the costs by production losses is not the only desirable objective when establishing a product manufacturing sequence. Indeed, in production environments that are governed by the Just-in-Time ideals [10, 11], the manufacturing sequences must have properties that are linked to the regularity of production. The incorporation of the regularity concept in manufacturing sequence problems can be characterized by an objective function that implies to maximize the constancy of the product manufacturing rates [12] and/or the component consumption rates with the purpose of minimizing the maximum stock levels of the latter [13, 14].

In this work, we will add to the MMSP-Cost problem the regularity concept in manufacturing to achieve sequences with minimum Total Operational Cost and with some properties that propitiate the Regularity of production. The characteristics of other previous works referenced (*Ref.* column) are summarized in Table 1.

The remaining text has the following structure. Section 2 is dedicated to concepts about the costs by production losses and regular sequences in production lines. In Sect. 3, we present the problem under study, which we call MMSP-Cost/Δ_{Q_x} with

Table 1. Characteristics of some previous works on the mixed model sequencing problem

MMSP	Ref.	Optimizing	Interruption	Regularity	Procedure
W	[6]	Work overload	Forced	No	BDP
W	[8]	Work overload	Free	Quota property	MILP
W/U	[9]	Work overload	Free	Quota property	GRASP-LP
W/ΔQx	[7]	Work overload	Forced	Quota property	GRASP
Cost	[2, 3]	Operational cost	Free	No	MILP

forced interruption of operations. In Sect. 4, we describe the GRASP algorithm that was designed. In Sect. 5, we present a case study with its data, the procedures used and their results. Finally, in Sect. 6, we offer some conclusions about this work.

2 Operational Costs and Regularity of Production in a Manufacturing Sequence

We define Operational Costs of an assembly line as those that correspond to unproductive times [3].

Given a manufacturing sequence $\pi(T)$, composed of T units of products, the unproductive time can be of two types: (i) Total work overload $W(\pi(T))$, that corresponds to the work not completed during the time that has been granted to the assembly line to make a demand plan, and (ii) Total idle time $U(\pi(T))$, that corresponds to the time during which the workers have been inactive in the completion of the demand plan.

The Total Operational Cost $\Gamma(\pi(T))$ is calculated according to (1).

$$\Gamma(\pi(T)) \equiv \Gamma_W(\pi(T)) + \Gamma_U(\pi(T)) = \gamma_W W(\pi(T)) + \gamma_U U(\pi(T)) \tag{1}$$

In formula (1), $\Gamma_W(\pi(T))$ and $\Gamma_U(\pi(T))$ are respectively the total work overload cost and the total idle time cost. Also in (1), γ_W is the cost per work overload unit (associated with the production drop), and γ_U is the cost per unit of idle time (associated with the cost per unit of time of a processor).

On the other hand, the characterization of regularity by restrictions [8] is not sufficient to discriminate between a set of regular sequences in production. Therefore, it is convenient to adopt a characterization by objective function both to measure the regularity of the sequences. To accomplish this goal, we will refer to the functions used in the Just-in-Time ideology [10].

To measure the non-regularity of a sequence $\pi_\varepsilon(T) = (\pi_{1,\varepsilon}, \ldots, \pi_{T,\varepsilon})$, which is composed of T units of products and associated with the demand plan ε, we will use the sum of the quadratic discrepancies between the ideal production $(\lambda_{i,\varepsilon} t)$ and the real productions $(X_{i,t,\varepsilon})$ in every partial sequence $\pi_\varepsilon(t) = (\pi_{1,\varepsilon}, \ldots, \pi_{t,\varepsilon}) \subseteq \pi_\varepsilon(T)$. As in [7, 9], we adopt the metric $\Delta_{Qx}(\pi_\varepsilon(T))$:

$$\Delta_{Q_x}(\pi_\varepsilon(T)) = \sum_{t=1}^{T} \sum_{i=1}^{|I|} (X_{i,t,\varepsilon} - \lambda_{i,\varepsilon} t)^2 \quad \forall \varepsilon \in \mathrm{E} \tag{2}$$

Where:

- $I, \mathrm{E}, \mathcal{T}$: Set of product types, $i = 1, \ldots, |I|$, set of demand plans, $\varepsilon = 1, \ldots, |\mathrm{E}|$, and set of manufacturing cycles in every demand plan, $t = 1, \ldots, |\mathcal{T}|$; $T \equiv |\mathcal{T}|$
- $d_{i,\varepsilon}, \lambda_{i,\varepsilon}$: Demand for units of type $i \in I$ in plan $\varepsilon \in \mathrm{E}$, and proportion of units of type $i \in I$ in plan $\varepsilon \in \mathrm{E}$: $\lambda_{i,\varepsilon} = d_{i,\varepsilon}/T \ \forall i \in I, \forall \varepsilon \in \mathrm{E}$
- $X_{i,t,\varepsilon}$: Actual production associated with the partial sequence $\pi_\varepsilon(t) \subseteq \pi_\varepsilon(T)$: number of units of type $i \in I$ in the partial sequence $\pi_\varepsilon(t) \subseteq \pi_\varepsilon(T)$ of plan $\varepsilon \in \mathrm{E}$.

Obviously, a minimum non-Regularity $\Delta_{Q_x}(\pi_\varepsilon(T))$ in a manufacturing sequence can contradict the minimization of the Total Operational Cost $\Gamma(\pi_\varepsilon(T))$. In fact, a problem of sequences that accounts for both objectives should be interpreted as a bi-objective problem.

3   MMSP-Cost/Δ_{Q_x} with Forced Interruption of Operations

We present a model for a variant of the MMSP-Cost sequencing problem that accounts for two types of aspects:

a.1   Economic: objective function to minimize the Total Operational Cost $\Gamma(\pi(T))$, taking into account the work overload $W(\pi(T))$ and the idle time of the processors $U(\pi(T))$.

a.2   Technical-productive: objective function to minimize the non-regularity of the product manufacturing $\Delta_{Q_x}(\pi(T))$, and productive operations subject to forced interruption to simplify the management of the production line.

We will call this problem MMSP-Cost/Δ_{Q_x}/forced, and we will describe it as follows.

We are given the following:

- The set of product types $(I:i = 1, \ldots, |I|)$ and the set of workstations $(K:k = 1, \ldots, |K|)$;
- The cycle time c and the temporary windows $l_k (k \in K)$, which are granted to each processor to work on a product unit at its station, the number of processors assigned to each station $b_k (k \in K)$, and the processing times $p_{i,k} (i \in I \wedge k \in K)$ of the operations, which are measured at normal activity ($\alpha^N = 1$);
- The demand vectors $\vec{d} = (d_1, \ldots, d_{|I|})$ and production mix $\vec{\lambda} = (\lambda_1, \ldots, \lambda_{|I|})$, where d_i is the number of product units of type $i \in I$ contained in the production-demand plan, and λ_i is the proportion of the model $i \in I$ in the plan, which satisfy $\vec{\lambda} = \vec{d}/D$ and $T \equiv D = \sum_{\forall i} d_i$; and
- The cost per work overload unit γ_W (linked to the production drop), and the cost per unit of idle time γ_U (linked to the cost per unit of time of a processor).

The problem is finding a sequence of T products $\pi(T) = (\pi_1, \ldots, \pi_T)$ with minimum operational cost $\Gamma(\pi(T))$ and minimum non-regularity of manufacture $\Delta_{Qx}(\pi(T))$ that satisfies the demand plan represented by the vector $\vec{d}$, while forcing the interruption of operations between the limits established by cycle c and the temporary windows $l_k(k \in K)$. The formulation of the model is as follows:

$$min \, F(\pi(T)) \equiv \Gamma(\pi(T)) \prec \Delta_{Qx}(\pi(T)) \tag{3}$$

$$\Gamma(\pi(T)) \equiv \Gamma_W(\pi(T)) + \Gamma_U(\pi(T)) = \gamma_W W(\pi(T)) + \gamma_U U(\pi(T)) \tag{4}$$

$$W(\pi(T)) = \sum_{t=1}^{T} \sum_{k=1}^{|K|} b_k w_{k,t}(\pi_t) \tag{5}$$

$$U(\pi(T)) = \sum_{k=1}^{|K|} b_k U_k(\pi(T)) \tag{6}$$

$$U_k(\pi(T)) = (T-1)c + l_k - \sum_{t=1}^{T} \left(p_{\pi_t,k} - w_{k,t}(\pi_t) \right) \forall k \in K \tag{7}$$

$$\Delta_{Qx}(\pi(T)) = \sum_{t=1}^{T} \sum_{i=1}^{|I|} \left(X_{i,t} - \lambda_i t \right)^2 \tag{8}$$

$$w_{k,t}(\pi_t) = max\left(0, s_{k,t}(\pi_t) + p_{\pi_t,k} - (k+t-2)c - l_k\right) \\ \forall k \in K \, \forall t = 1, \ldots, T \tag{9}$$

$$u_{k,t}(\pi_t) = s_{k,t}(\pi_t) - e_{k,t-1}(\pi_{t-1}) \\ \forall k \in K \, \forall t = 1, \ldots, T \tag{10}$$

$$X_{i,t} = |\{\pi_\tau \in \pi(t) = (\pi_1, \ldots, \pi_t) \subseteq \pi(T) : \pi_\tau = i \in I\}| \\ \forall i \in I \, \forall t = 1, \ldots, T \tag{11}$$

$$s_{k,t}(\pi_t) = max\left(e_{k,t-1}(\pi_{t-1}), e_{k-1,t}(\pi_t), (k+t-2)c\right) \\ k \in K \, \forall t = 1, \ldots, T \tag{12}$$

$$e_{k,t-1}(\pi_{t-1}) = s_{k,t-1}(\pi_{t-1}) + p_{\pi_{t-1},k} - w_{k,t-1}(\pi_{t-1}) \\ \forall k \in K \, \forall t = 2, \ldots, T \tag{13}$$

$$e_{k-1,t}(\pi_t) = s_{k-1,t}(\pi_t) + p_{\pi_t,k-1} - w_{k-1,t}(\pi_t) \\ \forall k \in K - \{1\} \, \forall t = 1, \ldots, T \tag{14}$$

$$e_{k,t}(\pi_t) = min\left(s_{k,t}(\pi_t) + p_{\pi_t,k}, (k+t-2)c + l_k\right) \\ \forall k \in K \, \forall t = 1, \ldots, T \tag{15}$$

$$X_{i,T} = d_i \, \forall i \in I \tag{16}$$

In the model, the identity (3) expresses the minimization of the objective function $\mathcal{F}(\pi(T))$ that attends to two hierarchical criteria: the first $\Gamma(\pi(T))$ corresponds to the global unproductive cost generated by the sequence, which is determined according to (4), and the second $\Delta_{Qx}(\pi(T))$ is associated with the non-regularity of the production of the sequence, which is determined according to (8). The expressions (5), (6) and (7) are used to calculate the global work overload $W(\pi(T))$, the global idle time of the processors $U(\pi(T))$ and the partial idle time that is generated in each workstation $U_k(\pi(T))$, respectively. The equalities (9) allow us to determine the partial work overloads of each station k and of each period t, with forced interruption of the operations, triggering it when the upper limit of the temporary window is reached: $(k + t - 2)c + l_k$. On the other hand, the equalities (10) define the partial idle time that is generated in each station and in each manufacturing period as a function of $\pi(T)$, while (11) serves to count the number of products of type $i \in I$ in the partial sequence $\pi(t) \subseteq \pi(T)$. The equalities (12), on one side, and (13), (14) and (15), on the other, determine the minimum starting, $s_{k,t}$, and completion, $e_{k,t}$, instances of the $|K| \times D$ operations. The equalities (16) impose the satisfaction of the demand plan $(d_i \forall i \in I)$. Finally, if also $\lfloor \lambda_i t \rfloor \leq X_{i,t} \leq \lceil \lambda_i t \rceil (\forall i \in I \, \forall t = 1, \ldots, T)$ is fulfilled, we will say that the sequence $\pi(T)$ satisfies the *Quota Property* or simply that $\pi(T)$ is a *Quota sequence*.

4   Algorithm for MMSP-Cost/Δ_{Q_X} with Forced Interruption of Operations

GRASP is a multi-start metaheuristic [15, 16] with two phases: (i) the constructive phase, which provides an initial solution through a randomized greedy procedure; and (ii) the improvement phase, which uses local search procedures to reach the local optima in one or more specific neighborhoods.

After a prefixed number of iterations (construction plus improvement), GRASP obtains a manufacturing sequence $\pi(T) = (\pi_1, \ldots, \pi_T)$, with forced interruption of the operations, and attending in this scheme, to the hierarchical bi-objective: minimum operational cost in the assembly line and minimum non-regularity of production.

Similar to [7, 10], we construct a sequence of models $\pi(T) = (\pi_1, \ldots, \pi_T)$, which assign progressively at each stage $t(t = 1, \ldots, T)$ a product from the $CL(t)$ list of candidates that can be selected for the t-th launch to line. Consequently, when stage t is reached, it is added to the partial sequence what is consolidated in the previous stage, $\pi(t - 1) = (\pi_1, \ldots, \pi_{t-1})$, a product $i \in CL(t)$.

For a product type $i \in I$ to enter the list $CL(t)$ of stage t, it must meet three conditions: (c1) the product does not have its demand fulfilled: $X_{i,t-1} < d_i$; (c2) it satisfies the Quota property [7] in cycle t: $\lfloor \lambda_i t \rfloor \leq X_{i,t-1} + 1 \leq \lceil \lambda_i t \rceil$; and (c3) to the greatest extent possible, the product that enters the list cannot randomly displace other critical products, which results in the violation of the Quota property in later stages (see details in [7]). If the list $CL(t)$ is empty by imposing the 3 conditions at the same time, then the list is opened for all products that do not have their demand fulfilled at stage t (condition c1).

Once the $CL(t)$ list is built, we order the candidate products $i \in CL(t)$ according to two hierarchical priority indexes: (p1) Operational Cost what is associated with the sequence $\pi_i(t) \equiv \pi(t - 1) \cup \{i\}$, which results from adding the product $i \in CL(t)$ to the consolidated partial sequence $\pi(t - 1)$; and (p2) Non-regularity of production what is associated with the sequence $\pi_i(t) \equiv \pi(t-1) \cup \{i\}$. The priority indexes p1 and p2, what are associated with the sequence $\pi_i(t)$, are calculated according to (17) and (18) respectively.

$$f_i^{(t)} \equiv \Gamma(\pi_i(t)) = \Gamma(\pi(t-1)) + \gamma_W \sum_{k=1}^{m} b_k w_{k,t}(i) + \gamma_U \sum_{k=1}^{m} b_k u_{k,t}(i) \tag{17}$$
$$\forall i \in CL(t) \wedge \forall t = 1, \ldots, T$$

$$g_i^{(t)} \equiv \Delta_{Qx}(\pi_i(t)) = \Delta_{Qx}(\pi(t-1)) + \sum_{j=1}^{|I|} (X_{j,t} - \lambda_j t)^2 \tag{18}$$
$$\forall i \in CL(t) \wedge \forall t = 1, \ldots, T$$

In (18) $\Delta_{Qx}(\pi(t-1))$ is the non-regularity that is associated with the partial sequence of products $\pi(t-1)$ of stage $t-1$, and the values $X_{j,t}$ are the productions in process of the various models $j \in I$ up to stage t. Obviously, it is true that $X_{i,t} = X_{i,t-1} + 1$ and $X_{j,t} = X_{j,t-1}, \forall j \neq i$.

In (17) $w_{k,t}(i)$ and $u_{k,t}(i)$ symbolize the partial work overload and the partial idle time supported by a processor of station $k \in K$ when the t-th product unit is of type i. If we consider the forced interruption of the operations, $w_{k,t}(i)$ and $u_{k,t}(i)$ these are calculated recursively while applying the formulas (19)–(23).

$$w_{k,t}(i) = max(0, s_{k,t}(i) + p_{i,k} - (k+t-2)c - l_k) \tag{19}$$

$$u_{k,t}(i) = s_{k,t}(i) - e_{k,t-1}(\pi_{t-1}) \tag{20}$$

$$s_{k,t}(i) = max(e_{k,t-1}(\pi_{t-1}), e_{k-1,t}(i), (k+t-2)c) \tag{21}$$

$$e_{k,t-1}(\pi_{t-1}) = s_{k,t-1}(\pi_{t-1}) + p_{\pi_{t-1},k} - w_{k,t-1}(\pi_{t-1}) \tag{22}$$

$$e_{k-1,t}(i) = s_{k-1,t}(i) + p_{i,k-1} - w_{k-1,t}(i) \tag{23}$$

In (19) and (20), $s_{k,t}(i)$ is the start time of the operation in station k when a model of type i occupies the t-th position in the sequence. Note that $s_{k,t}(i)$ depends on the start of the t-th manufacturing cycle in station $k((k+t-2)c)$ and the instants in which the operations in progress are considered to be completed in station k $(e_{k,t-1}(\pi_{t-1}))$ and in station $k-1$ $(e_{k-1,t}(i))$. To start these recursive calculations, we make $s_{1,1}(i) = 0 \forall i \in I$.

With $\Gamma(\pi_i(t))$ and $\Delta_{Qx}(\pi_i(t))$ indexes, we increasingly order the products from the list $CL(t)$ and convert it into the list $\overline{CL}(t)$. Considering that such ordering is applied hierarchically, the non-regularity $\Delta_{Qx}(\pi_i(t))$ intervenes only when there is a tie in the partial operational cost $\Gamma(\pi_i(t))$.

After this ordering, the list $\overline{CL}(t)$ is reduced according to the admission factor Λ (percentage of products that can be drawn among the best candidates). After this operation, we obtain the restricted list $\overline{RCL}(t, \Lambda)$, which coincides with $\overline{CL}(t)$ when $\Lambda = 100\%$. The construction phase concludes with obtaining the sequence $\hat{\pi}(T)$.

Once the construction phase is finished, as in [7, 9], we start with a *Quota* $\hat{\pi}(T)$ sequence and the local improvement phase, in which we proceed to consecutively and repetitively execute 4 descent algorithms in 4 neighborhoods until none of them improves the best solution that is achieved during the iteration. From two sequences, we select the one that offers the least total operational cost $\Gamma(\hat{\pi}(T))$ and, if the comparison is a tie, then the one with the least non-regularity of production $\Delta_{Q_x}(\hat{\pi}(T))$ is selected. The descent algorithms are based on the exchange and insertion of products, and they are oriented to the exploration of sequence cycles in both increasing and decreasing order.

Similar to [7, 9], the descent algorithms are: (i) Forward exchange, (ii) Backward exchange, (iii) Forward insertion and (iv) Backward insertion.

5   Case Study from Nissan's Barcelona Factory

In terms of the quality of the solutions and the CPU times, our computational experience is focused on analyzing the behavior of the GRASP algorithm proposed in this paper compared to BDP-1 algorithm [6]. The Quota property [7] has not been imposed on the sequences in either of the two procedures. As in [3, 7, 10], the analysis is conducted on a case study of the Nissan plant in Barcelona: an assembly line of 9 types of engines grouped into 3 families (Sport Utility Vehicle, Vans and Trucks).

In this production line, 42 operators work in shifts of 8 h. Briefly, the data that frames the case taken here are as follows: 21 workstations $|K| = 21$; 9 product types $|I| = 9$; cycle time $c = 175$ s; temporary window $l_k = 195$ s; 1 processor with 2 operators in each workstation $b_k = 1 \forall k \in K$; processing times $p_{i,k}(\forall i \in I, \forall k \in K)$ with values between 89 s and 185 s at normal activity; 23 engine demand plans $|E| = 23$, corresponding to the Nissan-9Eng.I instances; daily demand of 270 units for all demand plans $T \equiv D_\varepsilon = 270$ *units* $(\forall \varepsilon \in E)$; cost per work overload unit $\gamma_W = 2.2857$ and cost per unit of idle time $\gamma_U = 0.0111$ (euros per second).

The compiled codes of the procedures that we have selected in this work are BDP-1 (state of the art algorithm) and GRASP (this research). Table 2 shows the best results with respect to the work overload, idle time and operational cost from BDP-1 and GRASP, and for the 23 datasets of the problem $\varepsilon \in E$.

BDP-1 [6, 7] is an algorithm under the scheme of Bounded Dynamic Programming (BDP) which employs lower and upper bounds to reduce the space search of solutions. BDP-1 is focusing on minimizing the work overload, currently being the state of art algorithm for the mixed-model sequencing problem with work overload minimization and forced interruption of operations (MMSP-W/forced).

GRASP (this work) is a GRASP algorithm focused on hierarchically minimizing the operational cost $\Gamma(\pi(T)$ and the non-regularity of production $\Delta_{Q_x}(\pi(T))$. The code has been compiled and executed on an iMac (Intel Core 2 Duo 3.06 GHz, 3 GB RAM). The maximum number of iterations for each demand plan is equal to 10 with three candidate admission factors $\Lambda = (25\%, 50\%, 100\%)$, which generates in the constructive phase 690 solutions and 2906 improved solutions (improvement phase) in 69 executions. The average CPU time used per iteration is equal to 300.84 s, the average time required to obtain the best solution in each demand plan and each admission factor

Table 2. For each plan $\varepsilon \in E$, work overload W_{Proc} according to procedure (Bound, BDP1, GRASP). Unitary gain $\Delta GvB = (W_{BDP1} - W_{GRASP})/min(W_{BDP1}, W_{GRASP})$, best solutions work overload W_{BG}, idle time U_{BG} and operational cost Γ_{BG} and winning algorithm.

$\varepsilon \in E$	W_{Bound}	W_{BDP1}	W_{GRASP}	ΔGvB	W_{BG}	U_{BG}	Γ_{BG}	Winner
1	50	166	158	0.05	158	185408	2421.2	GRASP
2	241	464	340	0.36	340	185640	2839.8	GRASP
3	420	432	423	0.02	423	185833	3031.7	GRASP
4	235	440	444	-0.01	440	185610	3068.0	BDP
5	554	897	806	0.11	806	186101	3910.1	GRASP
6	285	663	678	-0.02	663	185828	3580.2	BDP
7	720	823	759	0.08	759	186164	3803.3	GRASP
8	72	129	76	0.70	76	185381	2233.5	GRASP
9	651	1149	1153	-0.00	1149	186204	4695.2	BDP
10	1208	1249	1228	0.02	1228	186763	4882.0	GRASP
11	43	50	48	0.04	48	185358	2169.2	GRASP
12	227	369	258	0.43	258	185568	2651.6	GRASP
13	162	379	357	0.06	357	185600	2878.2	GRASP
14	287	578	500	0.16	500	185768	3206.9	GRASP
15	392	553	428	0.29	428	185763	3042.3	GRASP
16	96	223	199	0.12	199	185474	2515.7	GRASP
17	408	640	518	0.24	518	185843	3248.9	GRASP
18	456	962	981	-0.02	962	186097	4266.6	BDP
19	945	980	949	0.03	949	186434	4240.6	GRASP
20	50	104	125	-0.20	104	185359	2297.3	BDP
21	480	854	854	0.00	854	185939	4018.0	BDP/GRASP
22	983	1104	1045	0.06	1045	186500	4460.8	GRASP
23	100	107	**100**	0.07	100	185435	2289.0	GRASP
Avg.	-	-	-	**0.113**	-	-	-	-

is equal to 1465 s, and the average time use for each demand plan and each admission factor is equal to 3008.39 s.

From Table 2, we can conclude the following points about the work overload and the operational cost of the assembly line:

-   In terms of the number of solutions, BDP-1 achieves the best solution 6 times, while GRASP achieves the best solution 18 times out of 23. BDP-1 and GRASP tie in plan 21 and GRASP obtains the optimal solution in plan 23.
-   In terms of work overload, GRASP beats BDP-1 in 17 plans, loses in 5 and ties in one. The average unitary gain of GRASP over BDP-1 is 16.74% when GRASP is the winner, while that of BDP-1 over GRAP-3 is 5.14% when BDP-1 wins. On a global average, the unitary gain of GRASP over BDP-1 is 11.25%.
-   In terms of work overload, GRASP and BDP-1 obtain solutions with an average GAP respect to the lower bound that is equal to 58% and 71.6%, respectively.
-   BDP-1 and GRASP needed on average 7664.3 s and 4395.0 s, respectively, to confirm their best solution in each demand plan.

6 Conclusions

We proposed, in this work, a GRASP procedure for solving a mixed-model sequencing problem. Specifically, the studied approach focused on minimizing both the total operational costs of the assembly line (taking into account the work overload and the idle time of the processors), and the non-regularity of the product manufacturing. The productive operations are subject to forced interruption.

The procedure designed for the problem, GRASP, was compared with BDP-1 (state of the art algorithm) using instances of a case study of the Nissan engine manufacturing plant in Barcelona.

Results show that GRASP (versus BDP-1) is the best procedure with respect to the minimum work overload, idle time and the operational costs of the line. This fact makes GRASP the state of art algorithm for the problems MMSP-W/forced (MMSP with work overload minimization and forced interruption) and MMSP-Cost/forced (MMSP with operational cost minimization and forced interruption).

Acknowledgments. This work was funded by the *Ministerio de Economía y Empresa* (Government of Spain) through project FHI-SELM2 (Ref. TIN2014-57497-P).

References

1. Boysen, N., Fliedner, M., Scholl, A.: Sequencing mixed-model assembly lines: survey, classification and model critique. Eur. J. Oper. Res. **192**(2), 349–373 (2009)
2. Bautista, J., Alfaro-Pozo, R., Batalla-García, C.: Minimizing lost-work costs in a mixed-model assembly line. In: Viles, E., Ormazábal, M., Lleó, A. (eds.) Closing the Gap Between Practice and Research in Industrial Engineering. Lecture Notes in Management and Industrial Engineering, pp. 213–221. Springer, Cham (2018). https://doi.org/10.1007/978-3-319-58409-6_24
3. Bautista-Valhondo, J., Alfaro-Pozo, R.: An expert system to minimize operational costs in mixed-model sequencing problems with activity factor. Expert Syst. Appl. **104**(2018), 185–201 (2018)
4. Yano, C.A., Rachamadugu, R.: Sequencing to minimize work overload in assembly lines with product options. Manag. Sci. **37**(5), 572–586 (1991)
5. Bolat, A., Yano, C.A.: Scheduling algorithms to minimize utility work at a single station on a paced assembly line. Prod. Plan. Control **3**(4), 393–405 (1992)
6. Bautista, J., Cano, A.: Solving mixed model sequencing problem in assembly lines with serial workstations with work overload minimisation and interruption rules. Eur. J. Oper. Res. **210**(3), 495–513 (2011)
7. Bautista, J., Alfaro-Pozo, R.: A GRASP algorithm for Quota sequences with minimum work overload and forced interruption of operations in a mixed-product assembly line. Prog. Artif. Intell. 1–15 (2018). https://doi.org/10.1007/s13748-018-0144-x
8. Bautista, J., Cano, A., Alfaro, R.: Modeling and solving a variant of the mixed-model sequencing problem with work overload minimisation and regularity constraints. An application in Nissan's Barcelona Plant. Expert Syst. Appl. **39**(12), 11001–11010 (2012)
9. Bautista, J., Alfaro-Pozo, R.: Free and regular mixed-model sequences by a linear program-assisted hybrid algorithm GRASP-LP. Prog. Artif. Intell. **6**(2), 159–169 (2017)

10. Monden, Y.: Toyota production system: an integrated approach to just-in-time, 4th edn. Productivity Press, New York (2011)
11. Aigbedo, H., Monden, Y.: A parametric procedure for multicriterion sequence scheduling for just-in-time mixed-model assembly lines. Int. J. Prod. Res. **35**, 2543–2564 (1997)
12. Miltenburg, J.: Level schedules for mixed-model assembly lines in just-in-time production systems. Manag. Sci. **35**(2), 192–207 (1989)
13. Bautista, J., Companys, R., Corominas, A.: Heuristics and exact algorithms for solving the Monden problem. Eur. J. Oper. Res. **88**(1), 495–513 (1996)
14. Bautista, J., Cano, A., Alfaro, R., Batalla, C.: Impact of the production mix preservation on the ORV problem. In: Bielza, C., et al. (eds.) CAEPIA 2013. LNCS (LNAI), vol. 8109, pp. 250–259. Springer, Heidelberg (2013). https://doi.org/10.1007/978-3-642-40643-0_26
15. Feo, T.A., Resende, M.G.C.: Greedy randomized adaptive search procedures. J. Glob. Optim. **6**(2), 109–133 (1995)
16. Resende, M.G., Ribeiro, C.C.: Greedy randomized adaptive search procedures: advances, hybridizations, and applications. In: Gendreau, M., Potvin, J.Y. (eds.) Handbook of Metaheuristics. International Series in Operations Research & Management Science, vol. 146. Springer, Boston (2010). https://doi.org/10.1007/978-1-4419-1665-5_10

A Software Tool for Categorizing Violin Student Renditions by Comparison

Miguel Delgado, Waldo Fajardo, and Miguel Molina-Solana[(⊠)]

Department of Computer Science and Artificial Intelligence, Universidad de Granada,
Granada, Spain
{mdelgado,aragorn,miguelmolina}@ugr.es

Abstract. Music Education is a challenging domain for Artificial Intelligence because of the inherent complexity of music, an activity with no clear goals and no comprehensive set of well-defined methods for success. It involves complementary aspects from other disciplines such as psychology and acoustics; and it also requires creativity and eventually, some manual abilities. In this paper, we present an application of machine learning to the learning of music performance. Our devised system is able to discover the similarities and differences between a given performance and those from other musicians. Such a system would be of great value to music students when learning how to perform a certain piece of music.

1 Introduction

It is a fact that student musicians spend most of their time practicing. Countless music lessons, endless scales and arpeggios, nightly rehearsals, and recitals for friends, family and teachers are commonplace in their lives. Hours of practicing help them learn to interpret a piece of music as the composer imagine it, but also developing their own signature sound—one that is unique to each of them. In other words, what makes a piece of music come alive is also what distinguishes great artists from any other.

In this paper we propose a software tool, based on a machine learning algorithm, able to classify a performance according to its similarity to those of famous musicians. Such a system would help students to understand the musical resources some of the greatest performers use and how to imitate them.

We focus on the task of characterizing performers from their playing style using descriptors that are automatically extracted from commercial audio recordings by means of feature extraction tools. This learning process is done by employing the system described in [12]. This approach is quite different from others in the literature, in which a heavy human intervention is needed in order to manually annotate the music. Our software does not require any human intervention and the whole process is done in an automatic way.

Many software tools aimed to help students to better perform have the drawback that they are limited to a range of previously analyzed songs. That is not

F. Herrera et al. (Eds.): CAEPIA 2018, LNAI 11160, pp. 330–340, 2018.
https://doi.org/10.1007/978-3-030-00374-6_31

the case of our system, which is able to capture how the performer plays, regardless the piece being performed. This way, the student can be assessed of how to play a work even though the expert musician did not play it.

Our system does not require any special hardware (e.g. midi instruments) to appraise the student's unique way of performing. To do so, it only uses an audio recording of the student performing with their own instrument. Thus, the performance is no biased in any sense: students play with their own instrument and they don't even know they are being recorded (if that is the case, informed consent from student should be obtained in advanced to avoid any ethical issues). Recording student performances is a new tendency that many teachers are putting in practice in order to get advantage of technology. Because of that, it would be relatively easy to count with students' recordings.

This paper is organized as follows. Section 2 provides background about both Music Performance and Music Education. In Sect. 3, we explain the model used to represent the information about performances, and how we extract that information from the audio files. Section 4 presents the devised software and its usage; and Sect. 5 points out conclusions and further developments.

2   Background

In this section we describe what *Music Performance* is, and how computers can be used for *Music Education*. We also revise former works in those areas to provide the reader a proper framework to understand the rest of the paper.

2.1   Music Performance

Most people would judge the literal execution of a musical score to be significantly less interesting than a performance of that piece by even a moderately skilled musician. Why is that so? The answer is straightforward: what we hear is not a literal rendition of the score. Human musicians do not play a piece of music mechanically, with constant tempo or loudness, exactly as written in the printed music score. Rather, skilled performers slow down at some places, speed up at others and stress certain notes or passages. Timing (tempo variations) and dynamics (loudness variations) are the most important parameters available to a performer, being mainly the source of expression in music [13]. The way these parameters 'should be' varied is not precisely specified in the printed score. Hence, it is the performer the one in charge of using them properly.

Briefly, *expressive music performance* can be defined as the deliberate shaping of the music by the performer in the moment of playing, by means of continuous variations of parameters such as timing, loudness or articulation [18]. Changes in tempo (timing) are non-linear warping of the regular grid of beats that defines time in a score. Changes in loudness (or dynamics) are modifications of the intensity of notes with respect to the others and to the general energy of the fragment in consideration. Articulation consists in varying the gap between contiguous notes by, for instance, making the first one shorter or overlapping it with the next.

Music performance is a complex activity involving physical, acoustic, psychological, social and artistic issues. At the same time, it is also a deeply human activity, relating to emotional as well as cognitive and artistic categories. Furthermore, it is dramatically affected by performer's mood and physical condition. [9] contains a review of what can be achieved in this domain.

As previously mentioned, research in musical performance has a multidisciplinary character, with studies that range from understanding expressive behavior to modeling aspects of renditions in a formal quantitative and predictive way. Historically, research in expressive music performance has focused on finding general principles underlying the types of expressive 'deviations' from the musical score (e.g., in terms of timing, dynamics and phrasing) that are a sign of expressive interpretation. Works by De Poli [4] and Widmer and Goebl [18] contains recent overviews on expressive performance modeling. The reader should refer to them for further information.

One of the issues in this area is the representation of the way certain performers play by just analyzing some of their renditions—studies into the individual style of famous musicians. That information would enable us to identify performers by just listening to their renditions [12,16]. These studies are difficult because the same professional musician can perform the same score in very different ways (cf. commercial recordings by Sergei Rachmaninoff and Vladimir Horowitz). Among the methods for the recognition of music performers and their style, the most relevant are the fitting of performance parameters in rule-based performance models, and the application of machine learning methods for the identification of performing style of musicians. Recent results of specialized experiments show surprising artist recognition rates (for instance, see those from Saunders et al. [16] or Molina-Solana et al. [12]).

2.2 Computers in Music Education

According to Brown [2], there are three main roles in which a computer can take part in music education: they can act as a tool, medium or musical instrument. A tool for recording, editing, analyzing and sequencing sounds; a medium for music storage, indexing and distribution; and a musical instrument for synthesizing music in real time.

In the last decades there have been several attempts to use computers in Music Education [17]. The field is highly interdisciplinary, involving contributions from disciplines such as Music, Education, Artificial Intelligence, Psychology, Linguistics, Human Computer Interaction and many others. Advances in research, as well as new software tools for the analysis of data, open up a new area in the field of music education, as stated by Friberg and Battel [7].

Artificial Intelligence in Music Education is also a very diverse field, being the work by Holland [10] still an interesting review in the topic. Attempts in this area can be classified into four categories, according to Brandao et al. [1], which are intended to: teach fundamentals of music, teach musical composition skills, perform analysis of music, and teach musical performance skills. In this paper, we propose an example within the scope of the last group.

However, music performance is not a well-defined domain. There are no clear goals, correct answers or an only way of doing things [6], making it a hard domain researchers are not always willing to tackle. Persson *et al.* [14] pointed it out: relatively little time is dedicated to interpretative and emotional aspects of performance in comparison with that dedicated to learning notation—a domain that can be more easily represented.

So why using technology for such a task? Despite computers will hardly replace music teachers, they can complement them. The benefits of individualized instruction, assessment and motivation can be used to supplement the learning. Using technology, students can work at their own pace, focusing on those aspects they need to improve. Because the learning is individual, there is no peer competition that, in many cases, is counterproductive. Evaluation of responses can be totally accurate and impartial. These possibilities have been tested with promising results by Friberg and Battel [7].

New pedagogical trends suggest recording students' performances. This way, they can hear how they play and analyze critically their work. Listening to both professional and amateur recordings is also a valuable teaching opportunity for music students. On the one hand, this enables them to compare their own current performance to those of professional musicians, being a motivational way of helping students to set goals for practicing. On the other hand, students can identify improvements with respect to their previous recordings and be aware of their progression and achievements during the time.

Beside the role recordings play to identify errors, the use of recordings can also be useful for going further and focus on specific musical elements or sections, providing a deeper understanding of the performance. Furthermore, it gives students the possibility of record themselves in other places different than the classroom, releasing them of the pressure of a classroom situation and allowing them to better perform.

Once recordings are available, we can also take advantage of them by means of computational tools, like the one we are presenting in this paper. These tools might be able to point out the tempo of the work, to discover any irregularities in the rhythm, to stress some musical parameters, or check the sound quality (as in [3]). Of special interest is the possibility of representing the sound in a new meaningful way, like in the works by Lagner [11] or Saap [15]. In our case, we are able to offer a list of performers that played in a similar way of the input recording.

The analytical comparison between a natural performance and performances with particular expressive intentions also seems to possess a potential for music pedagogy. One of these tools is *Director Musices* [8] which is able to play a song applying some performance rules. This way, certain acoustic parameters can be exaggerated so that anybody, regardless of musical training, can detect the difference and concentrate on particular aspects of the performance.

3 Data Representation Model

Our approach for dealing with the characterization of performers is based on the acquisition of *trend models* [5], which represent each particular performance. To do so, we employ the learning algorithm described in Molina-Solana *et al.* [12], which is a quick and automatic method for collecting information about the performer of a musical piece from its audio. As state-of-the-art feature extraction tools are employed and the collected information is not manually processed, it is not feasible to learn a deep model about each performer, i.e. to learn rules mapping specific musical concepts to expressive resources. From one side, a confident musical analysis from the (approximate) score provided by the extraction tools cannot be performed; on the other side, without a high precision in identifying note attacks, we know that the precision of expressive features is not guaranteed.

Despite those apparent drawbacks, the alternative that *trend models* presents consists in taking a more global perspective of a performer by trying to capture his essential (and recurrent) expressive decisions. For instance, playing in a romantic articulation affects note durations and produces characteristic energy envelopes.

These expressive decisions are somehow related to the musical analysis each musician performs. Nevertheless, as the real score (that one on the music sheet) is not provided, only an approximated score derived from the audio is available. For that reason, a melodic contour segmentation is employed as a way to capture the musical structure of the piece. This segmentation enables an analysis in terms of melodic intervals, i.e. precision on pitch detection is not critical.

Specifically, a *trend model* characterizes, for a specific audio descriptor, the relationships a given performer is establishing among groups of neighbor musical events. A qualitative analysis of the variations of the audio descriptors is performed with a local perspective.

We will be using two *trend models* in this paper: energy and duration. The *trend model* for the energy descriptor relates, qualitatively, the energy variation for a given set of consecutive notes, and it is related to dynamics. On the other hand, the *trend model* for duration indicates, also qualitatively, how note duration changes for note sequences. Duration is related to articulation and timing.

Given an input musical recording of a piece, the trend analysis is performed by aggregating the qualitative variations on their small melody segments. Thus, in advance of building *trend models*, input streams are broken down into segments of three-note long. As most automatic melody segmentation approaches do, note grouping is performed according to a human perception model.

At the training stage, the goal of the system is to characterize performers by extracting expressive features and constructing *trend models*. Next, at the identification stage, the system analyzes the input performance and looks for the most similar previously learned model. The training process is composed of three main steps: (1) the extraction of audio descriptors and the division of a performance in segments; (2) the tagging of segments according to contour patterns; and (3) the calculus of probabilistic distributions for each contour pattern and descriptor (trend generation).

3.1 Feature Extraction and Segmentation

The first step consists in extracting audio features. At the moment, we consider fundamental frequency and energy, as these are the main low-level audio features related to melody. These features are then used to identify note boundaries and to generate melodic segments. This module provides a vector with instantaneous fundamental frequency and energy values calculated every 0.01 s.

A post-processing module for determining the possible notes played by the performers was developed. This module first converts fundamental frequencies into quantized pitch values, and then a pitch correction procedure is applied in order to eliminate noise and sudden changes. This correction is made by assigning to each sample the value given by the mode of their neighbors around a certain window of size σ.

After this process, a smooth vector of pitches is obtained. By knowing on which samples pitches are changing, a note-by-note segmentation of the whole recording is performed. For each note, its pitch, duration and energy are collected.

It is assumed that there might be some errors in this automatic segmentation, given the heterogeneity of recording conditions. The *trendmodel* algorithm proposes a more abstract representation (but still close to the melody) than the real notes for dealing with this problem. That is, instead of focusing on the absolute notes, we are interested in modeling the melodic surface.

To do that, a simple contour model that identifies some melodic patterns from the melody is used. These patterns are three-note long and are related with the direction of the two intervals that exist in them.

3.2 Classification

A nearest neighbor classifier is used to generate a list of similar performers for each input recording. *Trend models* acquired in the training stage, as described in the previous section, are used as class patterns. When the student's recording is presented to the system, its *trend model* is created and compared with the previously acquired ones. The system outputs a ranked list of performer candidates where distances determine the order, with the first being the most likely performer relative to the results of the training phase.

3.2.1 Distance Measure

The distance d_{ij} between two performances i and j, is defined as the weighted sum of distances between respective contour patterns:

$$d_{ij} = \sum_{n \in N} w_{ij}^n dist(n_i, n_j) \tag{1}$$

where N is the set of the different contour patterns considered; $dist(n_i, n_j)$ measures the distance between two probability distributions (see (3) below); and w_{ij}^n are the weights assigned to each contour pattern. Weights have been introduced for balancing the importance of the patterns with respect to the number of times they appear. Frequent patterns are considered more informative due to the fact that they come from more representative samples. Weights are defined as the mean of cardinalities of respective histograms for a given pattern n:

$$w_{ij}^n = (N_i^n + N_j^n)/2 \tag{2}$$

Mean value is used instead of just one of the cardinalities to assure a symmetric distance measure in which w_{ij}^n is equal to w_{ji}^n. Cardinalities could be different because recognized notes can vary from a performance to another, even though the score is supposed to be the same.

Finally, distance between two probability distributions is calculated by measuring the absolute distances between the respective patterns:

$$dist(s, r) = \sum_{k \in K} |s_k - r_k| \tag{3}$$

where s and r are two probability distributions for the same contour pattern; and K is the set of all possible values they can take.

3.3 Model Performance

The feasibility and accuracy of the system have already been tested in [12] using several works from Sonatas and Partitas by J. S. Bach. In those experiments, the model was employed to identify several violinists by using some already labeled performances. The results were quite promising: much better than those from a random chance classifier and probably better that those achievable by a human expert. It was also demonstrated that the information that *trend models* contain can be used to derive meaningful musicological data.

The system has only been tested with violin recordings, but the underlying machine learning model (*trendmodel* algorithm) is flexible enough to allow the employment of other monophonic instruments. The application itself is capable of dealing with any instrument or song as long as the information it receives is properly formatted.

4   Software

We have implemented in MATLAB all the tools for extracting the information from musical signals, representing those data, and using it for comparing: the whole *trendmodel* algorithm and an interface to use all the developed tools.

The software is very user-friendly. The procedure is as follow: (1) the user/student selects the file with their performance; (2) selects the files to compare with; (3) waits for the audio analysis at the learning stage; (4) launches the comparator.

The devised interface (see Fig. 1) enables the user to load a set of files to be used as comparing performances. On the left, we find a list of all the audio

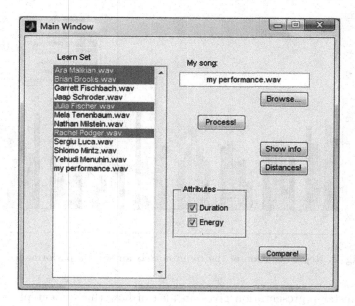

Fig. 1. Main window of the devised interface of the system

files with performances that we have in the folder. From those, we can select only those we are interested in being used. On the right, we select the audio file, which contains the student's performance, we want to compare with the others. The developed software uses audio files in *wav* format as input.

Once selected the files we want to use, the system needs to extract information from them, and represent them by means of *trend models*. That is done when the 'Process' button is pressed.

The interface offers the student the possibility to choose which (or all) attributes they want to use when comparing performances ('Attributes' section in the interface). Currently, only two possibilities are allowed as they are the attributes used in the learning stage: duration of notes and their energy. Those two attributes are the main ones in music performance, as previously explained in Sect. 2.1. Thus, students can compare themselves in terms of timing, in terms of energy, or in both together. Results may differ depending on the selection. However, any other musical attribute can be used as long as they are extracted from the audio and learned in the learning stage.

'Show info' offers a representation of *trend models*. An example of this can be seen in Fig. 2. Each bar indicates how a performer tends to play for a certain musical pattern (more details can be found in [12]).

'Distances' shows a distance matrix comparing the distances between each pair of performances (including the one by the student). An example can be seen in Fig. 3.

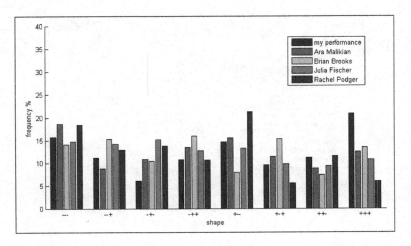

Fig. 2. Representation of the *trend models* for several performances.

This cluster representation gives an idea of how the student plays in comparison with others. However, it cannot tell where the differences are nor the changes to be made to their style in order get closer to a certain performer.

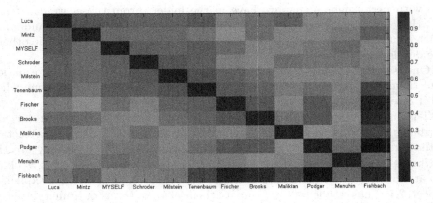

Fig. 3. Normalized similarity matrix between performances

Finally, we can obtain a list of sorted performers ('Compare' button), with the first one being the most similar to the student one, and the last being the most different.

5   Conclusions

We presented an application of machine learning to the teaching of music performance. Our developed software shows the similarities and differences between a given performance and those from other musicians. We think that such a system would be helpful to music students, and its application to the Music Education field would be of great interest, as it supports students to better perform music by showing them the similarities and differences with renditions of famous performers.

Technology applied to music education is a very promising area. New pedagogical tendencies suggest recording students' performances. Our system uses precisely those recordings to proceed, not needing any other special hardware to collect the information. Doing so, we get another advantage: all the data is gathered in a transparent way, not interfering with the performance.

Studies in music performance have a particular value in our time. Research on music performance can point out expressive resources that traditionally have been hiding in musicians' skill and musical intuition. When explicitly formulated, these resources will give the user the possibility to play music with different expressive coloring.

Acknowledgements. M. Molina-Solana is funded by the EU's Horizon 2020 research and innovation programme under the Marie Skłodowska-Curie grant agreement No 743623.

References

1. Brandao, M., Wiggins, G., Pain, H.: Computers in music education. In: Proceedings of the AISB 1999 Symposium on Musical Creativity, pp. 82–88 (1999)
2. Brown, A.R.: Computers in Music Education: Amplifying Musicality. Routledge, Abingdon (2007)
3. Chávez de la O, F., Fernández de Vega, F., Rodríguez Diaz, F.J.: Analyzing quality clarinet sound using deep learning. A preliminary study. In: 2017 IEEE Symposium Series on Computational Intelligence (SSCI), pp. 1–7 (2017)
4. De Poli, G.: Methodologies for expressiveness modelling of and for music performance. J. New Music. Res. **33**(3), 189–202 (2004)
5. Delgado, M., Fajardo, W., Molina-Solana, M.: Representation model and learning algorithm for uncertain and imprecise multivariate behaviors, based on correlated trends. Appl. Soft Comput. **36**, 589–598 (2015)
6. Dolan, D., et al.: The improvisational state of mind: a multidisciplinary study of an improvisatory approach to classical music repertoire performance. Front. Psychol. **9**, 1341 (2018). https://doi.org/10.3389/fpsyg.2018.01341
7. Friberg, A., Battel, G.U.: Structural communication. In: The Science and Psychology of Music Performance: Creative Strategies for Teaching and Learning, pp. 199–218. Oxford University Press, New York (2002)
8. Friberg, A., Colombo, V., Frydén, L., Sundberg, J.: Generating musical performances with Director Musices. Comput. Music J. **24**(3), 23–29 (2000)
9. Gabrielsson, A.: Music performance research at the millennium. Psychol. Music **31**(3), 221–272 (2003)
10. Holland, S.: Artificial intelligence in music education: a critical review. Readings in music and artificial intelligence. Contemp. Music Stud. **20**, 239–274 (2000)
11. Langner, J., Goebl, W.: Visualizing expressive performance in tempo-loudness space. Comput. Music J. **27**(4), 69–83 (2003)

12. Molina-Solana, M., Arcos, J.L., Gomez, E.: Identifying violin performers by their expressive trends. Intell. Data Anal. **14**(5), 555–571 (2010)
13. Palmer, C.: Anatomy of a performance: sources of musical expression. Music Percept. **13**(3), 433–453 (1996)
14. Persson, R.S., Pratt, G., Robson, C.: Motivational and influential components of musical performance: a qualitative analysis. In: Fostering the Growth of High Ability: European Perspectives, pp. 287–302. Ablex, Norwood (1996)
15. Sapp, C.: Comparative analysis of multiple musical performances. In: Proceeding of 8th International Conference on Music Information Retrieval (ISMIR 2007), Vienna, Austria, pp. 497–500 (2007)
16. Saunders, C., Hardoon, D., Shawe-Taylor, J., Widmer, G.: Using string kernels to identify famous performers from their playing style. Intell. Data Anal. **12**(4), 425–440 (2008)
17. Upitis, R.: Technology and music: an intertwining dance. Comput. Educ. **18**(1–3), 243–250 (1992)
18. Widmer, G., Goebl, W.: Computational models of expressive music performance: the state of the art. J. New Music. Res. **33**(3), 203–216 (2004)

Background Modeling for Video Sequences by Stacked Denoising Autoencoders

Jorge García-González(✉), Juan M. Ortiz-de-Lazcano-Lobato,
Rafael M. Luque-Baena, Miguel A. Molina-Cabello, and Ezequiel López-Rubio

Department of Computer Languages and Computer Science, University of Málaga,
Bulevar Louis Pasteur, 35, 29071 Málaga, Spain
{jorgegarcia,jmortiz,rmluque,miguelangel,ezeqlr}@lcc.uma.es

Abstract. Nowadays, the analysis and extraction of relevant information in visual data flows is of paramount importance. These images sequences can last for hours, which implies that the model must adapt to all kinds of circumstances so that the performance of the system does not decay over time. In this paper we propose a methodology for background modeling and foreground detection, whose main characteristic is its robustness against stationary noise. Thus, stacked denoising autoencoders are applied to generate a set of robust characteristics for each region or patch of the image, which will be the input of a probabilistic model to determine if that region is background or foreground. The evaluation of a set of heterogeneous sequences results in that, although our proposal is similar to the classical methods existing in the literature, the inclusion of noise in these sequences causes drastic performance drops in the competing methods, while in our case the performance stays or falls slightly.

Keywords: Background modeling · Deep learning · Autoencoders

1 Introduction

In today's society, the fact that it is necessary to automate the exploitation of visual information that we capture in the most reliable and efficient possible way is ever more present. In the field of artificial vision, video surveillance is still a very active field at research level, since not all the open fronts have been satisfactorily addressed in recent years. One of the main areas to improve resides in the background modeling, which consists of determining which pixels of the image correspond to the movement objects in the scene and which ones are part of the background of the scene.

Foreground detection algorithms must work 24 h a day, with robustness against background variations. This variability can be observed in outdoor environments where the weather can change and generate rain, hail or snow, or in indoor environments where changes in lighting compromise the reliability of

© Springer Nature Switzerland AG 2018
F. Herrera et al. (Eds.): CAEPIA 2018, LNAI 11160, pp. 341–350, 2018.
https://doi.org/10.1007/978-3-030-00374-6_32

detection. Not only is it necessary for a detection algorithm to behave correctly for a few hundred frames, but it is necessary to ensure that changes in the external conditions of the scene will not cause a drop in performance. These requirements are difficult to achieve in many of the works already published, in which changes in the scene could cause that the system stops working properly.

Most of the foreground detection algorithms work at pixel level, that is, modeling the intensity of each pixel and determining the probability of belonging to one of the two possible classes: foreground or background. Thus, there are many highly referenced proposals that achieve more than satisfactory results. The main differences between them reside in the underlying model that represents the intensity of color of each pixel over time. Wren et al. [11] uses a Gaussian distribution as a basis for the modeling of each pixel, while the GMM model [7] uses K distributions to manage multimodal funds. Zivkovic [13] uses an intermediate strategy, considering as many Gaussians as necessary up to a maximum value (K). On the other hand, Elgammal et al. [2] make use of kernel distributions, less restrictive than the previous ones statistically but more complex to update. Other more complex models go through modeling each pixel through self-organized maps, a type of unsupervised neural network. Both SOBS [4] and FSOM [3] models are based on the previous algorithm, in addition to combining the output of each pixel (probability of belonging to the background or foreground) with the output of their neighbors, which provides robustness to the model and makes it less sensitive to false positives.

The use of deep learning networks is not alien to this field. In this work we will use autoencoders, as an unsupervised learning technique, to minimize the impact of noise backgrounds in the modeling of the scene. Each image will be divided into patches whose noise will be eliminated by a previously trained autoencoder. Subsequently, and using the autoencoder information after the coding phase, a N dimensional Gaussian model will be used to estimate the probability of belonging to the background of each patch.

Autoencoders are well suited to information representation. Single layer linear autoencoders are proved to span the same subspace as a Principal Components Analysis (PCA) does when they attempt to learn an undercomplete representation of the input data, i.e., the number of neurons in the hidden layer is less than or equal to the input data dimension [1]. Therefore, the features that retain most of the input data variance will be kept. In stacked linear autoencoders, subsequent layers of the autoencoder will be used to condense that information gradually to the desired dimension of the reduced representation space. On the other hand, sparse autoencoders or autoencoders with layers made up of non-linear units will also obtain relevant features which can be expected to be easier to interpret and used by a classifier, though they will likely differ from those provided by the PCA technique, as it is discussed in [9].

The paper is divided in the following sections: Sect. 2 presents the object detection methodology based on the analysis of image patches to obtain a foreground mask from an input frame; Sect. 3 reports the experimental results over several public surveillance sequences and Sect. 4 concludes the article.

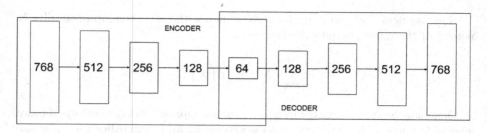

Fig. 1. Complete autoencoder structure with layers sizes.

2 Methodology

Most previous approaches to background modeling in video sequences model each pixel of the video frame separately. Our model intends to model small patches of size $N \times N$ pixels, so that for each incoming video frame an estimation is made in order to know whether each patch belongs to the background of the scene. The process is divided in two stages: firstly, a condensed representation of the patch, composed of significant features, is obtained by means of previously trained Stacked Denoising Autoencoder (SDA) [9]; secondly, a probabilistic model classifies the patch according to their computed set of relevant features.

2.1 Patch Feature Extraction

Let $\mathbf{X} \in \mathbb{R}^H$ be a patch of size H, where tristimulus pixel color values are assumed. The patch is processed by a stacked denoising autoencoder:

$$\tilde{\mathbf{X}} = g\left(f\left(\mathbf{X}\right)\right) \tag{1}$$

$$f : \mathbb{R}^H \to \mathbb{R}^L \qquad g : \mathbb{R}^L \to \mathbb{R}^H \tag{2}$$

where $\tilde{\mathbf{X}} \in \mathbb{R}^H$ is the reconstructed version of the input patch $\mathbf{X}$, f is the encoding part of the autoencoder, g is the decoding part of the autoencoder, and L is the number of neurons of the innermost layer of the neural architecture, i.e. the size of the last layer of the encoding part and the first layer of the decoding part (see Fig. 1). The goal of the autoencoder is to reduce the high dimensional input of size H to a low dimensional set of features of size L with $L < H$.

An autoencoder is usually trained to minimize the reconstruction error $\mathcal{E}$:

$$\mathcal{E} = \sum_{i=1}^{R} \left\| \mathbf{X} - \tilde{\mathbf{X}} \right\|^2 \tag{3}$$

where R is the number of patches in the training set.

However denoising autoencoders are trained with corrupted input samples $\hat{\mathbf{X}}$ instead of the input samples themselves $\mathbf{X}$.

$$\mathcal{E} = \sum_{i=1}^{R} \left\| \mathbf{X} - g\left(f\left(\hat{\mathbf{X}} \right) \right) \right\|^2 \tag{4}$$

Denoising autoencoders try to learn a robust representation made up of more general features which prevents from overtraining and diminishes the influence of scene factors such as illumination and local variation. In an attempt to enforce the invariance of the autoencoder to the diverse scene conditions, several authors [10,12] have used a training set that comprises not patches extracted from the frames corresponding to the video to process but a huge amount of generic natural image patches that may be corrupted. This approach is followed in our proposal, where the training set for our single autoencoder is generated from the Tiny Images dataset [8].

It turns out that stacked denoising autoencoders might find difficulties in modeling too small patches. Here we propose to overcome this limitation by augmenting the $N \times N$ pixel patch by M pixels in each direction (up, down, left and right), so that an augmented patch of size $(N + 2M) \times (N + 2M)$ is supplied to the autoencoder, while the estimation about the pertenence to the background only affects to the central $N \times N$ pixel section of the augmented patch. In this way, the augmented patches overlap with their neighbors, while the small patches do not. Therefore, the dimension of the samples the autoencoder processes is $H = 3\left(N + 2M\right)^2$.

2.2 Patch Classification

As the video sequence progresses, the features which are discovered by the autoencoder are extracted, and a probabilistic model is learned for them. (Figure 2) This model aims to capture the main characteristics of the probability distribution of the feature vector $\mathbf{v} \in \mathbb{R}^L$:

$$\mathbf{v} = f\left(\mathbf{X}\right) \tag{5}$$

To this end, the mean $\mu_j = E\left[v_j\right]$ and the variance $\sigma_j^2 = E\left[\left(v_j - \mu_j\right)^2\right]$ of each component of $\mathbf{v}$ are approximated by the Robbins-Monro stochastic approximation algorithm [5]:

$$\mu_{j,t+1} = (1 - \alpha)\,\mu_{j,t} + \alpha v_{j,t} \tag{6}$$

$$\sigma_{j,t+1}^2 = (1 - \alpha)\,\sigma_{j,t}^2 + \alpha\left(v_{j,t} - \mu_{j,t}\right)^2 \tag{7}$$

where t is the time instant (the frame index) and α is the step size.

Each small patch is declared to belong to the foreground whenever the number of components of the feature vector which are far from its estimated mean, as

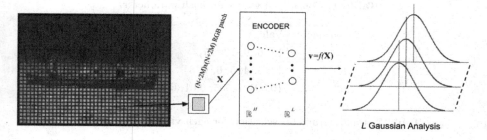

Fig. 2. Method overview scheme.

measured with respect to the estimated variance, is higher than a given threshold:

$$C < \sum_{j=1}^{L} \mathbb{I}\left(|v_{j,t} - \mu_{j,t}| > K\sigma_{j,t}\right) \tag{8}$$

where $\mathbb{I}$ stands for the indicator function, C is a tunable parameter which specifies the number of components which must be far from its estimated mean to declare that the small patch belongs to the foreground, and K is another tunable parameter which specifies how many standard deviations an observation must depart from its estimated mean to be considered to be far away.

3   Experimental Results

3.1   Methods

Five methods have been selected in order to make a performance comparison with our proposal: WrenGA [11], ZivkovicGMM [13], MaddalenaSOBS [4], ElgammalKDE [2] and Lopez-RubioFSOM [3].

Four of this methods are available on BGS library [6][1]. The version 1.3.0 of the BGS library has been implemented by using the C++ language and version 2.4.8 of the OpenCV[2] library. On the other hand, Lopez-RubioFSOM is written in Matlab, with MEX files written in C++ for the most time-consuming parts and Matlab scripts for the rest. The employed parameter values are those indicated as default by the authors.

Finally, our proposed approach has been implemented using Python version 2.7. For neural network implementation we have used TensorFlow[3] version 1.5.0 by means of Keras[4] version 2.1.3 as high-level API. All evaluation has been made using MATLAB R2017B.

---

[1] https://github.com/andrewssobral/bgslibrary.
[2] http://opencv.org/.
[3] https://www.tensorflow.org/.
[4] https://keras.io/.

Table 1. Considered values for each parameter.

Parameter	Values
C	{3, 6, 9, 12, 15}
K	{2, 3, 4, 5, 6, 7, 8}
α	{0.001, 0.005, 0.01, 0.05}

Table 2. Final parameter selection for each video an noise level.

Video	$\sigma = 0$	$\sigma = 0.1$	$\sigma = 0.2$
Canoe	$C = 3$, $K = 5$, $\alpha = 0.001$	$C = 6$, $K = 4$, $\alpha = 0.001$	$C = 3$, $K = 5$, $\alpha = 0.001$
Boats	$C = 3$, $K = 4$, $\alpha = 0.001$	$C = 3$, $K = 4$, $\alpha = 0.001$	$C = 3$, $K = 4$, $\alpha = 0.001$
Fountain02	$C = 15$, $K = 6$, $\alpha = 0.001$	$C = 12$, $K = 5$, $\alpha = 0.001$	$C = 12$, $K = 5$, $\alpha = 0.001$
Overpass	$C = 3$, $K = 6$, $\alpha = 0.001$	$C = 3$, $K = 6$, $\alpha = 0.001$	$C = 3$, $K = 6$, $\alpha = 0.001$
Port_0_17fps	$C = 15$, $K = 5$, $\alpha = 0.05$	$C = 12$, $K = 5$, $\alpha = 0.05$	$C = 12$, $K = 5$, $\alpha = 0.05$
Pedestrians	$C = 15$, $K = 7$, $\alpha = 0.001$	$C = 15$, $K = 6$, $\alpha = 0.001$	$C = 12$, $K = 7$, $\alpha = 0.001$

Our autoencoder implementation has been trained and tested using 100,000 random images from Tiny Images dataset [8][5]. Since each image has 32×32 pixels, we have divided each one to obtain four 16×16 images.

In order to be as fair as possible a random seed has been used to generate the input Gaussian noise for each noise level. The same videos with the applied noise are used in all the studied methods. We do not use any additional post processing in any of the methods.

3.2   Sequences

A set of video sequences have been selected from the 2014 dataset of the ChangeDetection.net website[6]. Four of the selected scenes are from Dynamic Background category, one from Low Frame Rate category and another one from the Baseline one. *Canoe* shows a river with water and forest background where a canoe goes across (320×240 pixels and 1189 frames). *Fountain02* shows a road behind a fountain that spits water out (432×288 pixels and 1499 frames). *Boats* shows a river next to a road. Two boats cross through the river while various vehicles move on the road (320×240 pixels and 7999 frames). *Overpass* shows a bridge traversed by a man with a river, forest and a road behind (320×240 pixels and 3000 frames). *Port_0_17fps* is a low frame rate video that shows a little dock with a lot of boats constantly moving, water and clouds as background and some persons and boats crossing from time to time as foreground (640×480 pixels and 3000 frames). *Pedestrians* is a baseline video where several people walk over a pavement next to grass with sun and shadows (360×240 pixels and 1099 frames).

---

[5] http://groups.csail.mit.edu/vision/TinyImages/.
[6] http://changedetection.net/.

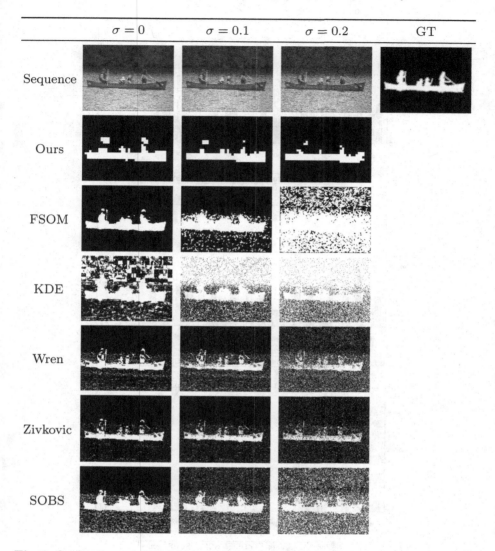

Fig. 3. Qualitative results for frame 960 from canoe dataset. From left to right: images with different amount of Gaussian noise with mean 0. First row is original dataset input image with different amounts of Gaussian noise and ground-truth. Other rows correspond to foreground segmentation performed for each method and each input image.

3.3 Parameter Selection

Our method needs three parameters to be selected (C, K and α). To get a good combination of parameter values, we have carried out and analyzed each possible combination from values in Table 1 on page 6 for images without noise

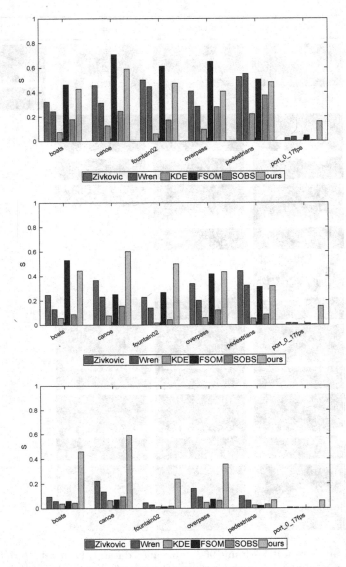

Fig. 4. Comparison between methods for each video with $\sigma = 0$, $\sigma = 0.1$ and $\sigma = 0.2$ Gaussian noises.

(140 combinations). We have used the parameter configuration that achieves best performance for each video without noise. The top 3 combinations have been tested for videos with Gaussian noise in order to select the combination which best performs. Table 2 on page 6 shows final parameter selection for each video and noise level.

3.4   Results

A well-known measure has been selected in order to compare the performance from a quantitative point of view. In this work we have considered the spatial accuracy S. This measure provides values in the interval $[0, 1]$, where higher is better, and represent the percentage of hits of the system.

The definition of this measure can be described as follows:

$$S = \frac{TP}{TP + FN + FP} \tag{9}$$

where TP refers to the foreground patches classified correctly(true positives) whereas FN and FP are the type II (false negatives) and type I (false positive) errors respectively.

S has been calculated for each binarized frame in Region of Interest (specified by ChangeDetection.net) generated using each previously mentioned method and we have obtained the mean for all frames with TP pixels in ground-truth.

Figure 4 on page 8 shows comparison between each method result for videos with different noises. We can observe our proposed method is able to deal with low level noise and even improve a bit its performance for some videos (canoe and boats). It is interesting to point that adding noise to other methods causes a lot more FP pixels while our method deals with it by increasing FN pixels as can be observed on Fig. 3 on page 7.

4   Conclusions

In this work we have proposed a methodology for the background modeling in video sequences, that uses autoencoders to filter the possible noise in the background and a multidimensional probabilistic model to determine the probability of belonging to one of the following two classes, background or foreground. Although our proposal works at region level, the comparative results with other techniques at pixel level where they take advantage of more information, leave us in good place for all the experiments carried out. To simulate more heterogeneous scenes, we have added Gaussian noise to each sequence, being our method much more robust than competitors to this increase in variability in the scene. In fact, the improvement is significant, being the best method on average in all the scenes analyzed. The greater the background noise, the greater the fall in performance of the method used. The results indicate that after introducing slight noise, the fall of our method is 4% on average, while the rest of the techniques have drops of over 30%. If the noise is magnified our performance goes down near 30% of its original value, but the performance of the rest falls 70% on average. These data corroborate the idea of robustness of our proposal, in addition to its usefulness for the processing and analysis of continuous data during uninterrupted periods of time.

Acknowledgements. This work is partially supported by the Ministry of Economy and Competitiveness of Spain under grant TIN2014-53465-R, project name Video surveillance by active search of anomalous events, besides for the projects with codes TIN2016-75097-P and PPIT.UMA.B1.2017. It is also partially supported by the Autonomous Government of Andalusia (Spain) under grant TIC-657, project name Self-organizing systems and robust estimators for video surveillance. All of them include funds from the European Regional Development Fund (ERDF). The authors thankfully acknowledge the computer resources, technical expertise and assistance provided by the SCBI (Supercomputing and Bioinformatics) center of the University of Málaga. They also gratefully acknowledge the support of NVIDIA Corporation with the donation of two Titan X GPUs used for this research. The authors would like to thank the grant of the Universidad de Malaga.

References

1. Baldi, P., Hornik, K.: Neural networks and principal component analysis: learning from examples without local minima. Neural Netw. **2**(1), 53–58 (1989)
2. Elgammal, A., Harwood, D., Davis, L.: Non-parametric model for background subtraction. In: Vernon, D. (ed.) ECCV 2000. LNCS, vol. 1843, pp. 751–767. Springer, Heidelberg (2000). https://doi.org/10.1007/3-540-45053-X_48
3. López-Rubio, E., Luque-Baena, R., Domínguez, E.: Foreground detection in video sequences with probabilistic self-organizing maps. Int. J. Neural Syst. **21**(3), 225–246 (2011)
4. Maddalena, L., Petrosino, A.: A self-organizing approach to background subtraction for visual surveillance applications. IEEE Trans. Image Process. **17**(7), 1168–1177 (2008)
5. Robbins, H., Monro, S.: A stochastic approximation method. Ann. Math. Stat. **22**(3), 400–407 (1951)
6. Sobral, A., Bouwmans, T.: BGS library: a library framework for algorithm's evaluation in foreground/background segmentation. In: Background Modeling and Foreground Detection for Video Surveillance. CRC Press, Taylor and Francis (2014)
7. Stauffer, C., Grimson, W.: Adaptive background mixture models for real-time tracking, vol. 2, pp. 246–252 (1999)
8. Torralba, A., Fergus, R., Freeman, W.: 80 million tiny images: a large data set for nonparametric object and scene recognition. IEEE Trans. Pattern Anal. Mach. Intell. **30**(11), 1958–1970 (2008)
9. Vincent, P., Larochelle, H., Lajoie, I., Bengio, Y., Manzagol, P.: Stacked denoising autoencoders: learning useful representations in a deep network with a local denoising criterion. J. Mach. Learn. Res. **11**, 3371–3408 (2010)
10. Wang, N., Yeung, D.: Learning a deep compact image representation for visual tracking. In: Advances in Neural Information Processing Systems 26, pp. 809–817 (2013)
11. Wren, C., Azarbayejani, A., Darrell, T., Pentl, A.: Pfinder: real-time tracking of the human body. IEEE Trans. Pattern Anal. Mach. Intell. **19**(7), 780–785 (1997)
12. Zhang, Y., Li, X., Zhang, Z., Wu, F., Zhao, L.: Deep learning driven blockwise moving object detection with binary scene modeling. Neurocomputing **168**, 454–463 (2015)
13. Zivkovic, Z., van der Heijden, F.: Efficient adaptive density estimation per image pixel for the task of background subtraction. Pattern Recognit. Lett. **27**(7), 773–780 (2006)

GraphDL: An Ontology for Linked Data Visualization

Juan Gómez-Romero[1,2(✉)] and Miguel Molina-Solana[2]

[1] Department of Computer Science and Artificial Intelligence,
Univ. de Granada, Granada, Spain
jgomez@decsai.ugr.es
[2] Data Science Institute, Imperial College London, London, UK
m.molina-solana@imperial.ac.uk

Abstract. Linked Data is an increasingly important source of information and contextual knowledge in Data Science, and its appropriate visualization is key to effectively exploit them. This work presents an ontology to generate graph-based visualizations of Linked Data in a flexible and efficient way. The ontology has been used to successfully visualize DrugBank and DBPedia datasets in a large visualization environment.

Keywords: Linked data · Ontologies · Visualization · Big data

1 Introduction

The late 80s saw the establishment of Visualization as a research area [24] aimed at providing solutions in three aspects of computer-aided Engineering [19]: (i) quick verification of simulation models; (ii) quick delivery of simulation results; and (iii) easier communication to the layperson. In Data Science, visualization is typically used at the beginning and at the end of the data analysis process [7]; i.e. to explore unknown datasets and to communicate results.

Linked Data is an important source of information in Data Science [30], given the availability of open knowledge resources that can be easily reused by relying on open standards. Linked Data visualizations are usually based on trees and graphs, which naturally represent the concept networks created with Linked Data languages such as RDF (Resource Description Framework) [32] and OWL (Ontology Web Language) [13]. Besides, graph-based representations have proven effective for data exploration [29,38]. Other visualizations, such as topic maps [27], geographic maps [28] and statistical charts [23] have also been considered in the literature when the semantics of the underlying data is appropriate.

Big Data has posed new challenges to visualization tools, in terms of data volume, velocity, and variety of sources. Accordingly, new hardware and software tools have been proposed to support large-scale visualizations on big screens and video-walls backed by distributed computation systems. Examples of such frameworks are the CAVE environment [8] at University of Illinois and the Data

© Springer Nature Switzerland AG 2018
F. Herrera et al. (Eds.): CAEPIA 2018, LNAI 11160, pp. 351–360, 2018.
https://doi.org/10.1007/978-3-030-00374-6_33

Observatory[1] at the Data Science Institute (DSI) of Imperial College London. Such approaches have successfully been used in domains in which large amounts of intensively-interrelated data are generated; e.g. Bitcoin transactions [25].

In the research work we present here, we studied how to generate in a flexible and efficient way graph-based visualizations of Linked Data to be displayed in a large screen. With this aim, we have developed a software package based on an ontology for graph modeling that implements a translation from Linked Data into customized visual graphs. This tool can run on a desktop computer and on a Spark cluster [41] —the state-of-the-art open source platform for Big Data— to process very large graphs.

The remainder of the paper is structured as follows. In the next section, we review some related works in the area of Linked Data and ontology visualization. In Sect. 3, we introduce the GraphDL ontology, which can be used to represent custom graphs and to export them to other common graph formats. In Sect. 4, we describe the use of GraphDL and the associated software to visualize data from the DrugBank and DBPedia datasets. The paper finishes with a summary of conclusions and prospective directions for future work.

2   Related Work

Knowledge graph visualization techniques date as early as ontology languages. Sowa's conceptual graphs, considered a predecessor of Description Logic ontologies, brought together the interpretability of visual semantic networks and the formal underpinnings of first-order logic [35]. Simpler models in the RDF language, not bound to a Description Logic, do not entail complex semantics, and therefore they can be directly depicted as directed graphs [32]. Linked Data has been typically visualized in this way. Conversely, the visualization of OWL models [13] is not straightforward. The TopBraid Composer [37] ontology development tool supports visualizing ontologies with a notation similar to UML (Universal Modeling Language). Protégé [26] has available different plugins to visualize different aspects of ontologies as graphs: Ontoviz [33], OWLGrEd [20] and VOWL [21] for complex concepts; Ontograf [6] and GrOWL [17] for interoperable representations; NavigOWL for automatic graph layout [15]; OWLViz [14] and Jambalaya [36] for taxonomies; and OWLPropViz [39] for object properties.

Visualization tools for dataset exploration are more focused on efficient browsing rather than depicting deeper semantics. Cytoscape, a network data toolbox widely used in the BioInformatics domain, includes a plugin to connect to SPARQL endpoints [34]. Gephi, a general-purpose graph visualization tool, can also pull data from Linked Data sources, lay out a graph, and measure different network parameters [16]. Focusing in heterogeneous datasets, in [22] the authors present and end-to-end exploration tool for integration, visualization, and interaction of multiple datasets. The main limitation of these approaches is that they only support straightforward graph visualization; that is, entities

---

[1] https://www.imperial.ac.uk/data-science/data-observatory/.

(resp. relationships) are always represented as vertices (resp. edges). In addition, they cannot manage large datasets.

Comparative analyses of visualization tools for knowledge graphs yield the conclusion that there is no one-size-fits-all solution, and the selection of tools will depend on the purpose [5,11]. Most approaches are however targeted to ontology modeling, and exploration tools often lack the capability of generating graphs with custom vertices and edges. Recently, it has also been emphasized that they have very limited capability to manage large amounts of data [2], since they do not address issues related to performance and scalability; e.g. memory overflow, use of external storage, or efficient automatic layout. Hence, we can find new proposals specifically aimed at these objectives: graphVizdb implements a graph partitioning schema supporting incremental navigation [1], and Memo Graph extracts relevant descriptors from the graph to generate a summarized model [9]. Nevertheless, both of them are implemented in a non-distributed fashion, which results in quite high execution times.

3   Representing Linked Data Graphs with GraphDL

3.1   Design

In the simplest case, we can visualize a RDF graph, encompassing several triples with form $<subject, predicate, object>$, by denoting entities (subject and objects) as vertices, and relationships (predicates) as edges. It is also possible to derive more complex graph structures from triples; for example, we can represent some elements as edges rather than vertices.

GraphDL is our proposal to represent any kind of graph with a simple OWL vocabulary denoting vertices, edges and attributes. GraphDL is strongly based on the Graph Markup Language (GraphML) conceptual model [3], an XML-based format for storing and exchanging graphs. GraphDL is available for download[2], along with several parsers to transform a GraphDL model into other formats (see Sect. 3.3). The main elements of GraphDL are the following:

1. Graph class: Graph encompassing vertices, edges, and nested graphs.
2. Node class: Graph vertex.
3. Edge class: Graph edge, connecting a source node and a target vertex.
4. Attribute class: Graph element attribute, associated to nodes or edges. Attribute instances are: id, label, color, node size, edge weight, and position x and y. More attributes can be added at convenience.
5. AttributeValue class: Instantiation of an attribute value. The instances of this class act as the reification of the relationships between graph elements and attribute values. AttributeValue instances are associated to Node and Edge instances by means of the property hasAttributeValue.
6. forAttribute property: Associates an AttributeValue with the Attribute for which it takes value.

---

[2] https://github.com/jgromero/graphdl.

7. val property: Associates an AttributeValue with a literal value. The type of this literal should be consistent with the type defined for the corresponding attribute (this is not forced by the ontology).

Figure 1 shows the GraphDL representation of a drug interaction obtained from DrugBank—see Sect. 4 for more details of this dataset. There are two instances of graphdl:Node, corresponding to the drugs DB00005 (*Etanercept*) and DB00072 (*Trastuzumab*), connected through an edge of type graphdl:Edge. Note that originally in DrugBank, drugs are represented as entities of the class Drug and drug interactions are entities of the class Drug-Drug-Interaction.

The vertex DB00005 has an attribute *color* with the value *yellow*. To do so, the attribute value is represented by a blank node of type graphdl:AttributeValue. This blank node has associated a datatype property (graphdl:val, with value yellow), and an object property (graphdl:forAttribute, with value graphdl:color).

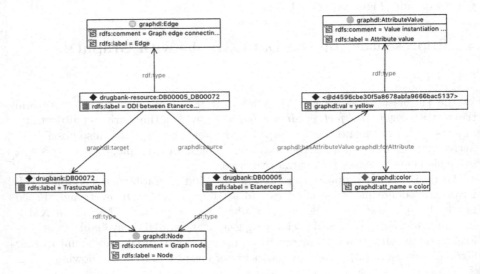

Fig. 1. Representation of vertices, edges and properties in GraphDL, example with DrugBank interaction *Etanercept–Trastuzumab* (created with TopBraid Composer™)

3.2 Graph Generation

The simplest way to create a GraphDL representation is to run a CONSTRUCT query over the triplestore that includes the bulk data such that it: (1) selects the data that we want to use to build our graph from the triplestore; and (2) uses GraphDL elements to define the new graph.

An example query on DrugBank linked data is presented in Listing 1. This query retrieves 10 drugs from the database with the inner SELECT, and then obtains their interacting drugs related via Drug-Drug-Interaction instances. It also retrieves drug categories. A *color* variable is bound to red if the category is Antithrombins; otherwise, it is set to darkGray.

Listing 1. SPARQL query to build a GraphDL from DrugBank (excerpt)

```
CONSTRUCT {
 ?d1 a graphdl:Node ;
     graphdl:hasAttributeValue [
         graphdl:forAttribute graphdl:id;
         graphdl:val ?d1_name ] ;
     graphdl:hasAttributeValue [
         graphdl:forAttribute graphdl:color;
         graphdl:val ?d1_color] ;
     graphdl:hasAttributeValue [
         graphdl:forAttribute viz:category ;
         graphdl:val ?d1_cat_label ] .
 ?d2 a graphdl:Node ;
     graphdl:hasAttributeValue [
         graphdl:forAttribute graphdl:id;
         graphdl:val ?d2_name ] ;
     graphdl:hasAttributeValue [
         graphdl:forAttribute graphdl:color;
         graphdl:val ?d2_color ] ;
     graphdl:hasAttributeValue [
         graphdl:forAttribute viz:category ;
         graphdl:val ?d2_cat_label ] .
 ?i  a graphdl:Edge ;
     graphdl:source ?d1 ; graphdl:target ?d2 ; rdfs:label ?i_label .
} WHERE {
   { SELECT ?d1 where { ?d1 a drugbank:Drug } LIMIT 10 }
   ?d1 a drugbank:Drug ; drugbank:ddi-interactor-in ?i ;
       dcterms:title ?d1_name ; drugbank:category ?d1_cat .
   ?d1_cat rdfs:label ?d1_cat_label .
   ?i  a drugbank:Drug-Drug-Interaction ;
       rdfs:label ?i_label .
   ?d2 a drugbank:Drug ; drugbank:ddi-interactor-in ?i ;
       dcterms:title ?d2_name ; drugbank:category ?d2_cat .
   ?d2_cat rdfs:label ?d2_cat_label .
   bind(if(contains(str(?d1_cat), "Antithrombins"),
        "red", "green" ) as ?d1_color) .
   bind(if(contains(str(?d2_cat), "Antithrombins"),
        "red", "green" ) as ?d2_color)
   bind(strafter(?i_label, "drugbank:") AS ?i_name) .
   filter(str(?d1) != str(?d2)) .
}
```

3.3 Tools

Triples in GraphDL format can be easily transformed into a graph structure in other format. Specifically, we offer Java parsers to translate GraphDL into Apache TinkerPop [31], a multi-language programming API implementing the Gremlin graph traversal language, and Gephi GraphML, which can be directly used for graph visualization and layout. We also provide a web tool[3] that can be used to reproduce the DrugBank example and extended for other datasets.

To be able to work with larger datasets, we have extended this software to run on Scala over Apache Spark, a distributed computing framework capable of processing larger datasets. Spark includes GraphX [40], a library that implements graph management over HDFS (Hadoop Distributed File System), as well as graph processing algorithms such as PageRank.

---

[3] http://sl.ugr.es/webgraphdl.

This implementation is based on the MapReduce paradigm. GraphDL triples, coming from a SPARQL query or from the HDFS, are mapped to cluster nodes that identify the kind of information conveyed by a triple —vertex, edge, property, property value, etc.— and tagged accordingly. Then, triples describing the same element of the graph —a vertex or an edge— are reduced together along their property value and inserted into a graph data structure. Finally, the graph is stored back in HDFS or rendered.

4 Examples

4.1 DrugBank

This example uses the Linked Data version of DrugBank created by the Bio2RDF initiative [4]. DrugBank is an online database containing biochemical and pharmacological information about drugs, including their structure, mechanisms, targets, and interactions. Here we are interested in drug to drug interactions, which mean a large deal of the information in the database and can be naturally visualized in the form of graphs. In total, DrugBank defines 8054 drugs, although only 1315 of them are categorized and have at least one interaction. The number of interactions between categorized drugs is 23172.

Figure 2 shows a drug interaction graph including 200 categorized drugs extracted with query 1. Red vertices correspond to drugs of the category *Antithrombins*. The graph has been translated into GraphML with our Java parser and laid out with the Yifan Hu algorithm through the Gephi API.

4.2 DBPedia

The previous workflow —encompassing triplestore query, parsing from GraphDL to GraphML and graph manipulation in Gephi— is not suitable for large graphs: the Java parser is too slow and requires a considerable amount of memory, while Gephi cannot work with graphs above a few thousands vertices. However, that is the general case with DBPedia [18], the Linked Data version of Wikipedia; e.g. if we want to extract a graph of connections between relevant people. Hence, we have used our Spark-based version of the parser to translate from GraphDL into GraphX.

Figure 3 shows an excerpt of a person-to-person connection graph, extracted from DBPedia, displayed on the screens of the Data Science Institute DO. Vertex colors are set according to the person category (given by YAGO2) and layout has been performed with a distributed version of the Fruchterman-Reingold layout algorithm (adapted from [12]). The processing pipeline has been executed in the Spark cluster backing the DO, which features 32 nodes connected to 64 46" high definition screens arranged in a 313° circle.

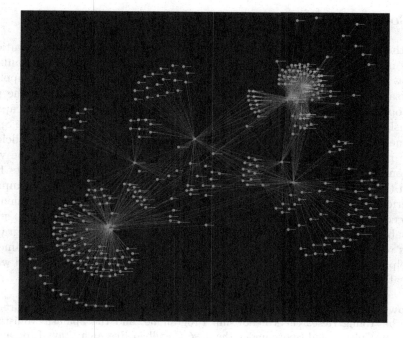

Fig. 2. Drug interaction sample graph, 200 drugs (Color figure online)

Fig. 3. Graphs displayed on the DO screens: DrugBank drug interactions [200 drugs] (left); DBPedia person-to-person relations [excerpt] (right)

5   Conclusions and Future Work

From the provided examples, we can conclude that graph-based representations, mostly used in the past for ontology modeling, can be created and manipulated at a low cost with state of the art Big Data technologies to generate compelling visualizations useful for data exploration. This can be done by using the provided open source software tool, based on the GraphDL ontology and capable of translating GraphDL representations into common graph formats. The Spark implementation shows that generating such visualizations can be very efficient.

We plan to continue our work by benchmarking the efficiency of the visualization stages in graphs with different topologies. Further experiments have proved that distributed force-based layout algorithms can significantly improve the performance of the graph visualization pipeline, even without implementing advanced graph partitioning techniques [10]. In addition, it is essential to study the usability of the resulting visualizations, in particular for very large graphs. In this regard, the DO, equipped with microphones and time-of-flight cameras, can help us to investigate how people interact with the visualizations and which new features are required.

Acknowledgements. Juan Gómez-Romero is supported by Universidad de Granada under the Young Researchers Fellowship Programme, and the Spanish Ministry of Education, Culture and Sport under the José Castillejo Research Stays Programme. Miguel Molina-Solana is supported by the EU's Horizon 2020 research and innovation programme under the Marie Skłodowska-Curie grant agreement No. 743623.

References

1. Bikakis, N., Papastefanatos, G., Skourla, M., Sellis, T.: A hierarchical aggregation framework for efficient multilevel visual exploration and analysis. Semant. Web **8**(1), 139–179 (2017)
2. Bikakis, N., Sellis, T.K.: Exploration and visualization in the web of big linked data: a survey of the state of the art. In: Proceedings of the Workshops of the EDBT/ICDT 2016 Joint Conference, EDBT/ICDT Workshops (2016)
3. Brandes, U., Eiglsperger, M., Lerner, J., Pich, C.: Graph markup language (GraphML). In: Handbook of Graph Drawing and Visualization, pp. 517–541. CRC Press (2013)
4. Callahan, A., Cruz-Toledo, J., Dumontier, M.: Ontology-based querying with Bio2RDF's linked open data. J. Biomed. Semant. **4**(1), S1 (2013)
5. Dudáš, M., Zamazal, O., Svátek, V.: Roadmapping and navigating in the ontology visualization landscape. In: Janowicz, K., Schlobach, S., Lambrix, P., Hyvönen, E. (eds.) EKAW 2014. LNCS (LNAI), vol. 8876, pp. 137–152. Springer, Cham (2014). https://doi.org/10.1007/978-3-319-13704-9_11
6. Falconer, S.: OntoGraf (2016). http://protegewiki.stanford.edu/wiki/OntoGraf
7. Fayyad, U., Grinstein, G.G., Wierse, A. (eds.): Information Visualization in Data Mining and Knowledge Discovery. Morgan Kaufmann, San Francisco (2002)
8. Febretti, A., et al.: CAVE2: a hybrid reality environment for immersive simulation and information analysis. In: Proceedings of the IS&T/SPIE Electronic Imaging, the Engineering Reality of Virtual Reality 2013, San Francisco, USA (2013)

9. Ghorbel, F., Hamdi, F., Ellouze, N., Métais, E., Gargouri, F.: Visualizing large-scale linked data with memo graph. Procedia Comput. Sci. **112**, 854–863 (2017)
10. Gómez-Romero, J., Molina-Solana, M., Oehmichen, A., Guo, Y.: Visualizing large knowledge graphs: a performance analysis. Future Gener. Comput. Syst. **89**, 224–238 (2018)
11. Haag, F., Lohmann, S., Negru, S., Ertl, T.: OntoViBe 2: advancing the ontology visualization benchmark. In: Lambrix, P. (ed.) EKAW 2014. LNCS (LNAI), vol. 8982, pp. 83–98. Springer, Cham (2015). https://doi.org/10.1007/978-3-319-17966-7_9
12. Hinge, A., Auber, D.: Distributed graph layout with spark. In: Proceedings of the IEEE 19th International Conference on Information Visualisation (iV 2015), pp. 271–276 (2015)
13. Hitzler, P., Krótzsch, M., Parsia, B., Patel-Schneider, P.F., Rudolph, S.: OWL2 Web Ontology Language Primer, 2nd edn. (2012). https://www.w3.org/TR/owl2-primer/
14. Horridge, M.: OWLViz (2013). http://protegewiki.stanford.edu/wiki/OWLViz
15. Hussain, A., Latif, K., Rextin, A.T., Hayat, A., Alam, M.: Scalable visualization of semantic nets using power-law graphs. Appl. Math. Inf. Sci. **8**(1), 355–367 (2014)
16. Jacomy, M., Venturini, T., Heymann, S., Bastian, M.: ForceAtlas2, a continuous graph layout algorithm for handy network visualization designed for the Gephi software. PLoS ONE **9**(6), 1–12 (2014)
17. Krivov, S., Williams, R., Villa, F.: GrOWL: a tool for visualization and editing of OWL ontologies. J. Web Semant. **5**(2), 54–57 (2007)
18. Lehmann, J., et al.: DBpedia - a large-scale, multilingual knowledge base extracted from Wikipedia. Semant. Web **6**(2), 167–195 (2015)
19. Leigh, J., et al.: Scalable resolution display walls. Proc. IEEE **101**(1), 115–129 (2013)
20. Liepinš, R., Grasmanis, M., Bojārs, U.: OWLGrEd ontology visualizer. In: Proceedings of the International Semantic Web Conference Workshop ISWC-DEV, vol. 1268, pp. 37–42 (2014)
21. Lohmann, S., Negru, S., Haag, F., Ertl, T.: Visualizing ontologies with VOWL. Semant. Web **7**(4), 399–419 (2016)
22. Mai, G., Janowicz, K., Hu, Y., McKenzie, G.: A linked data driven visual interface for the multi-perspective exploration of data across repositories. In: Proceedings of the 2nd International Workshop on Visualization and Interaction for Ontologies and Linked Data (VOILA2016), pp. 93–101 (2016)
23. Martin, M., Abicht, K., Stadler, C., Ngonga Ngomo, A.C., Soru, T., Auer, S.: CubeViz - exploration and visualization of statistical linked data. In: Proceedings of the 24th International Conference on World Wide Web (WWW 2015), pp. 219–222 (2015)
24. McCormick, B.H.: Visualization in scientific computing. ACM SIGBIO Newslett. **10**(1), 15–21 (1988)
25. McGinn, D., Birch, D., Akroyd, D., Molina-Solana, M., Guo, Y., Knottenbelt, W.J.: Visualizing dynamic bitcoin transaction patterns. Big Data **4**(2), 109–119 (2016)
26. Musen, M.A.: The Protégé project: a look back and a look forward. AI Matters **1**(4), 4–12 (2015)
27. Newman, D., et al.: Visualizing search results and document collections using topic maps. J. Web Semant. **8**(2–3), 169–175 (2010)
28. Nikolaou, C., et al.: Sextant: visualizing time-evolving linked geospatial data. J. Web Semant. **35**, 35–52 (2015)

29. Pienta, R., Abello, J., Kahng, M., Chau, D.H.: Scalable graph exploration and visualization: sensemaking challenges and opportunities. In: Proceedings of the 2015 International Conference on Big Data and Smart Computing (BIGCOMP), pp. 271–278, Februrary 2015

30. Ristoski, P., Paulheim, H.: Semantic web in data mining and knowledge discovery: a comprehensive survey. J. Web Semant. **36**, 1–22 (2016)

31. Rodriguez, M.A.: The Gremlin graph traversal machine and language. In: Proceedings of the 15th ACM Symposium on Database Programming Languages (DBLP 2015), pp. 1–10 (2015)

32. Schreiber, G., Raimond, Y.: RDF 1.1 Primer (2014). https://www.w3.org/TR/rdf11-primer/

33. Sintek, M.: OntoViz (2007). http://protegewiki.stanford.edu/wiki/OntoViz

34. Smoot, M.E., Ono, K., Ruscheinski, J., Wang, P.L., Ideker, T.: Cytoscape 2.8: new features for data integration and network visualization. Bioinformatics **27**(3), 431–432 (2011)

35. Sowa, J.F.: Conceptual graphs. In: van Harmelen, F., Porter, B., Lifschitz, V. (eds.) Handbook of Knowledge Representation, vol. 3, pp. 213–237. Elsevier, Amsterdam (2008). Chap. 5

36. Storey, M.A., Noy, N.F., Musen, M., Best, C., Fergerson, R., Ernst, N.: Jambalaya: an interactive environment for exploring ontologies. In: Proceedings of the 7th International Conference on Intelligent User Interfaces, p. 239 (2002)

37. TopQuadrant Inc.: TopBraid Composer (2016). http://www.topquadrant.com/tools/IDE-topbraid-composer-maestro-edition/

38. Von Landesberger, T., et al.: Visual analysis of large graphs: state-of-the-art and future research challenges. Comput. Graph. Forum **30**(6), 1719–1749 (2011)

39. Wachsmann, L.: OWLPropViz (2008). http://protegewiki.stanford.edu/wiki/OWLPropViz

40. Xin, R.S., Crankshaw, D., Dave, A., Gonzalez, J.E., Franklin, M.J., Stoica, I.: GraphX: unifying data-parallel and graph-parallel analytics. In: Proceedings of the USENIX Symposium on Operating Systems Design and Implementation (OSDI 2014) (2014)

41. Zaharia, M., Chowdhury, M., Franklin, M.J., Shenker, S., Stoica, I.: Spark: cluster computing with working sets. In: Proceedings of the 2nd USENIX Conference on Hot topics in cloud computing (HotCloud) (2010)

A First Step to Accelerating Fingerprint Matching Based on Deformable Minutiae Clustering

Andres Jesus Sanchez[1](✉) ⓘ, Luis Felipe Romero[1] ⓘ, Siham Tabik[2] ⓘ,
Miguel Angel Medina-Pérez[3] ⓘ, and Francisco Herrera[2] ⓘ

[1] Department of Computer Architecture, University of Malaga, Málaga, Spain
`ajsanchez@ac.uma.es, felipe@uma.es`
[2] Department of Computer Science and Artificial Intelligence,
University of Granada, Granada, Spain
`siham@ugr.es, herrera@decsai.ugr.es`
[3] Tecnologico de Monterrey, Campus Estado de México,
Carretera al Lago de Guadalupe Km. 3.5, 52926
Atizapán de Zaragoza, Estado de México, Mexico
`migue@itesm.mx`

Abstract. Fingerprint recognition is one of the most used biometric methods for authentication. The identification of a query fingerprint requires matching its minutiae against every minutiae of all the fingerprints of the database. The state-of-the-art matching algorithms are costly, from a computational point of view, and inefficient on large datasets. In this work, we include faster methods to accelerating DMC (the most accurate fingerprint matching algorithm based only on minutiae). In particular, we translate into C++ the functions of the algorithm which represent the most costly tasks of the code; we create a library with the new code and we link the library to the original C# code using a CLR Class Library project by means of a C++/CLI Wrapper. Our solution re-implements critical functions, e.g., the bit population count including a fast C++ PopCount library and the use of the squared Euclidean distance for calculating the minutiae neighborhood. The experimental results show a significant reduction of the execution time in the optimized functions of the matching algorithm. Finally, a novel approach to improve the matching algorithm, considering cache memory blocking and parallel data processing, is presented as future work.

Keywords: Fingerprint recognition · Cache optimization
Language interoperability

This work was partly supported by the National Council of Science and Technology of Mexico (CONACYT) under the grant PN-720 and by the TIN2016-80920R project (Spanish Ministry of Science and Technology) and the University of Malaga (Campus de Excelencia Internacional Andalucia Tech).

F. Herrera et al. (Eds.): CAEPIA 2018, LNAI 11160, pp. 361–371, 2018.
https://doi.org/10.1007/978-3-030-00374-6_34

1   Introduction

The task of fingerprint recognition has been widely studied in the recent years. The reason why this human characteristic is very popular to identify people lies in the fact that fingerprints are unique and do not change over time. That is, there are not two people with the same fingerprint, making them a perfect recognition feature. The major characteristics of fingerprints, based on ridge features, are known as minutiae and represent ridge endings and bifurcations as standard. Thereby, in almost all most accurate matching algorithms, the comparison method relies on the minutiae details computation carried out in the CPU [1], while other approaches perform the calculations by using GPUs [2,3].

Fingerprint matching algorithms are usually implemented using high-level programming languages such as C#, Java or Python [4]. This choice is made due to the large amount of programming tools that are available to make the programming work easier for the researcher. Nevertheless, the mayor drawback from this languages and similar ones consists in the lower execution speed of the code, in contrast with lower-level programming languages. For example, in C# programming language, when the file is compiled, the code is firstly translated into an intermediate managed language named *Intermediate Language* (IL) and then into machine code, during runtime, by the *Common Language Runtime* (CLR) tool from the .NET framework. This is why this type of code is also known as managed language, causing execution overhead.

This paper focuses on currently one of the most accurate fingerprint matching algorithms, based on minutiae processing, patented by Medina-Pérez et al. [5,6] and will be referred as "DMC" in the rest of the paper. This matching method was initially developed using C# for its implementation and uses the following three algorithms:

1. The *Minutia Cylinder-Code* method [7] which is based on 3D data structures, called Binary Cylinder Codes, built from minutiae positions and angles from merging local structures [8]. DMC uses this minutia representation to perform the matching processing.
2. The *Minutiae Discrimination* method [9] that calculates a quality value for each minutia based on the minutiae direction consistency inside its minutiae neighborhood.
3. The *Deformable Minutiae Clustering* method [5,10] which is used to avoid data loss due to fingerprint deformation. Similarity results are obtained by adding the weights of each matched minutiae pair from merging the clusters that are similar. In addition, the weight of each minutiae pair (q, p) is calculated by evaluating the minutiae from the matched cluster in which (q, p) is the centre. Statistical outcomes are obtained based on the performance evaluation proposed by Cappelli et al. [11].

We propose two techniques to accelerate the DMC matching algorithm in CPU. First, the profile of the matching algorithm is obtained to determine the highest computational cost functions. Afterwards, we apply novel optimizations by including a CLR Class Library project to work as a linker between the C# code, and a C++ Library where the DMC matching algorithm is actually executed. Focusing on the improvements, we optimize the bit population count operation performed within *Minutia Cylinder-Code* algorithm and the minutiae direction consistency algorithm from the *Minutiae Discrimination* method. Moreover, a future parallel approach to optimize the DMC algorithm is also proposed taking into account matching decomposition [12].

This paper is organized as follows. Section 2 introduces the structure of the DMC matching algorithm along with its profiling outcomes. Section 3 describes the optimization techniques performed on the original DMC matching algorithm. Section 4 presents the experimental results. Section 5 proposes a future approach to enhance matching algorithm execution. Finally, Sect. 6 discusses the result of this paper and future work.

2 DMC Structure and Profiling

The DMC fingerprint matching algorithm was developed from merging three independent matching methods whose procedures can be summarized as follows:

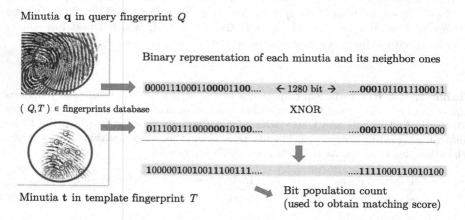

Fig. 1. Example of the bit population count operation used in the first step of the matching algorithm. The set bits, resulted from the XNOR operation performed on two minutiae descriptors, are counted and used to calculate the minutiae matching score.

1. Let Q and T be the minutiae sets of query and template fingerprints from a particular database. Each minutia $q \in Q$ is compared to all minutiae $t \in T$ based on their minutia descriptor. In our case study, the minutia descriptor is an array of 1280 bits which stores the positions and angles of all minutiae inside a circumference of radius r, centered on the selected minutia. Thus, two similar minutiae should have a large amount of bits with same position

and value inside both minutia descriptors, so that if the XNOR operation is done between them as it is shown in Fig. 1, it will result in a new vector with one bits in the concurring positions. Therefore, a function cited as *PopCount* is necessary to calculate the number of set bits, which is then used to obtain a similarity value between minutiae. Each matching minutiae pairs $(\mathbf{q}, \mathbf{t})$ is then included inside a set of local matching minutiae pairs (A) and sorted by its similarity value.

2. For each minutia $\mathbf{q} \in Q$, a quality value is computed relying on the minutiae direction consistency of all minutiae inside its neighborhood, and prior to achieve this goal, it is necessary to compute the three minutiae closest to the selected one by means of a function named as *UpdateNearest*. Moreover, the same procedure is performed on each minutia $\mathbf{t} \in T$. Finally, two sets containing all minutiae from each fingerprint and their particular quality value are obtained.

3. From the sorted A list, clusters of matching minutiae pairs are determined for each pair $(\mathbf{q}, \mathbf{t})$, along with a weight value which is obtained taking into account the matched minutiae pairs inside the matched cluster in which $(\mathbf{q}, \mathbf{t})$ is located in the centre. Then, all clusters inside both fingerprints are merged to find global matching minutiae pairs. Afterwards, a new search is performed, focusing on each minutia $\mathbf{q} \in Q$ and $\mathbf{t} \in T$, using the Thin Plate Spline method to find new matching minutiae pairs [13], which could have been discarded in the previous algorithms due to deformations.

Table 1. Call tree performance profiling of the original DMC where the functions are listed by the collected CPU exclusives samples in descending order. BBC denotes Binary Cylinder-Code and DMC-BCC. Match the main function in which the DMC matching algorithm is executed. The functions in bold are located within DMC whereas the other functions are native tools from .NET Framework.

Function name	Inclusive samples	Exclusive samples
clr.dll	2094	2094
mscorlib.ni.dll	1685	1187
Map	1242	358
UpdateNearest	315	231
DMC-BCC.Match	6193	209
MtiaPairComparer.Compare	381	205
BinaryCylinderCode.Match	309	189
DMC-BCC.MatchMinutiae	281	107
BCC.PopCount	95	95

Once the matching algorithm process has been studied, the next target consists in determining which functions are the most computationally expensive inside the four steps already mentioned. Thus, a thorough analysis of the original DMC implementation needs to be made, so the Performance Profiler Tools from Visual Studio 2017 is used to obtain two types of outcomes: inclusive and exclusive samples from the CPU. In our case, the exclusive samples are used to identify the functions with the highest computing lines of code because it only takes account of the samples obtained inside the function and not the ones that are called within it, representing the bottlenecks of the matching algorithm. Table 1 shows the functions with the highest computing load based on the CPU exclusives samples. Highlighted function names are functions which are situated inside the matching algorithm, while the rest of functions are tools used by the .NET Framework. From this list, our research will be focused on the *UpdateNearest* and *PopCount* functions due to the high number of exclusive samples that they have obtained from the profiling in contrast with the few amount of lines of code that they contains. In the next section, novel optimizations performed over this two functions will be presented.

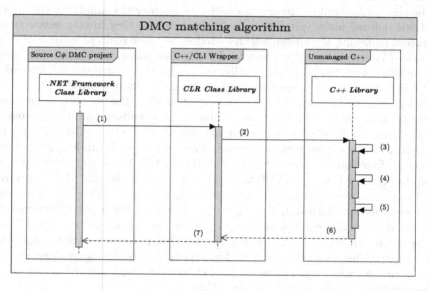

Fig. 2. Schematic structure of the modified DMC matching algorithm which includes a CLR Class Library and an unmanaged C++ Library where the fingerprint matching algorithm is performed.

3   Software Optimizations

Regarding the DMC algorithm, it was implemented using C# along .NET Framework 4 and hence, the first step constitutes the translation of the matching algorithm, which includes the functions previously determined from the profiling, to an unmanaged and lower-level programming language preserving the structure of the original solution. Therefore, C++ was chosen to replace C# as the programming language owing to its better performance and the fact that it is also an Object-Oriented Programming (OOP) language which results in less compatibility issues. Then, a connection between the original C# project and the new C++ library needed to be established in order to share the necessary data between them and hence, two .NET framework tools of interoperability were tested: Platform Invoke using extern "C" methods [14] and a C++/CLI Wrapper [15]. Both strategies speed up the part of the code which is translated into C++, however, the second method enables objects creation, name-spaces usage, and passing values, and it also enables the assembly along with managed codes.

Figure 2 shows the modified DMC algorithm using the C++/CLI Wrapper, where a CLR Class Library project and a C++ Library are included. The first one is written using C++/CLI code which provides several .NET Framework tools for working with managed and unmanaged memory. Thus, all the fingerprints data loaded inside the C# managed code can be passed as arguments and stored within the CLR Class Library with few transformations (1). Then, this data can be easily exported to the unmanaged C++ Library (2) simply by calling the functions defined inside the last one. Afterwards, the C++ library executes the optimized matching algorithm, which includes the three main processes: *Minutia Cylinder-Code Creation* (3), *Minutia Quality Computation* (4) and *Clustering Computation and Merging* (5). Finally, all the results achieved are returned to the CLR Class Library project first (6) and then, to the C# project (7).

Regarding the first step of the matching algorithm, the bit population count procedure is needed to compute the similarity value between minutiae. This bit operation is also known as pop count and define the action of calculating the number of set bits from a certain word. This technique is implemented in the *PopCount* function by using a SWAR (SIMD Within A Register) method. This technique deals with counting bits of duos to afterwards, add the duo-counts to a four-bit aggregation and then bytes inside a 64-bit register, to finally sum all bytes together. Although this algorithm is known for a good performance, we wanted to implement the best approach for the bit population count based on the computer characteristics where the DMC fingerprint matching algorithm was running. Therefore, the fast C++ bit population count library developed by Muła et al. [16] was implemented, which chooses the best algorithm depending on the computer features and the size of the bit array.

On the other hand, focusing on the second step of the matching algorithm, the minutiae neighborhood of each minutia is calculated within all minutiae that set the fingerprint, based on the Euclidean distance. The square root results from this operation are obtained by using two methods: accessing to a pre-established table with the 1024 former square root results or, if the value is not contained in this table, performing the square root operation using the C# Math Class. To avoid the overhead of this two methods in the translated matching algorithm, the square Euclidean distance is used instead.

4 Experimental Results

The experiments were executed using an Intel i5-8600K processor running at 3.60 GHz. The machine has six L1 and L2 cache memory of 32 and 256 kBytes, respectively, and a single L3 cache of 9 MBytes. Also, it has a 8 GB of RAM (DDR4 with dual-channel). Time results were obtained by using the Stopwatch Class from .NET Framework in C# and the C++11 Chrono Library in the C++ library. The FVC2004_DB1A database, which contains 800 hundred latent fingerprints with perturbations deliberately introduced, has been used to perform the experiments. Regarding the optimization of the bit population count, a fast C++ library (*libpopcnt*) was used inside the *PopCount* function in spite of the original SWAR algorithm from C#. The results achieved from the execution of both methods are shown in Table 2. The C# method runtime was reduced from 90.176 to 0.440 microseconds in C++ per iteration of counting the set bits from an array which takes about 160 bytes of memory (20 values of 8 bytes). Thus, since the array size is less than 512 bytes, an unrolled pop count algorithm is selected by the library to perform the operation. On the other hand, an AVX2 algorithm would be executed if the array size were over 512 bytes, however, this situation does not take place in the original DMC algorithm but will be addressed in a future work to improve the accuracy without increasing the execution time. In speed-up terms, the function acceleration achieved in C++ is up to 204.8 times faster than the original C# one.

Table 2. Time, in microseconds, and speed-up results per bit population count iteration inside the *PopCount* function, where the performances of the SWAR algorithm from C# and the fast C++ *libpopcnt* library are compared.

	C# SWAR algorithm	Fast C++ library
Time per iteration	90.176 µs	0.440 µs
Speed-up	1	204.8

With respect to the *UpdateNearest* function, the use of the Euclidean distance was replaced by the squared Euclidean distance (SED), eliminating the square root operation and improving time results per iteration as shown in the Table 3. This function was accelerated up to 2.83 times using the squared Euclidean distance, in contrast with the 2.33 achieved with the standard Euclidean distance.

5 Proposed Improvements

Fingerprint matching algorithms are useful methods which commonly perform the comparison between one latent fingerprint and all the fingerprints stored in a particular database. This technique reduces the number of fingerprints to analyze to a reduced set of similar ones, however, the time necessary to perform this

Table 3. Time, in microseconds, and speed-up results per iteration of the *UpdateNearest* function using three approaches: original C# method, this method translated into C++ and the final method implementing the squared Euclidean distance (SED) in C++.

	C#	C++	C++ (SED)
Time per iteration	25.47 μs	10.93 μs	9.00 μs
Speed-up	1	2.33	2.83

operation depends heavily on the size of the database. In addition, every fingerprint, either rolled, latent or plain, do not necessarily contain a fixed number of minutiae, owing to the high complexity of the fingerprint sample taking, which involves several aspect such as pressure variations, finger area (partial or total sample), large nonlinear distortion, among others [17]. Thus, if the query fingerprint and the selected template one contain a large amount of minutiae each, the total data used to perform the comparison will take more memory space than the available inside the L1 cache memory, increasing the cache misses and the execution time thereby [18]. This situation is shown in Fig. 3 where the average size of a fingerprints pair, obtained from executing the DMC matching algorithm on the FVC2004_DB1A fingerprints database, is close to the maximum L1 memory cache capacity.

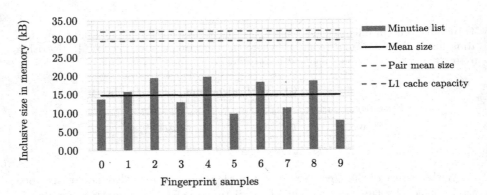

Fig. 3. Minutiae lists sizes results, in managed memory, from a sample of 10 fingerprints from the FVC2004_DB1A database. The average size and the mean pair size of the minutiae lists sample are presented along the L1 cache memory capacity.

In order to address this issue, the implementation of a parallel program structure for the DMC matching algorithm is proposed. A diagram of this structure is presented in Fig. 4, where each core from the CPU will have a copy of the query fingerprint and will perform the matching algorithm between this one and a block of fingerprints from the database. This block will contain entire fingerprints or sections, depending on its size in memory, e.g., F5-S1 and F5-S2 from the previous figure represent a division of the fifth fingerprint from the database in two sets of minutiae to be processed in two different cores. This processing block size will be calculated so that all minutiae data would fit properly in the L1 memory cache when processing the first two steps of the matching algorithm. If a match occurs between minutiae, the similarity value will be stored and then computed to obtain a global similarity outcome. Hence, this proposal tries to solve two common problems of fingerprint matching algorithms: poor scalability for large fingerprints databases and low efficiency of cache memory usage.

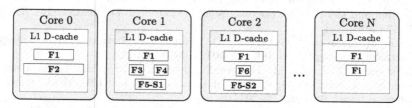

Fig. 4. Structure of the proposed parallel DMC matching algorithm where each thread has a copy of the query fingerprint ($F1$) to be compared with a block of selected fingerprints from the database whose data fit properly inside the L1 D-cache memory.

6 Conclusions

Two approaches to improve the speed of the fingerprint matching algorithm DMC proposed by Medina-Pérez et al. [5] has been presented in this work. The use of C++ as the programming language for executing the DMC matching algorithm has proved to be the critical step to reduce runtime, owing to the computational overhead that C# causes. To achieve this goal, a CLR Class Library was implemented inside the original DMC code to work as a linker between the original C# project and the unmanaged C++ file, enabling the pass of the fingerprints data in both directions. The unmanaged C++ was deployed into a static C++ Library (.lib) which contains the optimized matching algorithm written in this low-level programming language. Regarding more specific optimizations, the implementation of a fast C++ library for the *PopCount* function within the matching algorithm improves the acceleration of the set bits counting operation up to 204.8 times, in contrast to the SWAR C# method. Concerning the *Update-Nearest* function, the same algorithm as C# was implemented in the C++ library, obtaining an enhancement up to 2.33 times faster by translating this method into C++. Then, the use of the squared Euclidean distance inside the *UpdateNearest* function is addressed for calculating the minutiae neighborhood, proving to be up to 2.83 times faster than the no-squared approach in C#. To conclude, a new approach to enhancement the DMC matching algorithm performance has been proposed using parallel programming techniques where each thread compares the copied query fingerprint with blocks of template fingerprints which would fit properly in the L1 D-cache memory for the two first steps of the matching algorithm. It has been shown that, focusing on the DMC matching algorithm, the total data size in managed memory from two fingerprints is, in some cases, bigger than the L1 cache memory capacity. This situation could slow down the calculation performance owing to cache memory misses and the poor scalability for large databases, issues that could be addressed by using our novel suggestion.

References

1. Peralta, D., et al.: A survey on fingerprint minutiae-based local matching for verification and identification: taxonomy and experimental evaluation. Inf. Sci. **315**, 67–87 (2015)
2. Lastra, M., Carabao, J., Gutierrez, P.D., Benitez, J.M., Herrera, F.: Fast fingerprint identification using GPUs. Inf. Sci. **301**, 195–214 (2015)
3. Gutierrez, P.D., Lastra, M., Herrera, F., Benitez, J.M.: A high performance fingerprint matching system for large databases based on GPU. IEEE Trans. Inf. Forensics Secur. **9**(1), 62–71 (2014)
4. Medina-Pérez, M.A., Loyola-González, O., Gutierrez-Rodríguez, A.E., García-Borroto, M., Altamirano-Robles, L.: Introducing an experimental framework in C# for fingerprint recognition. In: Martínez-Trinidad, J.F., Carrasco-Ochoa, J.A., Olvera-Lopez, J.A., Salas-Rodríguez, J., Suen, C.Y. (eds.) MCPR 2014. LNCS, vol. 8495, pp. 132–141. Springer, Cham (2014). https://doi.org/10.1007/978-3-319-07491-7_14
5. Medina-Perez, M.A., et al.: Latent fingerprint identification using deformable minutiae clustering. Neurocomputing **175**, 851–865 (2016)
6. Medina-Perez, M.A., et al.: Sistema y método para la comparación de huellas dactilares y palmares basada en múltiples clústeres deformables de minucias coincidentes, No. ES2581593, Spain (2018). (in Spanish)
7. Cappelli, R., Ferrara, M., Maltoni, D.: Minutia cylinder-code: a new representation and matching technique for fingerprint recognition. IEEE Trans. Pattern Anal. Mach. Intell. **32**(12), 2128–2141 (2010)
8. Jiang, X., Wei-Yun, Y.: Fingerprint minutiae matching based on the local and global structures. Pattern Recognit. **2000**(2), 1038–1041 (2000)
9. Cao, K., Liu, E., Pang, L., Liang, J., Tian, J.: Fingerprint matching by incorporating minutiae discriminability. In: 2011 International Joint Conference on Biometrics (IJCB), pp. 1–6. IEEE (2011). https://doi.org/10.1109/IJCB.2011.6117537
10. Medina-Perez, M.A., et al.: Improving fingerprint verification using minutiae triplets. Sensors **12**(3), 3418–3437 (2012)
11. Cappelli, R., et al.: Performance evaluation of fingerprint verification systems. IEEE Trans. Pattern Anal. Mach. Intell. **28**(1), 3–18 (2006)
12. Peralta, D., Garcia, S., Benitez, J.M., Herrera, F.: Minutiae-based fingerprint matching decomposition: methodology for big data frameworks. Inf. Sci. **408**, 198–212 (2017)
13. Bazen, A.M., Sabih, H.G.: Fingerprint matching by thin-plate spline modelling of elastic deformations. Pattern Recognit. **36**(8), 1859–1867 (2003)
14. Using C++ Interop (Implicit PInvoke). https://msdn.microsoft.com/en-gb/library/2x8kf7zx.aspx. Accessed 16 Feb 2018
15. .NET Programming with C++/CLI (Visual C++). https://msdn.microsoft.com/en-gb/library/68td296t. Accessed 22 May 2018

16. Muła, W., Kurz, N., Lemire, D.: Faster population counts using AVX2 instructions. Comput. J. **61**(1), 111–120 (2017)
17. Jain, A.K., Feng, J.: Latent fingerprint matching. IEEE Trans. Pattern Anal. Mach. Intell. **33**(1), 88–100 (2011)
18. Kowarschik, M., Weiß, C.: An overview of cache optimization techniques and cache-aware numerical algorithms. In: Meyer, U., Sanders, P., Sibeyn, J. (eds.) Algorithms for Memory Hierarchies. LNCS, vol. 2625, pp. 213–232. Springer, Heidelberg (2003). https://doi.org/10.1007/3-540-36574-5_10

Learning Planning Action Models with Numerical Information and Logic Relationships Using Classification Techniques

José Á. Segura-Muros(✉) ⓘ, Raúl Pérezⓘ, and Juan Fernández-Olivaresⓘ

University of Granada, Granada, Spain
{josesegmur,fgr,faro}@decsai.ugr.es

Abstract. The task of constructing a planning domain is difficult and requires time and vast knowledge about the problem to be solved. This paper describes PlanMiner-O3 a planning domain learner designed to alleviate this problem, based on the use of a classification algorithm, able to learn planning action models from noisy and partially observed logic states. PlanMiner-O3 is able to learn continuous numerical fluents as well as simple logical relations between them. Testing was realized with benchmark domains obtained from the International Planning Competition and the results demonstrate PlanMiner-O3's capabilities of learning planning domains.

1 Introduction

When designing a planning domain able to solve correctly planning problems, a domain designer must face several obstacles during the designing process. These obstacles come from the complexity of the world to be modelled by the planning domain: the more complex worlds lead the more restrictions to be taken into account. This provokes that the task of designing a planning domain to be tiring and requires a lot of time and effort, as well as an important amount of expert knowledge about the problem to be solved using AI automated planning. Trying to lessen those obstacles and help the planning domain designers planning domain learners can be used. Planning domain learners can automatically create a planning domain from input data extracted from existing processes. This leads to a new problem: Planning domain learners must deal with incomplete or noisy data obtained from those processes, as well as they need to produce planning domains with enough expressivity to model the relations between the different elements of the world.

This research is being developed and partially funded by the Spanish MINECO R&D Project PLAN MINER TIN2015-71618-R.

F. Herrera et al. (Eds.): CAEPIA 2018, LNAI 11160, pp. 372–382, 2018.
https://doi.org/10.1007/978-3-030-00374-6_35

In this paper, we describe the PlanMiner-O3 planning domain learner. This domain learner is based on the use of an state-of-the-art classification algorithm to learn planning domains from an incomplete and noisy input data. Planning domains learned by PlanMiner-O3 are complex enough to model continuous numerical information, simple numerical and logic relations, as well as manage basic resources (action costs). PlanMiner-O3 creates a series of action models that represent the different actions of a planning domain to be learned by obtaining a representation of the fluents that form the previous (prestate) or subsequent (poststate) states of each different action of the domain. When creating those action models PlanMiner-O3 relies only on the input data provided knowing no information beforehand about the ontology of the domain to be learned.

Trying to overcome state-of-the-art approaches, PlanMiner-O3 was designed to be a planning domain learner robust to incompleteness and noise and able to learn action models with continuous numerical predicates and relations between its fluents. In order to achieve this, an array of techniques designed to extract correctly the planning information given as input and to preprocess it are going to be presented in the next pages. By considering the domain learning problem as a classification problem we reduce the problem to find a collection of hypothesis that model a collection of states of the world. When testing PlanMiner-O3, benchmark domains of the International Planning Community were used to obtain a series of execution traces of solved plans. In order to assure the results facing low quality data, noise and incompleteness were included and the learned domains were evaluated by measuring the difference between them and the benchmark domains but also by testing its problem-solving capabilities using tested tools from the International Planning Community.

Next section will cover previous works related to the one presented here. Section 3 will explain every background concept needed to understand how our solution works. In Sect. 4 our domain learner algorithm will be explained. Then, Sect. 5 will contain the information about our experiments and its results. Finally, in Sect. 6 the conclusions drawn from the results will be discussed together with possible improvements of our solution in the near future.

2   Related Work

The problem of planning domain learning has been faced several times in the last few years [1]. In the literature, we can find ARMS [2] or LAMP [3], both domain learners obtain action models by generating a set of logic formulas and solve them. Both ARMS and LAMP can handle incomplete logic states, but cannot deal with any type of noise. On the other hand, there are some other planning domain learners that use noisy information as input [4,5]. The solutions presented by those approaches range from considering every input plan trace as noisy and fits a collection of models from them or fitting an initial model and readjust it when new examples contradict it. In contraposition with the approaches presented earlier, those domain learners cannot deal with any data incompleteness. Finally, we can find state-of-the-art domains learners like [6] that handle both incompleteness and noise. This domain learner relies on fitting a series of kernels and extract action models from them. Very few of these solutions can handle incompleteness or noise, and even, less can deal with both. This issue burdens the implementation of planning domain learning techniques in real-world applications. Those solutions that use machine learning techniques are based on black-box models and can't deal with any type of numerical information lowering the expressivity of the models obtained and hindering the export of the information learned into other types of problems.

Planning domain learners that work with numerical information are getting very important in the last years. NLOCM [7] creates a series of linear equations using the plan's actions as variables and the cost as the coefficient. When solved the system of linear equation that forms the sum of every linear equation created from the input plan the cost of each action is the value of its associated variable. NLOCM requires of a very low amount of data to work, but it can not deal with any incompleteness nor noise in the input data and cannot infer relations between the variables involved in the problem.

3   The Learning Problem

In automated planning the world is represented as a conjunction of fluents. A fluent is a statement in the form of $p(arg_1, arg_2, \ldots, arg_n)$ where p is a logic predicate and arg_x an object of the world. Objects may have a type associated, and those types may have a hierarchical relationship with other types. Each fluent has a value associated: *True* or *False* in the case of literal fluents or a numerical value (discrete or continuous) in the case of function fluents.

A STRIPS [8] planning domain can be seen as a tuple $<Ont, Act>$ where Ont is the ontology of the world, the definition of the predicates and objects of the world, and Act is a collection of STRIPS actions. In the same way, a STRIPS planning action is a tuple $<Name, Par, Pre, Eff>$, where $Name$ is the action's name, Par are the parameters of the action, Pre the preconditions that must be true to allow the execution of the action and Eff the effects of the action in the world after being executed. An action whose Par, Pre and Eff are not instantiated with world's objects is called Action Model. This paper focuses on the learning of deterministic action models. In these action models, its preconditions and effects are unique among the rest of the domain's preconditions and effects. Finally, a *plan* is an ordered sequence of instantiated actions whose execution modify the world to achieve a given goal.

A *Plan Trace* (PT) can be defined as:

$$<S_0, A_0, S_1, A_1, \ldots, S_n, A_n, S_{n+1}>$$

where A_x are the actions that form the plan and S_x are interleaved states. S_0 is the initial state of the problem solved by this plan, S_{n+1} is the goal state of that problem and the rest of states are snapshots of the world in a given point during the execution of the plan. Each action has an associated prestate and poststate. The state S_x of an action A_x is the prestate associated with the action and can be seen as the world just before executing the action. In the same way, the state S_{x+1} is the poststate associated with A_x and is the state of the world just after executing the action. Figure 1 shows an example of a PT.

$$S_0:(at\ a1\ c2) \wedge (= (fuel\text{-}level\ a1)\ 100) \wedge (at\ p1\ c2)$$
$$A_0:(board\ p1\ -\ person\ a1\ -\ aircraft\ c2\ -\ city)$$
$$S_1:(at\ a1\ c2) \wedge (= (fuel\text{-}level\ a1)\ 100) \wedge (in\ p1\ a1)$$
$$A_1:(fly\ a1\ -\ aircraft\ c2\ -\ city\ c1\ -\ city)$$
$$S_2:(= (fuel\text{-}level\ a1)\ 0) \wedge (in\ p1\ a1) \wedge (at\ a1\ c1)$$

Fig. 1. Extract of a PT from a Zeno Travel problem.

The world's states of a PT are usually observed during the execution of a given plan. This can lead to have partially (incomplete) or wrongly (noisy) states observed. Incompleteness occurs when some fluents of the state (or the whole state) are not observed. Noise, on the other hand, is a problem where the value of a fluent is different of the value of the observed fluent.

4   PlanMiner-O3

PlanMiner-O3, the algorithm presented in this paper, first extracts the actions of an input set of plan traces with its associated prestates and poststates grouping them by the action's names. Then, the information of the states associated with a given action is included in a dataset, and after a preprocessing stage, is sent to a classification algorithm in order to obtain a classification model. The process is repeated until every action of the domain has been covered. The classification models contain the information needed to discern between the prestates and poststates of an action and disassociates them. This disassociation helps to extract the maximum information by obtaining the tuples $<attribute, value>$ that form both the prestates and poststates of the action. In the final stages of PlanMiner-O3 this disassociation can be reversed easily in order to create the PDDL action models of the domain. Algorithm 1 shows an overview of PlanMiner-O3 where PTs is the collection of input plan traces.

Algorithm 1. Plan Miner algorithm overview.

Input: A collection of Plan Traces. **Output:** A set of learned action models.

PlanMiner-O3(PTs)

1. ActM ← {}
2. **Foreach** *action* **in** PTs, **Do**
   (a) *Dataset* ← EXTRACT_INFO(*action*, PTs)
   (b) *Dataset* ← PREPROCESS(*Dataset*)
   (c) *rules* ← LEARN_RULES(*Dataset*)
   (d) ActM ← ActM + COMBINE(*rules*)
3. **Return** ActM

4.1   EXTRACT_INFO

In order to fit the input plan traces to a more adequate input model for a classification algorithm, PlanMiner-O3 creates a dataset for each different *action* in *PTs*. Two actions are equal in the *PTs* if they share names and argument types in its headers, regardless of the objects' instances of those arguments. Given an *action* of the domain to be learned it extracts from the set of plan traces *PTs* every pair <*prestate, poststate*> associated with it. Then, PlanMiner-O3 calculates the schema form of every pair by taking each argument <$arg_1, arg_2, \ldots, arg_n$> of the action and replacing the i-th argument in every prestate's and poststate's fluent in which it appears with a $Param_i$ token that represents a variable. Finally, every fluent in the state that has not undergone at least one substitution is erased from the state following a criterion of *relevance* [2]. A fluent is relevant if it shares anyone of its parameters with the associated action's parameters or as no parameters at all (0-arity predicates). Figure 2 shows an example of the schema form of the action board presented in Figure 1 and its associated states.

S_0: $(at\ ?Y\ ?Z) \wedge (= (fuel - level\ ?Y)\ 100) \wedge (at\ ?X\ ?Z)$
T_0: $(board\ ?X - person\ ?Y - aircraft\ ?Z - city)$
S_1: $(at\ ?Y\ ?Z) \wedge (= (fuel - level\ ?Y)\ 100) \wedge (in\ ?X\ ?Y)$

Fig. 2. Schema form of an action and its associated states.

Once the states are extracted PlanMiner-O3 forms a dataset using them. Those datasets are common in machine learning and are described as size $n * m$ matrices where n is the number of examples of the dataset and m the number of attributes. Attributes correspond with the fluents which define extracted states, and the values of the i-th example of the dataset correspond with the fluent's values of the i-th state extracted. Values depend on the type of the fluents. Literal fluents values can be True or False, while function fluents values are a numerical value. Dataset class are *prestate* or *poststate* depending on the relation of the state with the given action. When representing states with a different number of attributes, the set of all attributes is calculated as the union of the different sets of attributes of each example. If a fluent doesn't appear in an state its value is set as a Missing Value. Dataset's Missing Values (MV) are treated depending on the world assumption made in the planning domain. When interpreting the incompleteness of a state PlanMiner-O3 follows the Open World Assumption (OWA) interpretation. OWA considers that unobserved fluents are missing, nor true or false, and can't be evaluated.

4.2   PREPROCESS

Before beginning the learning process and adjusting a classification model, the dataset is preprocessed in order to reduce noise and incompleteness and to include extra attributes to it. On the first hand, dataset's cleaning will help the learning process increasing the robustness of the models generated by the classification algorithm. On the other hand, the addition of new attributes allows us to learn new information such as relations between the original attributes of the dataset. This is realized during the PREPROCESS step and carries a huge impact on the whole performance of PlanMiner-O3.

During the preprocessing step, new attributes are added to each example of the dataset. For each pair <*prestate, poststate*> included in the dataset new information is added by calculating the difference between the value of a certain attribute in the prestate and its value in the poststate. This is realized for each attribute with numerical values. Logic relations between fluents are calculated too by taking an example of the dataset and comparing every attribute with each other attribute of the dataset with the same type of values. Comparisons are realized with the logic relations $=, \neq, <, >, \leq$ and $\geq$. During this process, if there is an MV in any fluent involved in the process the new attributes cannot be calculated and left as an MV for the given example.

Noise is treated differently when facing numerical, logical or nominal noisy attributes. Noisy values cleaning process for logical attributes is realized by erasing noisy values and substituting them by missing values. Noisy values are detected by calculating the Conditional Relative Frequency of each attribute's value of the dataset with the classes of the dataset. If the frequency of a certain value is below a threshold it is replaced by an MV in the dataset. Setting the threshold value wrong will lead to the learning algorithm to detect low appearance rate values as noise, this leading to discard important fluents for the action models.

Noise reduction to numerical values is realized by applying discretization techniques to each attribute. Discretization [9] is a process where numerical attributes are converted into nominal attributes. The main advantages of use discretization techniques range from a reduced computing time from better resistance to noisy values. As PlanMiner-O3 doesn't know anything about the data is using we preferred to use an unsupervised learning algorithm to find the best sets to group the different attributes' values.

The technique used to fit the sets is the k-means clustering algorithm. As k-means needs as input a number k of clusters which is unknown a priori, we use an incremental approach to find the best clusters to a given collection of data. Starting from a given k, k-means fits k-clusters, measures them and repeat the process with k+1 clusters. This lasts as long as k-means produce better clusters. The measure used to compare the results of k-means in each iteration is the Calinski-Harabasz (CH) Index [10]. This measure is calculated as:

$$CH(K) = \sum_{k \in K} [B(k)/W(k)] \times [(N - |K|)/(|K| - 1)]$$

where N is the total number of data points, K the clusters, $W(k)$ the within-cluster variation of a cluster k and $B(k)$ its between-cluster variation. The higher the CH Index value, the "better" are the clusters. Higher CH Index values mean that clusters are more compact (elements in the cluster closer to each other) and separated (cluster's centers far from each other).

Once the best number of clusters is selected and its centers calculated, PlanMiner-O3 extracts the information needed to create a collection of sets from them. This is realized by taking each cluster and calculating the density function of the data points assigned to it. These density functions help the learning process by adjusting the sets to how the data are distributed in each cluster, allowing a more accurate discretization of the data.

Finally, to end the preprocessing step, those examples of the dataset whose attributes only contains MVs are erased. This ensures that every example in the dataset had at least one attribute with useful information.

4.3 LEARN_RULES

Within this step, PlanMiner-O3 tries to fit a set of models from an input dataset. Those models define a collection of examples of the given dataset and are used to create a planning action model. We are interested in the models created by the classifier rather the classification accuracy, so highly interpretable models will help in the translation process from a classification model to an action model.

As a result, Inductive rule learning algorithms were chosen. Among the collection of different inductive rule learning algorithms, the NSLV (New SLaVe) [11] algorithm was selected. The rules learned by NSLV follow the next structure:

$$\textbf{IF } C_1 \text{ and } C_2 \text{ and} \ldots \text{and } C_m \textbf{ THEN } Class \text{ is } B$$

where a condition C_i is a sentence X_n is A, with A a label (or a set of labels) of the domain of the variable X_n. X_n is an attribute of the problem's dataset. Each variable domain's label corresponds with a set that englobes a collection of elements of the given variable. PlanMiner-O3 uses the discrete sets calculated previously as sets for NSLV.

NSLV uses a genetic algorithm (GA) to select which tuples $<attribute,$ $value>$ define the antecedent of the rule that best fits a set of examples of the dataset. Starting from an empty ruleset, the GA extracts and adds a new rule to it in each iteration. NSLV is able to output descriptive rules. Descriptive rules contain every $<attribute, value>$ tuple relevant to the class description of the set of examples covered by the rule.

4.4 COMBINE

This last step separates the rules obtained using NSLV by its class and creates a new ruleset for each class. Then, it combines the rules of each ruleset in a single rule. In a noise-free dataset, the number of rules learned by NSLV is always 2, one for each different class, so rules combination is not necessary.

Starting from an empty *rule* and a ruleset, COMBINE extracts the tuples $<attribute, value>$ from the rule's antecedents of the ruleset. These tuples, then, are sorted by counting the number of examples covered by its original rule.

Once every tuple has been extracted and sorted, they are taken in order and COMBINE tries to insert in *rule*, if there's no conflict between the tuple and the rule's antecedent the tuple is added to it. A conflict arises when the antecedent contains a tuple with the same attribute but a different value than the tuple that COMBINE tries to insert in the rule. If a conflict is found, COMBINE creates a new rule starting from *rule* and swapping the conflicting tuples. COMBINE then calculates the performance of both rules, computing the number of examples covered by both of them. If there's a significant statistical difference between the performance of both rules, the conflictive tuples are erased.

Once every ruleset have been combined in a single rule, COMBINE converts the DNF rules to a PDDL action following a straightforward process: First, action's preconditions are taken directly from the prestate class' rule antecedent. Antecedent's attributes are translated directly into predicates whose value is the attribute value. And second, action's effects are extracted from the difference $\Delta(pre, post)$ between the prestate rule's antecedent and the poststate rule's antecedent. $\Delta(pre, post)$ is defined as the set of changes that must be done over *pre* in order to make it equal to *post*.

Numerical fluents are converted to a numerical PDDL numerical precondition or effect by taking the a representative value associated with the set assigned during the LEARN_RULES procedure. COMBINE differentiates between the original numerical attributes of the dataset and the new ones added artificially.

5   Experiments and Results

PlanMiner-O3 was tested using a collection of plan traces obtained from domains from the International Planning Competition IPC. The objective of these experiments is to demonstrate that PlanMiner-O3 is able to learn planning domain's action models with numerical information and logical relations between fluents even in presence of high levels of missing states' information and some levels of noise. The details of the domains used can be seen in Table 1. Domains can be sorted by its characteristics: STRIPS only representation, constant numeric information and boolean relations between state's fluents.

Table 1. Domain characteristics

Domain	Actions		Predicates		Caracteristics
	No	Max arity	No	Max. arity	
BlocksWorld	4	2	5	2	STRIPS
Depots	5	4	6	2	STRIPS
Satellite	5	4	8	2	STRIPS, functions
Parking	4	3	6	2	STRIPS, functions
DriverLog	6	4	6	2	STRIPS, functions, relations
ZenoTravel	5	6	8	2	STRIPS, functions, relations

From each domain, 200 PTs were used as input. In order to ensure the results, a 5 fold cross-validation was used. The results of the experiments were calculated as the average result of every run. Noise and incompleteness were included randomly in the PTs, affecting only the states defined in them. Noise in the states was included by changing the value of randomly selected fluents by a valid value, or by adding feasible non-existent fluents to them. Incompleteness, on the other hand, was included by erasing a given number of fluents. Both incompleteness and noise affect a given percentage of fluents from the total sum of fluents in the states of a PT. The threshold value used to discern between noise or not during the preprocessing stage is set to 0.05%.

Learned domains performance is measured using 2 different criteria: Learned domain's error rate and learned domain's validation rate with test problems. On the first hand, domain's error rate [3] is measured by comparing the learned domain with the original handmade one. Domain's error is computed by counting the number of missing or extra fluents in the learned domain action's preconditions and effects and dividing it between the number of possible fluents in those preconditions and effects.

The results showed in Fig. 3 demonstrate that our solution can model planning domains close to the original handmade planning domains: error rates fall below 4% even with high levels of incompleteness. In the worst cases, domains' error rate doesn't rise beyond 7% with high levels of incompleteness and some levels of noise. In those cases where with noiseless and complete plan traces PlanMiner-O3 learns domains with some errors (Satellite and Parking domains), the error is a product of the frame axiom: PlanMiner-O3 does not know in advance any information about how the fluents relate between themselves, so it can overfit some models by learning new useless information. This information is useless has it can be inferred using other fluents of the state or maybe PlanMiner-O3 found an irrelevant relation between fluents.

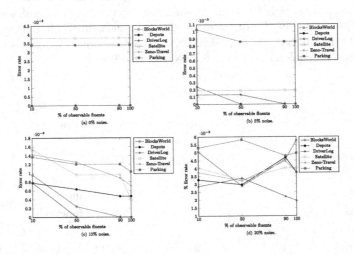

Fig. 3. Learned domains error rates.

On the second hand, domain validation is realized using VAL [12], an automatic validation tool used in the IPC. Roughly, VAL takes a problem, a plan and a planning domain and executes the plan's actions in order over the initial state defined in the problem using the action's definition contained in the planning domain. A domain is valid if the resultant state of applying every plan's action is equal to the problem's goal state. Once a new learned domain was obtained, it was used to solve a set of test problems. The resultant plans were used as input of VAL with the original handmade planning domain. If every plan created by the new domain validated the original domain, the learned domain was valid. The results of PlanMiner-O3 learned domains' validity can be seen in Table 2. Validity criterion is the hardest one to meet because a single error can make a whole domain not valid: a wrong fluent in the action's effects makes useless the whole action by producing wrong plans and therefore the whole domain.

Table 2. Domains validity matrix.

Noise %	0%				5%				10%				20%			
Incompleteness %	0%	10%	50%	90%	0%	10%	50%	90%	0%	10%	50%	90%	0%	10%	50%	90%
BlocksWorld	✓	✓	✓	✓	✓	✓	✓	✓	✓	✓	✓	✓	X	X	X	X
Depots	✓	✓	✓	✓	✓	✓	✓	✓	X	X	X	X	X	X	X	X
DriverLog	✓	✓	✓	✓	✓	✓	✓	✓	✓	✓	X	X	X	X	X	X
Satellite	✓	✓	✓	✓	✓	✓	✓	✓	X	X	X	X	X	X	X	X
ZenoTravel	✓	✓	✓	✓	✓	✓	✓	✓	X	X	X	X	X	X	X	X
Parking	✓	✓	✓	✓	✓	✓	✓	✓	X	X	X	X	X	X	X	X

Comparing our solution with other state-of-the-art solutions is very difficult: There's a wide variety of experimental settings that cannot be emulated such as input data types, planning domains used and domain learners expressivity. We have selected two of the most successful domains learners among the international community: ARMS [2] and the kernels based approach presented in [6]. Comparisons were realized by using the same testing domains. If a domain learner could not deal with a characteristic of a given domain this characteristic was ignored during the comparison. Comparing with [2], PlanMiner-O3 learned domains show lower rates even with some levels of noise in the input data: with a 90% of missing fluents and no noise (this solution only deals with incomplete input data) the error rates of the domains learned by ARMS range from above 0.2 to less than 0.6. The results presented in [6] show similar results. Error rates presented ranged from above 0.04 to 0.1 in the worst cases, while our learned domains' error rates maintain below 0.016 in every domain even with higher levels of noise.

6   Conclusions and Future Work

In this paper, we have presented a new planning domain learner (PlanMiner-O3) that encapsulates a machine learning classification algorithm and by applying various data preprocessing and postprocessing techniques is able to output planning domains. Those domains can be classical STRIPS, contain fixed numerical information and/or logic relations between fluents. The results obtained showed that PlanMiner-O3 is able to deal with incomplete and noise information. In fact, experiments demonstrate that incompleteness affects little to the results of PlanMiner-O3, even with some levels of noise.

Our new work is going to focus on the improvement of the expressivity of the models that PlanMiner-O3 outputs. We want to add a new preprocessing technique to allow PlanMiner-O3 to learn arithmetic relations between numerical fluents. The actual strategy to generate logic relations (add every possible relation as a new attribute) is impossible to carry out when dealing with complex arithmetic expression because of the size of the search space. This leads us to try to implement a more intelligent solution when facing this problem.

References

1. Jiménez, S., Rosa, T.D.L., Fernández, S., Fernández, F., Borrajo, D.: A review of machine learning for automated planning. Knowl. Eng. Rev. **27**, 433–467 (2012)
2. Yang, Q., Wu, K., Jiang, Y.: Learning action models from plan examples using weighted MAX-SAT. Artif. Intell. J. **171**, 107–143 (2007)
3. Zhuo, H.H., Yang, Q., Hu, D.H., Li, L.: Learning complex action models with quantifiers and logical implications. Artif. Intell. **174**, 1540–1569 (2010)
4. Rodrigues, C., Gérard, P., Rouveirol, C.: Incremental learning of relational action models in noisy environments. In: Frasconi, P., Lisi, F.A. (eds.) ILP 2010. LNCS (LNAI), vol. 6489, pp. 206–213. Springer, Heidelberg (2011). https://doi.org/10.1007/978-3-642-21295-6_24
5. Pasula, H.M., Zettlemoyer, L.S., Kaelbling, L.P.: Learning symbolic models of stochastic domains. J. Artif. Intell. Res. **29**, 309–352 (2007)
6. Mourao, K., Zettlemoyer, L.S., Petrick, R.P.A., Steedman, M.: Learning STRIPS operators from noisy and incomplete observations. In: Proceedings of the Twenty-Eighth Conference on Uncertainty in Artificial Intelligence (2012)
7. Gregory, P., Lindsay, A.: Domain model acquisition in domains with action costs. In: Proceedings of ICAPS (2016)
8. Fikes, R.E., Nilsson, N.J.: STRIPS: a new approach to the application of theorem proving to problem solving. Artif. Intell. **2**, 189–208 (1971)
9. Fayyad, U.M., Irani, K.B.: Multi-interval discretization of continuous-valued attributes for classification learning. In: IJCAI (1993)
10. Liu, Y., Li, Z., Xiong, H., Gao, X., Wu, J.: Understanding of internal clustering validation measures. In: 2010 IEEE 10th International Conference on Data Mining (ICDM) (2010)
11. González, A., Pérez, R.: Improving the genetic algorithm of SLAVE. Mathw. Soft Comput. **16**, 59–70 (2009)
12. Howey, R., Long, D.: VAL's Progress: the automatic validation tool for PDDL2.1 used in the international planning competition. In: Proceedings of ICAPS (2003)

An Intelligent Advisor for City Traffic Policies

Daniel H. Stolfi[✉][iD], Christian Cintrano[iD], Francisco Chicano[iD],
and Enrique Alba[iD]

Departamento de Lenguajes y Ciencias de la Computación, University of Malaga,
Malaga, Spain
{dhstolfi,cintrano,chicano,eat}@lcc.uma.es

Abstract. Nowadays, city streets are populated not only by private vehicles but also by public transport, fleets of workers, and deliveries. Since each vehicle class has a maximum cargo capacity, we study in this article how authorities could improve the road traffic by endorsing long term policies to change the different vehicle proportions: sedans, minivans, full size vans, trucks, and motorbikes, without losing the ability of moving cargo throughout the city. We have performed our study in a realistic scenario (map, road traffic characteristics, and number of vehicles) of the city of Malaga and captured the many details into the SUMO microsimulator. After analyzing the relationship between travel times, emissions, and fuel consumption, we have defined a multiobjective optimization problem to be solved, so as to minimize these city metrics. Our results provide a scientific evidence that we can improve the delivery of goods in the city by reducing the number of heavy duty vehicles and fostering the use of vans instead.

Keywords: Application · Evolutionary algorithm · Road traffic
City policy · Real world · Smart mobility

1 Introduction

Cities have been growing in number of inhabitants along the history, as many people are leaving the countryside to settle in urban areas [14]. Consequently, there is a noticeable increment in the number of trips that citizens have to take nowadays and their duration [18], especially because urban infrastructures are not scaling properly. Furthermore, the need of cargo space makes those services to use small trucks to perform their commercial activities [19], which represents an increment not only in the street space usage but also in the pollutant emissions.

Road traffic is a well-known source of air pollution in urban areas. The delivery of goods in a city affects air pollution, then the citizens quality of life. Local authorities are responsible for taking care of this matter despite the importance of negative or zero cost options when formulating their climate policy [10].

Some cities implement fixed speed policies where all the vehicles must observe a reduced maximum speed in the city's streets [2], while others have more green

© Springer Nature Switzerland AG 2018
F. Herrera et al. (Eds.): CAEPIA 2018, LNAI 11160, pp. 383–393, 2018.
https://doi.org/10.1007/978-3-030-00374-6_36

buildings, focus on pedestrian and bicycle infrastructure, or have implemented more programs to divert waste from methane-generating landfills [13].

Another type of pollution analysis is the one carried out in [8], where authors model the gas emissions of delivery trucks in urban logistics. The authors analyze the effect on greenhouse gas emissions of different trips and conclude that traveled distance and vehicles' weight have a capital impact on the pollution levels emitted. They conclude that replacing trucks by another less polluting vehicle may be a solution for reducing greenhouse gas emissions. In [12] the authors proposed banning Heavy Duty Vehicles (HDV) by 1.5 Light Duty Vehicles (LDV) among others strategies to keep the same cargo capacity while reducing gas emissions. Also, the total demand was increased by 1.035 as more vehicles were needed. We wished to go further in our study and analyze not only the optimal vehicle proportion (instead of banning some types), but also how this proportion affects travel times, gas emissions, and fuel consumption.

We have taken some small starting steps in [17] which have led us to the proposal in this article. It consists of studying different configurations of road traffic in a realistic scenario of Malaga, Spain, to better know how travel times, greenhouse gas emissions, and fuel consumption change. Common sense would suggest that reducing the number of HDV in the city's streets and incrementing the LDV should be the right thing to do. We wish to check if it is so, when the cargo capacity is a constraint to be kept and possible traffic jams are taken into account. After this study, city authorities would be capable of deciding the best strategy to apply when the pollutant levels are high or if they want to foster fuel saving or shorter travel times.

The rest of this paper is organized as follows. The problem description and our proposal are discussed in Sects. 2 and 3, respectively. In Sect. 4 we present the characteristics of the real scenarios analyzed. Section 5 focuses on the numerical study and the discussion of the results. And, finally, in Sect. 6, conclusions and future work are outlined.

2 Problem Description

In this article we present a new strategy to reduce travel times, gas emissions, and fuel consumption in the city by using a multi-objective evolutionary strategy. We start from the real number of vehicles measured in the city, their proportions and routes, and calculate new vehicle proportions according to an evolutionary algorithm, to improve metrics, without losing the observed cargo capacity.

Formally, let $v = (v_s, v_{mv}, v_{fsv}, v_t, v_m)$ be a vector containing the number of vehicles in the actual city (sedans, minivans, full-size vans, trucks and motorbikes) obtained from the proportions sampled during one hour. We assumed that only the 20% of sedans ($v_{s\prime} = 0.2 \cdot v_s$) and motorbikes ($v_{m\prime} = 0.2 \cdot v_m$) are used for delivering goods so that $v' = (v_{s\prime}, v_{mv}, v_{fsv}, v_t, v_{m\prime})$.

According to the number of cargo vehicles v' and its average cargo capacity $t = (t_{s\prime}, t_{mv}, t_{fsv}, t_t, t_{m\prime})$, we can calculate the cargo capacity available in the real city during our study time as $T = v' \cdot t^\mathsf{T}$.

Our objective is to obtain a Pareto set [5] of N vectors $vr^\star_j = (vr^\star_{s_j}, v^\star_{mv_j}, v^\star_{fsv_j}, v^\star_{t_j}, vr^\star_{m_j}), j \in N$ which contains different optimal solutions, minimizing travel times, emissions, and fuel consumption in the city subject to:

$$T^\star_j = \sum_i t_{ij} \cdot vr^\star_{ij} \geq T, \quad \forall j \in N \tag{1}$$

$$v^\star_{wj} \geq \tau \cdot v_w, \quad \forall j \in N \text{ and } w \in \{sedan, motorbike\} \tag{2}$$

The set of vectors $vr^\star_j$ represents the number of vehicles delivering goods, while $v^\star_j$ represents the total number of vehicles in the city (deliver and cargo trips). Thus, we have set $\tau = 0.8$, according to our cargo use estimation.

A set of solutions would represent different key indicators for new policies (restrictions, tax reductions, etc.) to be applied to the road traffic in the city to foster shorter travel times, less gas emissions, and saving fuel.

3 Solving the Problem

In this work, we are looking for the best proportion of vehicle types in the city so as to optimize different aspects of the entire road traffic in the city. Modifying the number and type of vehicles used in transportation of goods while keeping the total demanded cargo will allow us to change the whole traffic characteristics to improve the quality of life of drivers and citizens. We use a realistic scenario of Malaga featuring different vehicle proportions (Table 1) to be optimized. Using each new configuration we calculate the amount of cargo $T^\star$ for this new scenario (subject to the restriction in Eq. 1).

Figure 1 shows the architecture of our proposal. The algorithm calculates the optimal proportions of vehicles evaluating its individuals by using the SUMO traffic simulator [11]. This evaluation comprises a realistic city model made of data measured *in situ*, open data published by the local city council, and the city map obtained from OpenStreetMap [15] (see our case study in Sect. 4). The different parts of the architecture are explained as follows.

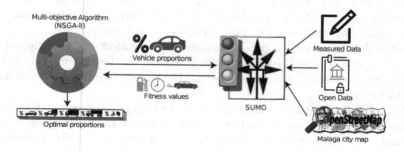

Fig. 1. System architecture.

3.1   Solution Encoding

We have encoded each solution as a vector of real numbers $x = (x_s, x_{mv}, x_{fsv}, x_t, x_m)$ in which each component is the proportion of sedans, minivans, full-size vans, trucks, and motorcycles intended to transport goods. However, a gas oil engine does not have the same fuel consumption as a gasoline engine, nor the same amount of gas emissions. Because of these differences, and to get closer to reality, before evaluating a solution x we transform it into more detailed proportions $x^\star$ according to the volume of engine types (fuel and gas emissions characteristics) existing in the car fleet. These proportions will be used by our evaluation function. We have two possibilities for fuel: gas oil and gasoline, and we have different gas emission classes according to the European emission standard [6]. Below we describe the steps to obtain $x^\star$:

Step 1. Calculate the amount of cargo vehicles $N_c = T/(\sum_i^{|x|} x_i \cdot t_i)$ that we need to supply the tonnage demand T (constraint).
Step 2. Compute the total amount of vehicles $N^\star = N_p + N_c$, being $N_p = v_s + v_m$ the number of vehicles intended for private use (constraint).
Step 3. Get the correction factor ϕ in the number of vehicles $\phi = N^\star/N$, being N the total amount of vehicles in the base solution.
Step 4. For each proportion, we correct its value for the new total of vehicles: $x_i' = x_i \cdot N_c/N^\star$, obtaining the new vector x'.
Step 5. Calculate the new proportion of private use vehicles (sedans and motorcycles) and add it to the solution $x_j' = x_j' + v_j/N^\star, j \in \{s, m\}$.
Step 6. According to the proportions of each vehicle classes in Table 1 we calculate the extended solution $x^\star = \{x' \cdot f \cdot e | \forall x' \in x', f \in Fuels, e \in Emissions\}$ where $Fuels$ is the proportion of gasoline and gas oil of each vehicle, and $Emissions$ the proportion of engine according to the data published in [4].
Step 7. Return the extended solution $x^\star$ and the factor ϕ.

3.2   Evaluation Function

We evaluate the quality of each solution making use of the SUMO traffic microsimulator [11]. SUMO is a free and open traffic simulation suite developed by the German Aerospace Center[1]. It allows modeling of intermodal traffic systems including road vehicles, public transport and pedestrians.

Each solution to be evaluated consists of a solution $x^\star$. Additionally, an increment factor ϕ is used to increase (or decrease) the total number of vehicles N (measured amount of vehicles). Note that the realistic scenario to be optimized has a factor $\phi = 1.00$.

Also, since Malaga has a large number of vehicles, solutions which notably increase this number could end in several traffic jams with many vehicles enqueued, waiting to enter the city after the analysis ends. To prevent that, we have calculated the proportion of vehicles entering the city in our realistic scenario (84%) and slightly penalized with the term k, those configurations that

---

[1] http://dlr.de/ts/sumo.

are under this threshold. Concretely, we have set a threshold $\theta = 0.8$ to allow up to 20% of vehicles waiting in the queue when the analysis time ends.

We show in Eq. 3 the fitness function used in the optimization process and in Eq. 4 the penalty term.

$$\mathbf{f}(x) = \left(k + \frac{1}{n}\right) \cdot \sum_{i=1}^{n} \left(travel\ time(x_i), emissions(x_i), fuel(x_i)\right), \qquad (3)$$

$$k = \begin{cases} 0 & \text{if } \frac{n}{N} \geq \theta, \\ \frac{100}{n} \cdot \left(\frac{n}{N} - \theta\right) & \text{otherwise.} \end{cases} \qquad (4)$$

3.3 Algorithm

In order to solve our optimization problem, we use a well-known multi-objective metaheuristic algorithm: the NSGA-II algorithm proposed by Deb et al. [3]. Our goal is not to research in the algorithm itself, but to make a first model of this real problem and see if we can create an intelligent advisor for city managers. As described above, each individual is represented as a vector of real numbers. This allows us to use simple and fast operators as the following ones:

– Crossover: Simulated binary crossover with probability 0.9.
– Mutation: Polynomial mutation with probability 0.25.
– Selection: Same selection applied in [3].
– Replacement: Elitist without including repeated individuals.

To avoid unfeasible solutions, before evaluating each solution, we normalize it and then calculate $x^\star$ and ϕ. We also perform 200 evaluations of the algorithm with a population of 48 individuals (the initial population was randomly generated). These values were selected after a preliminary study in which we tested two population sizes: 24 and 48, and two maximum number of evaluations: 100 and 200. Then, we selected the best configuration in order to maximize the diversity of the calculate Pareto set, taking into account the limitation of time (the mobility scenarios require long computation times).

4 Case Study

We have chosen as our case study an area comprising the East side of the city of Malaga (Spain), including those zones where traffic jams are common. The geographical area studied encompasses an area of about $32\,km^2$.

The city map (shown in Fig. 2) was imported from OpenStreetMap into the SUMO traffic microsimulator. This allows us to work with a real scenario, e.g., streets, traffic lights, left turns, and roundabouts. From our observations, we have defined five types of vehicle for representing the road traffic in the streets of Malaga. The average characteristics of vehicles according to the manufacturers are shown in Table 1. The vehicle distribution was obtained by counting and classifying the type of vehicles at four different locations in the city (blue labels in Fig. 2). We observed that 68.9% of vehicles are sedans, 6.4% are minivans, 7.0% are full-size vans, 2.9% are trucks, and 14.9% are motorcycles. Some types are divided into gasoline and gas oil variants (Table 1) as stated by the data for Andalusia [4], and into their equivalent emission classes in SUMO (HBEFA3 [9]).

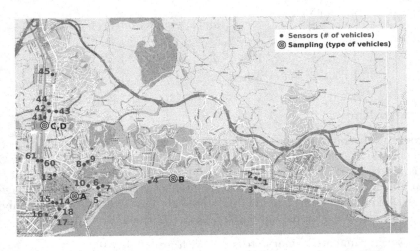

Fig. 2. Case study: east malaga. The measurement points are the red numbers and the sampling points are the blue letters. (Color figure online)

Using the data published by the local council of Malaga [1] corresponding to 23 measurement points (red numbers in Fig. 2) and the Flow Generator Algorithm (FGA) [16], we have obtained the average traffic per hour for working days in the third quarter of 2015. The FGA assigns vehicles to the traffic flows generated by the program DUARUTER included in the SUMO software package, and adjusts its number and routes in the simulation map to the values measured by the 23 real sensors in the city. After this process, we ended up with a scenario, consisting of 10,438 vehicles (7,193 sedans, 616 minivans, 768 full-size vans, 297 trucks, and 1,564 motorbikes), to be optimized with the aim of finding the best policies for reducing travel times, gas emissions and fuel consumption.

Each vehicle type has an assigned capacity t obtained from standard commercial models of vehicles as shown in Table 1. By multiplying the number of vehicles n by their capacity t, we obtained a total cargo of 4,101 tons, which is the lower bound T of the problem (Eq. 1). Each solution generated by our algorithm must be able to provide the total of vehicles needed to delivery this tonnes of goods, so that the city is not losing cargo capacity.

5   Results

After presenting our problem we are going to study the different solutions obtained by the algorithm and compare our improvements on the city with other strategies carried out by traffic managers for reducing pollution.

Table 1. Characteristics of the vehicles in our case study, the observed proportions, the cargo availability, and the individual and total cargo capacity.

Type	Accel. (m/s^2)	Decel. (m/s^2)	Length (m)	Max.Spd. (m/s)	Gasoline (%)	Gas Oil (%)	Rate (%)	Cargo (%)	Capacity t (ton)	Total T (ton)
Sedan	0.720	12.341	4.500	25.25	44.86	55.15	68.9	20	0.20	288
Minivan	0.720	12.341	4.500	25.25	10.88	89.12	6.4	100	1.00	616
Full-size van	0.720	12.341	4.878	25.25	10.88	89.12	7.0	100	2.00	1,536
Truck	0.263	3.838	7.820	16.67	0.00	100.00	2.9	100	5.50	1,633
Motorcycle	0.460	8.147	2.200	16.67	100.00	0.00	14.9	20	0.09	28

5.1 Pollutants Correlations

Since SUMO provides different types of pollutants (CO_2, CO, HC, etc.), we wished to select the ones with the lowest correlations to travel time and fuel consumption. After carrying out the simulation and measuring these emissions we calculated the Pearson correlation coefficients where hydrocarbons (HC) presented the lowest correlation with travel times (0.23) and fuel consumption (0.40). Hence, we have chosen HC as a gas emission measure in our experiments. The rest of metrics will be also reduced since there also exist relations between HC and CO_2 (0.38), CO (0.67), NO_x (0.25), PM_x (0.30), and fuel (0.40).

5.2 Solution Analysis

After performing 30 independent runs the average hypervolume is 95,736, the standard deviation is 6,550, their average ϵ-indicator is 12.6 and the deviation is 6.3. From these indicators, we assume the fronts were similar to each other.

We studied the different solutions using the attainment surfaces [7] technique. We selected the 25%-attainment surface to ensure the quality of the selected solutions (Fig. 3). By analyzing the individuals in this Pareto front, we were able to spot the differences in the number of vehicles in each class (Fig. 4(a)).

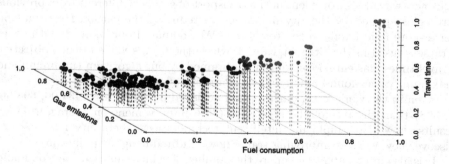

Fig. 3. 25%-attainment surface. Fitness values are scaled to the [0,1] range.

The most used cargo vehicles were vans: 1,419 minivans (median) and 1,332 full-size vans. Their speed and cargo capacity thus are an ideal choice for delivery goods throughout the city. However, HDV, like trucks, and also motorcycles were not good options, as the former are too slow and pollutant and the latter have too little cargo capacity. Sedans had a greater presence than motorcycles and trucks, but were still overshadowed by the utility of vans for delivering goods.

5.3   Solution Improvement

Next, after discussing the solutions obtained by the algorithm, we will analyze the fitness values of each one of them. Figure 4(b) shows the percentage improvement of each solution of the front with respect to the current situation (scenario) in the city of Malaga. Although travel time was improved by 10%, fuel and all the emissions (not only HC) were considerably better than the ones measured in Malaga. This is good news for the city's environment, as not only there were fewer emissions of polluting gases, but also fewer fossil fuels were consumed.

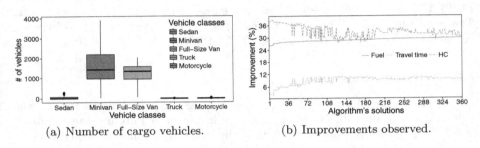

(a) Number of cargo vehicles.   (b) Improvements observed.

Fig. 4. 25%-attainment surface analysis.

5.4   Comparing with Other Strategies

This new way of improvement in the transport of goods is different from previous strategies applied by the city managers for reducing the carbon footprint and get a more fluid traffic in the roads [12]. We compared our solution with other strategies used in the cities in Table 2, where the fitness value of each objective (travel times, gas emissions, and fuel) obtained by our algorithm (minimum and maximum values found among all the Pareto sets) are shown.

Even in the worst case, our proposal achieved better fitness (434.06, 698.66, 523.08) than the rest of the strategies (30% improvement on average). These results show that employing intelligent techniques can help city managers to discover new ways of improving traffic flows and reducing air pollution.

It is also interesting to compare the number of vehicles in each solution found with those currently moving in Malaga. Figure 5 shows the number of vehicles of each algorithm's run. The red line marks how many vehicles (10,438) were in Malaga according to our case study. We can see that the total number of vehicles is lower than in Malaga (74.09% of solutions are under the red line). But, this

Table 2. Fitness values in the different strategies.

Strategies	Travel time	Gas emissions	Fuel
NSGA-II Min	**633.77**	**434.15**	**406.20**
NSGA-II Max	**698.66**	**523.08**	**434.06**
Malaga	711.12	708.68	579.19
Limit 30 km/h	2048.74	1412.43	1089.75
Limit 70 km/h	709.47	621.21	499.71

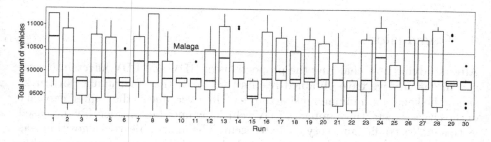

Fig. 5. Total number of vehicles in each run. (Color figure online)

does not mean that more trucks need to be added instead of LDV, as we saw in the 25%-attainment surface. The original, realistic scenario (Malaga), is close to the maximum capacity of vehicles that fits into the analyzed area of the city without producing traffic jams. Consequently, our solutions do not only have less emissions while improving the efficiency of transport operators, but also citizens and drivers are enjoying less congestion on the roads.

5.5   City Improvements

The global improvements achieved on Malaga are shown in Table 3. Being Malaga our realistic case study, we have improved all the metrics using the algorithm, although we report only the best solutions for each objective. We can see that the shortest travel times are obtained when there are 1,734 full-size vans as they are faster than trucks an their emissions are lower. The less emitting solution is the one that uses 3,193 minivans for delivering cargo instead of the other vehicles. Finally, citizens would save more fuel in a more equilibrate distribution of both van types. Note that the algorithm has definitely banned trucks from the city as they are large, slow, and pollutant. Additionally, we present the metrics for 30 and 70 km/h, and conclude that their improvements are marginal despite being commonly used by city councils when the emissions are high.

Table 3. Improvements of metrics when using the analyzed strategies in Malaga and the number of vehicles of each type.

Strategy	T.Time (s)	G. Emiss. (g HC)	Fuel (l)	Distance (m)	Sedan	Minivan	Full-size van	Truck	Motorcycles	Total
Malaga	696.3	626.7	513.8	5552.0	6095	521	649	254	1307	8826
Limit 30 km/h	912.7	629.2	485.5	5261.4	5655	484	600	237	1224	8200
Limit 70 km/h	709.5	621.2	499.7	5527.5	6004	512	643	246	1302	8707
NSGA-II T.Time	**633.8**	504.8	412.9	5605.5	5118	109	**1734**	1	1084	**8046**
NSGA-II G. Emiss.	694.5	**434.1**	425.6	**5532.9**	5034	**3193**	76	1	1011	9315
NSGA-II Fuel	642.9	491.7	**406.2**	5593.7	5027	**1274**	**1153**	0	1093	8547

6   Conclusions

After studying how changing the proportions of HDV and LDV affects the metrics in the city we have obtained results that show that the number of trucks should be kept at a minimum inside the city. However, due to the limited capacity of the city's streets, the number of LDV vehicles cannot be considerably increased as this makes traffic jams very likely to occur. The multi-objective algorithm was capable of identifying this restriction and obtained solutions where using minivans and full-size vans for delivering goods is advisable.

It does not mean that companies have to sell all their trucks, but HDVs should be used as freight transport by highways and then use vans for local delivery. All this information would be extremely useful for city managers and these results would serve as goals for the creation of municipal strategies to promote the well-being of drivers, workers and citizens.

Acknowledgements. This research has been partially funded by the Spanish MINECO and FEDER projects TIN2014-57341-R, TIN2016-81766-REDT, and TIN2017-88213-R. University of Malaga, Andalucia TECH. Daniel H. Stolfi is supported by a FPU grant (FPU13/00954) from the Spanish MECD. Christian Cintrano is supported by a FPI grant (BES-2015-074805) from Spanish MINECO.

References

1. Ayuntamiento de Malaga: Area de Movilidad (2017). http://movilidad.malaga.eu
2. Bel, G., Bolancé, C., Guillén, M., Rosell, J.: The environmental effects of changing speed limits: a quantile regression approach. Transp. Res. Part D: Transp. Environ. **36**, 76–85 (2015)
3. Deb, K., Pratap, A., Agarwal, S., Meyarivan, T.: A fast and elitist multiobjective genetic algorithm: NSGA-II. IEEE Trans. Evolut. Comput. **6**(2), 182–197 (2002)
4. DGT: Tablas estadisticas (2017). http://www.dgt.es/es/seguridad-vial/estadisticas-e-indicadores/parque-vehiculos/tablas-estadisticas/
5. Ehrgott, M.: Multicriteria Optimization, vol. 491. Springer, Berlin (2005). https://doi.org/10.1007/3-540-27659-9
6. European Commission: Transport & Environment (2018). https://web.archive.org/web/20121111033537/http://ec.europa.eu/environment/air/transport/road.htm

7. Fonseca, C.M., Fleming, P.J.: On the performance assessment and comparison of stochastic multiobjective optimizers. In: Voigt, H.-M., Ebeling, W., Rechenberg, I., Schwefel, H.-P. (eds.) PPSN 1996. LNCS, vol. 1141, pp. 584–593. Springer, Heidelberg (1996). https://doi.org/10.1007/3-540-61723-X_1022
8. Gan, M., Liu, X., Chen, S., Yan, Y., Li, D.: The identification of truck-related greenhouse gas emissions and critical impact factors in an urban logistics network. J. Clean. Prod. **178**, 561–571 (2018)
9. Hausberger, S., Rexeis, M., Zallinger, M., Luz, R.: Emission factors from the model PHEM for the HBEFA Version 3. Technical report I (2009)
10. Kousky, C., Schneider, S.H.: Global climate policy: will cities lead the way? Clim. Policy **3**(4), 359–372 (2003)
11. Krajzewicz, D., Erdmann, J., Behrisch, M., Bieker, L.: Recent development and applications of SUMO - simulation of urban MObility. Int. J. Adv. Syst. Meas. **5**(3), 128–138 (2012)
12. Mahmod, M., van Arem, B., Pueboobpaphan, R., de Lange, R.: Reducing local traffic emissions at urban intersection using ITS countermeasures. Intell. Trans. Syst. IET **7**(1), 78–86 (2013)
13. Millard-Ball, A.: Do city climate plans reduce emissions? J. Urban Econ. **71**(3), 289–311 (2012)
14. United Nations: World urbanization prospects the 2011 revision. Technical report (2011)
15. OSM Fundation: OpenStreetMap (2017). http://www.openstreetmap.org
16. Stolfi, D.H., Alba, E.: An evolutionary algorithm to generate real urban traffic flows. In: Puerta, J.M. (ed.) CAEPIA 2015. LNCS (LNAI), vol. 9422, pp. 332–343. Springer, Cham (2015). https://doi.org/10.1007/978-3-319-24598-0_30
17. Stolfi, D.H., Cintrano, C., Chicano, F., Alba, E.: Natural evolution tells us how to best make goods delivery: use vans. In: Proceedings of the Genetic and Evolutionary Computation Conference Companion, GECCO 2018, pp. 308–309. ACM, NY, USA (2018)
18. TNS Opinion & Social: Attitudes of Europeans towards urban mobility. Technical report, June 2013
19. Visser, J., Nemoto, T., Browne, M.: Home delivery and the impacts on urban freight transport: a review. Procedia. Soc. Behav. Sci. **125**, 15–27 (2014)

Author Index

Printed in the United States
By Bookmasters